Forms of Luminosity: Epistemic Modality and Hyperintensionality in Mathematics

David Elohim

© Hasen Joseph Khudairi 2017, 2025
All rights reserved.

Abstract

This book concerns the foundations of epistemic modality and hyperintensionality and applications of the latter to the philosophy of mathematics. David Elohim examines the nature of epistemic modality, when the modal operator is interpreted as concerning both apriority and conceivability, as well as states of knowledge and belief. The book demonstrates how epistemic modality and hyperintensionality relate to the computational theory of mind; metaphysical modality and hyperintensionality; the types of mathematical modality and hyperintensionality; to the epistemic status of large cardinal axioms, undecidable propositions, and abstraction principles in the philosophy of mathematics; to the modal and hyperintensional profiles of the logic of rational intuition; and to the types of intention, when the latter is interpreted as a hyperintensional mental state. Chapter **2** argues for a novel type of expressivism based on the duality between the categories of coalgebras and algebras, and argues that the duality permits of the reconciliation between modal and hyperintensional cognitivism and modal and hyperintensional expressivism. Elohim develops a novel, topic-sensitive truthmaker semantics for dynamic epistemic logic, and develops a novel, dynamic two-dimensional semantics grounded in two-dimensional hyperintensional Turing machines. Chapter **3** provides an abstraction principle for two-dimensional (hyper-)intensions. Chapter **4** advances a topic-sensitive two-dimensional truthmaker semantics, and provides three novel interpretations of the framework along with the epistemic and metasemantic. Chapter **5** applies the fixed points of the modal μ-calculus in order to account for the iteration of epistemic states in a single agent, by contrast to availing of modal axiom 4 (i.e. the KK principle). The fixed point operators in the modal μ-calculus are rendered hyperintensional, which yields the first hyperintensional construal of the modal μ-calculus in the literature and the first application of the calculus to the iteration of epistemic states in a single agent instead of the common knowledge of a group of agents. Chapter **6** advances a solution to the Julius Caesar problem based on Fine's 'criterial' identity conditions which incorporate conditions on essentiality and grounding. Chapter **7** provides a ground-theoretic regimentation of the proposals in the metaphysics of consciousness and examines its bearing on the two-dimensional conceivability argument against physicalism. The topic-sensitive epistemic two-dimensional truthmaker semantics developed in chapters **2** and **4** is availed of in order for epistemic states to be a guide to metaphysical states in the hyperintensional setting.

Chapters **8-12** provide cases demonstrating how the two-dimensional hyperintensions of hyperintensional, i.e. topic-sensitive epistemic two-dimensional truthmaker, semantics, solve the access problem in the epistemology of mathematics. Chapter **8** examines the interaction between Elohim's hyperintensional semantics and the axioms of epistemic set theory, large cardinal axioms, the Epistemic Church-Turing Thesis, the modal axioms governing the modal profile of Ω-logic, Orey sentences such as the Generalized Continuum Hypothesis, and absolute decidability. These results yield inter alia the first hyperintensional Epistemic Church-Turing Thesis and hyperintensional epistemic set theories in the literature. Chapter **9** examines the modal and hyperintensional commitments of abstractionism, in particular necessitism, and epistemic hyperintensionality, epistemic utility theory, and the epistemology of abstraction. Elohim countenances a hyperintensional semantics for novel epistemic abstractionist modalities. Elohim suggests, too, that higher observational type theory can be applied to first-order abstraction principles in order to make first-order abstraction principles recursively enumerable, i.e. Turing machine computable, and that the truth of the first-order abstraction principle for two-dimensional hyperintensions is grounded in its being possibly recursively enumerable and the machine being physically implementable. Chapter **10** examines the philosophical significance of hyperintensional Ω-logic in set theory and discusses the hyperintensionality of metamathematics. Chapter **11** provides a modal logic for rational intuition and provides a hyperintensional semantics. Chapter **12** avails of modal coalgebras to interpret the defining properties of indefinite extensibility, and avails of hyperintensional epistemic two-dimensional semantics in order to account for the interaction between interpretational and objective modalities and the truthmakers thereof. This yields the first hyperintensional category theory in the literature. Elohim invents a new mathematical trick in which first-order structures are treated as categories, and Vopenka's principle can be satisfied because of the elementary embeddings between the categories and generate Vopenka cardinals in the category of Set in category theory. Chapter **13** examines modal responses to the alethic paradoxes. Elohim provides a counter-example to epistemic closure for logical deduction. Chapter **14** examines, finally, the modal and hyperintensional semantics for the different types of intention and the relation of the latter to evidential decision theory.

Acknowledgements

Please note that I have changed my name from Hasen Joseph Khudairi to David Hyrulaolmec Elohim, effective April, 2024, owing to my being a religious Jew since 2018, and the influence of Jewish philosophy. I changed my name, from March, 2023 to April, 2024, to Timothy Alison Bowen, owing to my partner, Alison.

From 2014 to 2017, I was a Ph.D. Student at the Arché Philosophical Research Centre for Logic, Language, Metaphysics, and Epistemology at the University of St Andrews. St Andrews is an ideal place to live and work. At Arché, I was supported by a St Leonard's College (SASP) Research Scholarship, for which I record my gratitude.

This book is a revised version of my dissertation. Most of the novel ideas in the dissertation were developed and written between early 2014 and the end of 2016. I had conceived of using epistemic two-dimensional semantics in the philosophy of mathematics (namely, applied to abstraction principles, large cardinal axioms, and undecidable sentences such as CH) in that time, anticipating Linnebo (2018b), Hale (2020), and Waxman (ms_1), and had a chapter at the time on two-dimensional truthmaker semantics, although the idea of rendering the semantics topic-sensitive occurred subsequently in the spring of 2022. Rossi and Özgün's paper on hyperintensional positive epistemic possibility was published in July, 2023, which is when I was first apprised of it.

I conceived of dynamic two-dimensional semantics, with updates to the first parameter of a sentence ranging over epistemic states determining the update to the second parameter of the sentence ranging over metaphysical states, in May, 2022. I decided, in 2022, to use epistemic actions as the update for the epistemic space in the manner of Baltag (2016)'s Logic of Epistemic Dependency, and Hannes Leitgeb suggested that interventions in Pearl (2009)'s structural equation models were the analogue of updates in metaphysical space. Peter Hawke suggested Hodges (1997)'s Team Semantics to me, in May, 2024, where Team Semantics allows valuation assignments to a set of variables. In this book, I avail of the Logic of Epistemic Dependency and structural equation models, in my novel, dynamic two-dimensional semantics.

The idea of accounting for a novel type of modal expressivism based on the duality between the categories of coalgebras and algebras, and arguing that the duality permits of the reconciliation between modal cognitivism and modal expressivism occurred to me in October, 2016. The extension to hyperintensional cognitivism and hyperintensional expressivism occurred to me in May, 2022. I was first apprised of Hawke's hyperintensional expressivism in January, 2024, and Hawke's paper was published in June, 2024.

The idea of accounting for indefinite extensibility with regard to category-theoretic Grothendieck Universes came to mind in January, 2016, and the corresponding chapter was written in that month. Hamkins and Linnebo (2022)'s paper on the topic was posted on Arxiv.org in 2017, which is when I was first apprised of it.

The idea of the conceptual engineering of (hyper-)intensions, which I write about in chapter **2**, and doing so via a novel topic-sensitive truthmaker semantics for dynamic epistemic logic and dynamic interpretational modalities, as well as a novel dynamic epistemic two-dimensional hyperintensional semantics, occurred to me in the spring of 2022.

Rosen and Yablo (2020) avail of real, or essential, definitions in their attempt to solve the Julius Caesar problem, although their real definitions do not target grounding-conditions. The need for a grounding-condition is mentioned in Wright (2020: 314, 318). The approach here developed, of solving the Caesar problem by availing of metaphysical definitions, was arrived at independently of Rosen and Yablo (op. cit.) and Wright (op. cit.), in 2016, although Rosen and Yablo's paper was archived at the time and I had not yet had the occasion to read it. The examination of the relation between abstraction principles and grounding, though not essence, has been pursued by Rosen (2010); Schwartzkopff (2011); Donaldson (2017); De Florio and Zanetti (2020); and deRosset and Linnebo (2023). Mount (2017: ch. 5) examines the relations between essence and number and grounding and number separately. The approach here developed is novel in examining metaphysical definitions which incorporate conditions on both essentiality and grounding.

The idea of specifying an abstraction principle for intensions was suggested to me by Dave Chalmers, in 2013, before I pursued my Ph.D. at

Arché in 2014. I specify a recursively enumerable abstraction principle for two-dimensional hyperintensions, in chapter **3**. See §**1.2**, for further discussion.

The idea of applying hyperintensional semantics – i.e. a novel epistemic two-dimensional semantics which I developed which involves hyperintensional truthmakers rather than worlds and is further topic-sensitive – and epistemically necessary truthmakers to the axioms of epistemic set theory, large cardinal axioms, the Epistemic Church-Turing Thesis, the modal axioms governing Ω-logical consequence in set theory, the modal profile of rational intuition, and the semantic profile of intentions also occurred to me between 2021 and 2022.

Otherwise, every major proposal in the dissertation was developed between 2014 and 2016.

Subsequent to writing the foregoing, Beddor and Goldstein (2022) have published on topic-sensitive intentions, although my proposal differs in being multi-hyperintensional (involving both truthmakers and topics), and targeting the semantic profiles of the various types of intention.

For comments on the structure of the book, I am grateful to Otávio Bueno.

For comments which lead to revisions to individual chapters, I am grateful to Francesco Berto for comments on chapter **4**, Josh Dever for comments on chapter **6**, Dave Chalmers, Sean Kelly, Kit Fine, and Shawn Standefer for comments on chapter **7**, Peter Milne for comments on chapter **9**, and Joel Hamkins, Peter Koellner, Gabriel Uzquiano, and Hugh Woodin for comments on chapter **12**.

Thanks to Anna Traverse, for helpful comments, while the work was being written, from 2014 to 2017, and edited, from 2017 till early 2026.

For her love and encouragement, over the years, I am grateful to Petunia Elohim. This book is dedicated to her.

I was recruited into the United States Intelligence Community, with credentials for the United States Central Intelligence Agency (the CIA), in March, 2017. I am grateful to King Charles III, President George W. Bush, Michael Traynor, Tamar Lando, Peter Milne, Toby Meadows, Liz Warren, President Don Trump, Mark Gormley, President Barack Obama, Karl Rove, Otávio Bueno, Melvin Fitting, Ned Block, Tim Williamson, James Turner, Kit Fine, Jason Kessel, Stephane Leblois, Ramsey, Gabriel Uzquiano, Sarah Moss, Jeremy Goodman, Bilgrami, Richard Pettigrew, Steinberger, and John Notruf, for their support during my transition from being a professional research academic in mathematical and philosophical logic to being a professional spy op.

Thanks to Hararabi, for positioning relations with the Skull and Bones at Yale.

Thanks sincerely to Peter Milne, Hararabi, Peter Hill, and Nelson, for their support of my professional ambitions.

I am grateful to Min and Alexandre Lefebvre, for enjoining me to mention that my two-dimensional hyperintensional Turing machines ground my novel, dynamic two-dimensional semantics, in the book abstract.

For productive conversations at Arché, I am grateful to Poppy Mankowitz.

For productive discussion of philosophy of mathematics at the Musical Offering Cafe, I am grateful to Branden Fitelson.

Thanks to Graham Priest, Michael Williams, Cian Dorr, Zee Perry, Evan Barrett, Huw Price, Claudia Schaer, and John Curran, for their good company over the years.

From 2005 to 2008, I was an honors undergraduate in philosophy at Johns Hopkins University. For their encouragement and example, I am grateful to Hent de Vries.

Thanks to Øystein Linnebo and Stephen Read, for their support of my research endeavors in philosophical logic and the philosophy of mathematics, in the 'Models, Modality, and Meaning' and 'History and Philosophy of Logic and Mathematics' research seminars, while I was a Ph.D. Student at Arché at the University of St Andrews.

Chapter **7** has been published as "Grounding, Conceivability, and the Mind-Body Problem" in *Synthese*, 195 (2):919-926 (2018), doi:10.1007/s11229-016-1254-2. Chapter **10** has been published as "Modal Ω-Logic" in Don Berkich and Matteo Vincenzo d'Alfonso (eds.), *On the Cognitive, Ethical, and Scientific Dimensions of Artificial Intelligence – Themes from IACAP 2016*, Springer (2019).

Table of Contents

1. Methodological Foreward (p. 15)

1.1 The Need for a New Approach

1.1.1 History

1.1.2 The Target Conception of Epistemic Modality and Hyperintensionality

1.1.3 The Literature

1.2 Chapter Summary

Part I: A Framework for Epistemic Modality and Hyperintensionality

2. Modal and Hyperintensional Cognitivism and Modal and Hyperintensional Expressivism (p. 48)

2.1 Introduction

2.2 The Hybrid Proposal

2.2.1 Epistemic Modal Algebra

2.2.1.1 Epistemic Two-dimensional Truthmaker Semantics

2.2.2 Modal Coalgebraic Automata

2.3 Material Adequacy

2.4 Precedent

2.5 Conceptual Engineering of Intensions and Hyperintensions

2.6 Two-dimensional Hyperintensionality and the Epistemic Church-Turing Thesis

2.7 Expressivist Semantics for Epistemic Possibility

2.8 Modal Expressivism and the Philosophy of Mathematics

2.9 Concluding Remarks

3. Cognitivism about Epistemic Modality and Hyperintensionality (p. 79)

3.1 Introduction

3.2 An Abstraction Principle for Two-dimensional (Hyper-)Intensions

3.3 Examples in Philosophy and Cognitive Science

3.4 Objections and Replies

3.5 Concluding Remarks

4. Topic-Sensitive Two-Dimensional Truthmaker Semantics (p. 94)

4.1. Introduction

4.2 Two-Dimensional Truthmaker Semantics

4.2.1 Intensional Semantics

4.2.2 Truthmaker Semantics

4.2.3 Two-Dimensional Truthmaker Semantics

4.3. Topic-sensitive Two-dimensional Truthmaker Semantics

4.4 New Interpretations

4.4.1 Fundamental and Derivative Truths

4.4.2 Decision Theory

4.4.3 Intentional Action

4.5 Concluding Remarks

5. Fixed Points in the Epistemic Hyperintensional μ-Calculus and the KK Principle (p. 113)

Part II: The Relation between Hyperintensional Conceivability and Metaphysical States

6. Conceivability, Haecceities, and Essence (p. 124)

6.1 Introduction

6.2 Super-Rigidity and Hyper-Rigidity

6.3 Two Dogmas of Semantic Rationalism

6.3.1 The First Dogma

6.3.2 The Second Dogma

6.3.2.1 The Julius Caesar Problem

6.3.3 Mereological Parthood

6.3.4 Summary

6.4 Determinacy and Consistency

6.5 Concluding Remarks

7. Grounding, Conceivability, and the Mind-Body Problem (p. 145)

Part III: Epistemic Modality and Hyperintensionality in the Philosophy of Mathematics

8. Epistemic Hyperintensionality and Absolute Decidability (p. 156)

8.1 Mathematical Modality

8.1.1 Metaphysical Mathematical Modality

8.1.2 Epistemic Mathematical Modality

8.1.3 Modal Axioms

8.1.4 Hyperintensional Epistemic Set Theory

8.1.5 Two-dimensional Hyperintensional Large Cardinals

8.2 Departures from Precedent

8.3 Knowledge of Absolute Decidability

8.4 Concluding Remarks

9. Hyperintensional Foundations of Mathematical Platonism (p. 177)

9.1 Introduction

9.2 The Abstractionist Foundations of Mathematics

9.3 Abstraction and Necessitism

9.3.1 Hale and Wright's Arguments against Necessitism

9.3.2 Hale on the Necessary Being of Purely General Properties and Objects

9.3.2.1 Objections

9.3.3 Cardinality and Intensionality

9.4 Epistemic Hyperintensionality, Epistemic Utility, and Entitlement

9.4.1 Epistemic Two-dimensional Truthmaker Semantics

9.5 Concluding Remarks

10. Hyperintensional Ω-Logic (p. 201)

10.1. Introduction

10.2 Definitions

10.2.1 Axioms

10.2.2 Large Cardinals

10.2.3 Ω-Logic

10.2.4 Two-dimensional Hyperintensionality and Ω-logic

10.3 Discussion

10.3.1 Ω-Logical Validity is Genuinely Logical

10.3.2 Hyperintensionality and the Concept of Set

10.4 Concluding Remarks

11. A Modal Logic and Hyperintensional Semantics for Gödelian Intuition (p. 216)

11.1 Introduction

11.2 Modalized Rational Intuition and Conceptual Elucidation

11.3 Hyperintensionality

11.4 Concluding Remarks

12. Hyperintensional Category Theory and Indefinite Extensibility (p. 233)

12.1 Introduction

12.2 Indefinite Extensibility in Set Theory: Modal and Extensional Approaches

12.3 Hyperintensional Coalgebras and Indefinite Extensibility

12.4 Concluding Remarks

13. Truth, Modality, and Paradox (p. 245)

13.1 Introduction

13.2 Scharp's Replacement Theory

13.2.1 Properties of Ascending and Descending Truth

13.2.2 Scharp's Theory: ADT

13.2.3 Semantics for ADT

13.3 New Extensions of ADT

13.3.1 First Extension: The Preface Paradox

13.3.2 Second Extension: Absolute Generality

13.3.3 Third Extension: Probabilistic Self-reference

13.3.4 Fourth Extension: The Sorites Paradox

13.4 Issues for ADT

13.4.1 Issue 1: Revenge Paradoxes

13.4.2 Issue 2: Validity

13.4.3 Issue 3: Hybrid Principles and Compositionality

13.4.4 Issue 4: ADT and Indeterminacy

13.4.5 Issue 5: Descending Truth, Ascending Truth, and Objectivity

13.4.6 Issue 6: Paradox, Sense, and Signification

13.4.7 Epistemic Closure and Logical Deduction

13.5 Concluding Remarks

14. Intention: Hyperintensional Semantics and Decision Theory (p. 264)

14.1 Introduction

14.2 The Modes of Intention

14.2.1 Intention-in-Action
14.2.2 Intention-as-Explanation
14.2.3 Intention-for-the-Future
14.3 Intention in Decision Theory
14.4 Concluding Remarks

Bibliography (p. 273)

Chapter 1

Methodological Foreward

This book concerns the foundations of epistemic modality and hyperintensionality and their applications to the philosophy of mathematics. The work aims to advance our present understanding of the defining contours of epistemic modal space. I endeavor, then, to develop the theory of epistemic modality and hyperintensionality, by accounting for its interaction with metaphysical modality and hyperintensionality; the types of mathematical modality and hyperintensionality;[1] the epistemic status of large cardinal axioms, undecidable propositions, and abstraction principles in the philosophy of mathematics; the modal and hyperintensional profiles of rational intuition; and the types of intention, when the latter is interpreted as a modal and hyperintensional mental state. In each chapter, I examine the philosophical significance of the foregoing, by demonstrating its import to a number of previously intransigent philosophical issues.

In brief summary of the novel theories here developed, I develop, inter alia, a novel topic-, i.e. subject matter, sensitive two-dimensional truthmaker semantics (chs. **2** and **4**). Atomic topics record the hyperintensional intentional content of atomic propositions, i.e. what the atomic propositions are about at a hyperintensional level (Berto (2022: ch. 2). The multi-hyperintensional semantics includes atomic topics, topic-sensitive truthmakers from epistemic and metaphysical state spaces, where the truthmakers are hyperintensional parts of whole possible worlds (Fine, 2017a,b),[2] subject matter similarity re-

---

[1]See §**2.1**, fn. 6, for further discussion.

[2]Evans (1979/1985) distinguishes between epistemic verification and metaphysical satisfaction. Evans writes: 'If this conclusion is correct, it follows that we must use the notion of *what makes a sentence true* with a great deal of care, for, if it is used without care it

lations between topic-sensitive truthmakers, and structured hyperintensions for subsentential expressions of propositions, where structured hyperintensions are functions from subsentential expressions verified by topic-sensitive truthmakers to extensions. A multi-hyperintensional two-dimensional hyperintension is defined in chapter **9** for numerical term-forming operators, N:x, in abstraction principles and for entire abstraction principles. Two-dimensional hyperintensions are functions from a term or sentence verified by a topic-sensitive epistemic hyperintensional state, which determines the value of the term or sentence as verified by a topic-sensitive metaphysical hyperintensional state, to the term's or sentence's extension. An abstraction principle for (hyper-)intensions modeled on Voevodsky's Univalence Axiom and function type equivalence is countenanced in chapters **2** and **3**. The truth of the first-order abstraction principle for two-dimensional hyperintensions is supposed to secure the existence of two-dimensional hyperintensions, and its truth is grounded in its possibly being recursively enumerable i.e. Turing machine computable, owing to results in higher observational type theory (see §**3.2**, for further discussion), and the machine being physically implementable. I focus on the epistemic interpretation of two-dimensional semantics (ch. **2**), and apply it (i) to conceivability arguments concerning hyperintensional phenomena (ch. **7**), and the epistemic and metaphysical profiles of (ii) abstraction principles (ch. **9**); (iii) set-theoretic axioms (including large cardinal axioms), the Epistemic Church-Turing Thesis, the modal axioms governing Ω-logical consequence, and Orey, i.e. undecidable, sentences such as the Continuum Hypothesis (ch. **8**); (iv) rational intuition, for which

---

will collapse the distinction between content and proposition which we have been at pains to describe. There are two ways in which one can use the notion of what makes a sentence true. One can follow traditional practice, as I shall do, and tie the notion to the content of a sentence, so that if two sentences are epistemically equivalent, they are verified by exactly the same state of affairs, and what one believes, in understanding the sentence and accepting it as true, is precisely that some verifying state of affairs obtains. On this conception, the same set of states of affairs makes the sentence "Julius is F" true as makes the sentence "The inventor of the zip is F" true.
[...]
'Alternatively, one can tie the notion of what makes a sentence true to the proposition it expresses by means of the principle: σ makes (Q) true iff, for any world w which comprises σ, (Q) is true$_w$.' (205-206). See Chalmers (2011: 70), for further discussion.

Chalmers (2006b: 3.5) countenances structured intensions, i.e. functions from subsentential expressions in worlds to extensions. See Stanley (2014) and Chalmers (2014c) for further discussion.

I develop a novel modal logic and hyperintensional semantics (ch. **11**); (v) indefinite extensibility. I invent a new mathematical trick in which first-order structures are treated as categories, and Vopenka's principle can be satisfied because of the elementary embeddings between the categories and generate Vopenka cardinals in the category of Set in category theory (ch. **12**); and (vi) the types of intention (ch. **14**).

Two-dimensional hyperintensions are functions from epistemic verifiers considered as actual for propositions to metaphysical verifiers to extensions. This is interpreted as conceivability being a guide to the relevant metaphysical profile. My topic-sensitive epistemic two-dimensional truthmaker semantics thus differs from Intuitionism in logic and mathematics by being governed by a classical logic and being committed to the reality of the classical continuum. Unlike Mathematical Intuitionism, which grounds object realism in constructions or proofs (see Linnebo, 2023), there are epistemic, non-maximally objective, and maximally objective i.e. metaphysical verifiers for (i)-(vi) above. Epistemic states which serve as verifiers for the propositions comprising (ii)-(iii) concern the conceivability thereof, rather than constructive provability as in Mathematical Intuitionism, ideal knowability as in Epistemic Arithmetic (Shapiro, 1985), or Grassmann (1844/1995)'s general theory of forms, and concern the possible reinterpretations of quantifiers with regard to (iv-v) above.[3] Similarly to Epistemic Arithmetic, however, epis-

---

[3]See Grassmann (1844/1995: Introduction), and Cantù (2020), for discussion of Grassmann's general theory of forms. Cantù (2022) explicates the general theory of forms thus: 'Defining mathematics as the science of (thought) forms, Grassmann claims that mathematics is about concepts, which are considered particulars generated by means of an act from some initial element.

'1. Thought forms are particulars that have "come to be through thought" (Grassmann 1844, Intro., §§2–3, 22–23).

'2. Thought forms might come to be by different types of generation and different ways to relate them to the initial element of the generation (Grassmann 1844, Intro., §5, 25).

'But he also characterizes mathematics by contrasting its peculiar method to the method followed in philosophy.

'3. The mathematical method goes from the particular to the general (Grassmann 1844, Intro., §13, 30).

'According to both characterizations of mathematics, which are not mutually exclusive but rather complementary, mathematics is conceived as the science of the particular.

'Mathematical thought forms (Denkformen) – or simply forms – are determined by their generating law: any form is a particular being that has come to be by some act of thought (it is the result of a particular act).

'Forms are abstract concepts that result from a generating thought from an initial

temic two-dimensional truthmaker semantics can capture the phenomenon of partial constructivity, e.g. a conditional mathematical claim which can be formalized neither in Heyting Arithmetic nor Peano Arithmetic, because the antecedent of the conditional concerns a property which can be effectively found, and the consequent concerns a property which cannot be effectively found (see e.g. Horsten, 1998: 7). Topic-sensitive epistemic two-dimensional truthmaker semantics answers, further, Horsten's problem concerning the construction of a 'theory of presentations of mathematical objects' in Epistemic Type Theory (op. cit.: 18). Horsten writes: '[T]he *first-order* variables range over the numbers, *given in a canonical way* (by a finite number of successor symbols prefixed to *0*, say). [O]ne cannot always substitute coextensive higher-order presentations for each other in epistemic contexts *salva veritate* [including the absolute provability operator - D.E.] – for first-order presentations there is no such problem ... In sum, in extensional contexts the associated presentations do no real work, but in intensional contexts it can make all the difference in the world how a mathematical object is presented' (17). Topic-sensitive epistemic two-dimensional truthmaker semantics might be one way to countenance hyperintensional construals of the epistemic rigidity, and thus super-rigidity, of concepts for mathematical objects such as num-

---

element, which is itself a particular concept. Mathematics is thus the science of the particular that is posited by thought and not the science of the general laws of thinking (logic)' (3.1, 3.1.1).

'Under the name of "general theory of forms" (GTF) Grassmann gathers the investigation of equality, difference, and the common properties of some connections that make their appearance in all branches of mathematics ... Forms are determined by their generating law, and are therefore equal if the same law from the same initial element generates them ... Given that forms are not given objects but the results of an act of thought that generates them according to a certain law, only the characteristics that depend on the specific way in which forms have been generated will be taken into account in the comparison.

'Mathematics, as we will see in section 3.1, is for Grassmann the science of the particular. GTF, on the contrary, investigates formal operations, which are necessarily underdetermined, because the nature of the forms and their generating law induce the properties of the operation, which might vary relative to the domain of application. Grassmann considers as more "general" the product relative to a variable domain—a domain that is not closed under the operation but rather a result of our carrying out the operation itself. It is more general in the sense that it is underdetermined, because the determination or particularization of the operation depends on further conditions dictated by the nature of mathematical objects and by generating rules. The refusal to admit a domain of elements given prior to, or independently from, the generation of the elements themselves is an idea that Grassmann never abandons, and a basic assumption of his "constructivism"' (2.3.1).

ber, against Benacerraf (1965)'s contention that, in the reduction of number theory to set theory, there must be, and is not, a principled reason for which to prefer the identification of natural numbers with, in this case, the presentation of von Neumann ordinals (e.g., 2 = {∅, ∅}), rather than with the presentation of Zermelo ordinals (e.g., 2 = {{∅}}). Super-rigid expressions are rigid in all epistemic worlds and in all metaphysical worlds. Hyper-rigid expressions are rigid in all epistemic states and in all metaphysial states.[4]

Horsten (1994) and (1998) countenances an interpretation of arithmetic, which he refers to as Modal-Epistemic Arithmetic. Modal-Epistemic Arith-

---

[4]See Zermelo (1908/1967) and von Neumann (1923/1967). Ordinals are transitive (a∈A iff a ⊆ A), and totally ordered sets: reflexive (a < a), anti-symmetric [if (a < b) ∧ (b < a), then a = b], logically transitive (a < b ∧ b < c → a < c), and satisfy the law of trichotomy, ∀a, b[a < b ∨ a = b ∨ a > b]. According to Chalmers (2012: 14th Excursus), the phrase, 'absolutely rigid', is introduced by Haas-Spohn (1995), and the phrase, 'super-rigidity', is introduced by Nida-Rümelin (2002; 2003), who credits Haas-Spohn (op. cit.) in Nida-Rümelin (2003). Super-rigidity is an intensional notion. See §**6.2** and chapter **7**.

Kripke (1980: 122, fn. 63)'s definition of analyticity corresponds to the concept of super-rigidity. See §**4.2.1**, fn. 1, for further discussion. Super-rigid expressions are rigid in all epistemic worlds and in all metaphysical worlds, and have constant two-dimensional intensions: 'Here, in effect, a term's subjunctive intension depends on which epistemic possibility turns out to be actual. / This can be seen as a mapping from scenarios to subjunctive intensions, or equivalently as a mapping from (scenario, world) pairs to extensions. We can say: the two-dimensional intension of a statement S is true at (V, W) if V verifies the claim that W satisfies S. If $[A]_1$ and $[A]_2$ are canonical descriptions of V and W, we say that the two-dimensional intension is true at (V, W) if $[A]_1$ epistemically necessitates that $[A]_2$ subjunctively necessitates S. A good heuristic here is to ask "If $[A]_1$ is the case, then if $[A]_2$ had been the case, would S have been the case?". Formally, we can say that the two-dimensional intension is true at (V, W) iff '$\Box_1([A]_1 \to \Box_2([A]_2 \to S))$' is true, where '$\Box_1$' and '$\Box_2$' express epistemic and subjunctive necessity respectively' (Chalmers, 2006).

Conditions (II) and (III) could further be added to the definition of super-rigidty, if one were to prefer to add conditions to the above.

(II) epistemic factivity, ■($\Box\phi \to \phi$) (Chalmers and Rabern, 2014: 210); or

(III) the apriority operator from (Chalmers and Rabern, op. cit.: §5), according to which ■ϕ iff there is a function from the 2-intension of the 2D-intension to the 2-intension to the extension, and the 1-intension is necessary, ■$\phi^{v,w} = 1$ iff

– (i) $\sec_w(\phi) = \sec_{v,w}(\phi)$ [i.e. the liveness constraint]

– (ii) for all v', $[\![\phi]\!]^{v',v'} = 1$ [i.e. necessary primary intension] (Rabern and Chalmers, op. cit.).

Hyper-rigid expressions are rigid in all epistemic states and in all metaphysical states. My notion of hyper-rigidity corresponds to the notion of super-rigidity, in the hyperintensional setting. See §**4.2.3**.

metic augments Peano Arithmetic with a complex operator, K, which combines a modal operator, $\Diamond$, interpreted as provability, and an epistemic operator, P, interpreted as 'some (not further specified) mathematician has a proof that ... ' (Horsten, 1994: 285; see, too, Horsten, 1998: 12). Modal-Epistemic complex operators can be captured in my multi-hyperintensional semantics by a fusion of necessary truthmakers $A(x) \sqcup A(x)$. See §2.2, for the semantic clauses for necessary truthmakers.

In chapter **2**, I also develop a novel topic-sensitive truthmaker semantics for dynamic epistemic logic, and develop a novel dynamic epistemic two-dimensional hyperintensional semantics. I apply the foregoing to account for conceptual engineering, and this application of formal methods to conceptual engineering is, as far as I know, novel in the literature.

The other novel moves in the book which concern epistemic modality and hyperintensionality are (vii) availing of modal algebra-coalgebra categorical duality in order to model a novel type of expressivism about epistemic modality and hyperintensionality (ch. **2**), and (viii) availing of fixed points in the hyperintensional μ-calculus in order to model iterated epistemic states, by contrast to epistemic modal axiom 4 i.e. the KK principle (ch. **5**). I also examine the role of modality and hyperintensionality in Ω-logic in set theory (ch. **10**, published by *Springer*); and provide an account of conceivability which is sensitive to essences and metaphysical haecceities (ch. **6**). On the metaphysics side, I provide a novel ground-theoretic regimentation of the proposals in the metaphysics of consciousness (ch. **7**, published in *Synthese*) and examine its bearing on conceivability arguments concerning hyperintensional phenomena; and apply Fine's 'criterial' identity conditions to solve the Julius Caesar problem (ch. **6**).

In more detail, in chapter **8**, I countenance hyperintensional versions of epistemic set theory and the Epistemic Church-Turing thesis. I countenance the hyperintensional Epistemic Church-Turing thesis two-dimensionally, where the formula describing a Turing machine is defined relative to two parameters, a topic-sensitive epistemic state space, and a topic-sensitive metaphysical state space.

In chapter **2**, I apply two-dimensional hyperintensional epistemic Turing Machines and the two-dimensional hyperintensional Epistemic Church-Turing thesis to provide mechanistic grounds for my novel dynamic two-dimensional semantics.

Two-dimensional sentences are updated as follows. The first parameter for sentences ranges over epistemic worlds or states and the update effects a reassignment of values to variables in the sentences in the class of epistemic states similar to Hodges (1997)'s Team Semantics and Baltag (2016)'s Logic of Epistemic Dependency, and updates thereby a second parameter ranging over metaphysical worlds or states via interventions in Pearl (2009)'s structural equation models for sentences in the class of metaphysical states. If a sentence is two-dimensional and the two parameters for the formula range over distinct spaces, then there won't be only one subject matter for the sentence, because total subject matters are construed as sets of verifiers and falsifiers and there will be distinct verifiers and falsifiers relative to each space over which each parameter ranges. This is especially clear if one space is interpreted epistemically and another is interpreted metaphysically. Availing of topics, however, and assigning the same topics to each of the states from the distinct spaces relative to which the sentence gets its value is one way of ensuring that the two-dimensional sentence has a single subject matter. This countenances the first topic-sensitive epistemic two-dimensional truthmaker semantics and the first dynamic two-dimensional semantics. This would answer Berto (2022)'s problem of 1C contents and 'topic-diverging (co-)necessities' (§2.3.3). According to 1C semantics, 'either topics reduce to truth conditions, or vice versa' (§2.3). A truthmaker for ϕ is exact when 'it can necessitate the sentence while being wholly relevant to its truth' (Fine and Jago, 2019: sect. 1). According to Berto, 'ϕ expresses a *semantic necessity* just in case it is true at unrestrictedly every circumstance. Correspondingly, say that ϕ and ψ express semantic co-necessities if they coincide in truth value at unrestrictedly every circumstance' (Berto, 2022: 42). Circumstances can be interpreted as possible worlds or hyperintensional states i.e. truthmakers. Making truthmakers topic-sensitive would, then, avoid the issues of topic-divergence which are advanced by Berto for truthmaker semantics.

In Part III, I apply the Epistemic Church-Turing thesis grounded account of my dynamic two-dimensional semantics to provide a solution to the access problem as witnessed by the axioms of ZFC (ch. **8**), abstraction principles (ch. **9**), large large cardinal axioms (ch. **10**), rational intuition (ch. **11**), and indefinite extensibility and the relation between possible reinterpretations of quantifiers and ontological expansion (ch. **12**).

I countenance an abstraction principle for two-dimensional hyperintensions via Voevodsky's Univalence Axiom and function type equivalence (ch. **3**), and use two-dimensional hyperintensions and modal epistemology to solve the access problem. Higher observational type theory can be applied to first-order abstraction principles in order to make first-order abstraction principles recursively enumerable, i.e. Turing machine computable, and the truth of the first-order abstraction principle for two-dimensional hyperintensions is grounded in its being possibly recursively enumerable and the machine being physically implementable. Indefinite extensibility is the best case of how epistemic hyperintensional two-dimensional semantics solves the access problem, because it provides a semantics grounded in the two-dimensional Epistemic Church-Turing Thesis, in which the possible reinterpretations of quantifier domains, rendered hyperintensional via availing of topic-sensitive epistemic states rather than worlds, determines possible ontological expansion, rendered hyperintensional via availing of topic-sensitive metaphysical states. I apply, further, modal rationalism in modal epistemology to solve the access problem witnessed by the topics in each chapter of Part III, including indefinite extensibility. Epistemic possibility and hyperintensionality can be a guide to metaphysical possibility and hyperintensionality, when (i) epistemic worlds or epistemic hyperintensional states are interpreted as being centered metaphysical worlds or hyperintensional states, i.e. indexed to an agent; when (ii) the epistemic (hyper-)intensions and metaphysical (hyper-)intensions for a sentence coincide, i.e. the (hyper-)intensions have the same value irrespective of whether the worlds or states in the argument of the functions are considered as epistemic or metaphysical; and when (iii) sentences are said to consist in super-rigid expressions. I argue that (i) and (ii) obtain in the case of the access problem.

I interpret the epistemic possibility and hyperintensionality at issue as concerning conceivability rather than ideal knowability as in Epistemic Arithmetic, intuitionistic constructability as in Intuitionism, or Grassmann (1844/1995)'s general theory of forms. Epistemic hyperintensionality is defined as concerning parts of epistemically possible worlds rather than the whole worlds themselves, and epistemically possible worlds are defined as concerning negative ideal conceivability, i.e. the dual of apriority, not apriori not, which is indefeasible.

In Section **1** of what follows, I examine the limits of competing proposals in the literature, and outline the need for a new approach. In Section **2**, I provide a more detailed summary of each of the chapters.

1.1 The Need for a New Approach

1.1.1 History

The proposal that mental representations can be defined as possibilities relative to states of information dates back at least to Wittgenstein (1921/1974), although there are a number of precursors to the literature in the twentieth century.[5] While novel, the limits of these incipient proposals consists in that they are laconic with regard to the explanatory foundations of the general approach.

Wittgenstein writes: 'Logical pictures can depict the world. / A picture has a logico-pictorial form in common with what it depicts. / A picture depicts reality by representing a possibility of existence and non-existence of states of affairs. / A picture represents a possible situation in logical space. / A picture contains the possibility of the situation that it represents ... A logical picture of facts is a thought. / 'A state of affairs is thinkable': what this means is that we can picture it to ourselves. / The totality of true thoughts is a picture of the world. / A thought contains the possibility of the situation of which it is the thought. What is thinkable is possible too' (op. cit.: 2.19-2.203, 3-3.02). Wittgenstein notes, further, that 'The theory of knowledge is the philosophy of psychology' (4.1121), and inquires: 'Does

---

[5]For an examination of epistemic logic in, e.g., the late medieval period, see Boh (1993).

For the role of logical, rather than epistemic, modality in defining the modes of judgment, see Aristotle's *Prior Analytics I* (1987b, ch. 3), Buridan's *Summulae de Dialectica* (2001: 5.6), Kant (1787/1998: A74/B99-A76/B101), and Bolzano (1810/2004: 15-16). For the synthetic apriori determination of which possible predicates ought to be applied to objects see Kant (op. cit: A53/B77-A57/B81; A571/B599-A574/B602). Anticipating Kripke (1980: 56), Husserl (1929/1999: §6) refers, in a section heading and the discussion therein, to transcendental logic as pertaining to conditions on the 'contingent apriori'. Martha Kneale also anticipates Kripke's notion of aposteriori or metaphysical necessity in her 1938 paper, "Logical and Metaphysical Necessity". [See Leech (2019) for further discussion.]

For a discussion of epistemic logic, see Lewis (1932); von Wright (1951); and Hintikka (1962).

For the role of possibilities in accounting for the nature of subjective probability, i.e., partial belief, see Bernoulli (1713/2006: 211), Wittgenstein (op. cit.: 4.464, 5.15-5.152), and Carnap (1945). Bernoulli (op. cit.) writes: 'Something is possible if it has even a very small part of certainty, impossible if it has none or infinitely little. Thus something that has 1/20 or 1/30 of certainty is possible'. For subjective interpretations of probability, see Pascal (1654/1959), Laplace (1774/1986), Boole (1854), Ramsey (1926/1960), de Finetti (1937/1964), and Koopman (1940). For the history of the development of the theory of subjective probability, see Daston (1988; 1994) and Joyce (2011).

not my study of sign-language correspond to the study of thought processes which philosophers held to be so essential to the philosophy of logic? Only they got entangled for the most part in unessential psychological investigations, and there is an analogous danger for my method' (op. cit.).[6] Despite Wittgenstein's reluctance to accept the bearing of cognitive psychology on thought, chapters 2 and 3 endeavor to argue that epistemic modality and hyperintensionality comprise a materially adequate fragment of the language of thought, i.e., the computational structure and semantic values of the mental representations countenanced in philosophy and cognitive science.

Modal analyses of the notions of apriority and of states of information broadly construed are further proffered in Russell (1919), Lewis (1923), and Peirce (1933).

Russell (1919: 345-346) contrasts the possible truth-value of a propositional function given an assignment of values to the variables therein with an epistemic – what he refers to as the 'ordinary' – interpretation of the modal according to which 'when you say of a proposition that it is possible, you mean something like this: first of all it is implied that you do not know whether it is true or false, and I think it is implied; secondly, that it is one of a class of propositions, some of which are known to be true. When I say, e.g., 'It is possible that it may rain to-morrow' ... We mean partly that we do not know whether it will rain or whether it will not, but also that we do know that that is the sort of proposition that is quite apt to be true, that it is a value of a propositional function of which we know some value to be true' (op. cit.: 346). Russell (1905: 30-31) writes: 'In the first place, since the discussion belongs to symbolic logic, which already possesses technical names for the ideas we require, it is desirable to compare Dr. Hobson's terms with those in current use. What he calls a *norm* is what I call a *propositional function*. A *propositional function of x* is any expression $\phi!x$ whose value, for every value of x, is a proposition; such is "x is a man" or "$\sin x = 1$." Similarly we write $\phi!(x, y)$ for a propositional function of two variables; and so on. / In this paper I shall use the words *norm*, *property*, and *propositional function* as synonyms. / The word *aggregate* is used sometimes with an implication of order, sometimes without; I shall use *class* where there is no implication of order, and where there is order I shall consider the *relation* of *before* and *after* which generates the order. This last is necessary because

---

[6]The remarks are anticipated in Wittgenstein [1979: 21/10/14, 5/11/14, 10/11/14, 12/11/14 (pp. 16, 24-29)].

every class which can be ordered at all can be ordered in many ways; so that only the ordering relation, not the class, determines what the order is to be. A *relation* will be used in an extensional sense, *i.e.*, so that two relations are identical provided each holds whenever the other holds. We shall find that a propositional function $\phi!x$ may be perfectly definite, in the sense that, for every value of x, $\phi!x$ is determinably true or determinably false, while yet the values of x for which $\phi!x$ is true do not form a class. And, similarly, we shall find that a propositional function $\phi!x$ may be in the same sense definite, without there being any relation R which holds between x and y when and only when $\phi!(x, y)$ is true'. Quine (1967: 151) writes: 'From his hierarchy of types of propositions Russell derives a hierarchy of types of propositional functions. He speaks here of substitution, in a way that suggests that his functions also are notational in character; they seem simply to be open sentences, sentences with free variables. Still, he assigns them types and lets them be values of quantified variables. Insofar, they should be viewed not as open sentences, but as *attributes*, or, when they are functions of two or more arguments, *relations*. Failure to distinguish thus between open sentences on the one hand and attributes and relations on the other had grave consequences for this paper[, "Mathematical Logic as based on the Theory of Types", - D.E.] and equally for *Principia mathematica*, for which this paper sets the style'. See Russell (1908a: §II). For a construction of propositional functions which takes them to correspond to intensions i.e. functions from possible worlds to extensions, see Mares (2019: §9).

Lewis (1923.: 172) defines the apriority of the laws of mathematical languages as consisting in their being 'true in all possible worlds'.

Peirce (1933: §65) writes:

'[L]et me say that I use the word information to mean a state of knowledge, which may range from total ignorance of everything except the meanings of words up to omniscience; and by informational I mean relative to such a state of knowledge. Thus, by "informationally possible," I mean possible so far as we, or the persons considered, know. Then, the informationally possible is that which in a given information is not perfectly known not to be true. The informationally necessary is that which is perfectly known to be true. The informationally contingent, which in the given information remains uncertain, that is, at once possible and unnecessary'.

The notion of epistemic modality was, finally, stipulated independently by Moore (c.1941-1942/1962) in his commonplace book. According to Moore, 'epistemic' possibilities include that 'It's possible that [for some individual, a: a] is [glad] right now [iff] [a] may be [glad]', where 'I know that he's not' contradicts' the foregoing sentence (op. cit.: 187). Another instance of an epistemic possibility is advanced – 'It's possible that I'm not sitting down right now' – and analyzed as: 'It's not certain that I am' or 'I don't know that I am' (184).

1.1.2 The Target Conception of Epistemic Modality and Hyperintensionality

The conception of epistemically possible worlds which I will avail of throughout the course of this book is as follows.

Epistemically possible worlds or scenarios can be thought of, following Chalmers, as 'maximally specific ways things might be' (Chalmers, 2011: 60).

One can define epistemic possibility as a for all one knows operator following, informally, Chalmers (op. cit.)[7] and, formally, MacFarlane (2011: 164). Chalmers refers to this type of epistemic possibility as 'strict' epistemic possibility (2011: 62). Following MacFarlane, FAK(Φ) (read: for all I know, Φ), relative to an agent α and time τ is true at $\langle c, w, i, a \rangle$ iff Φ is true at $\langle c, w', i', a \rangle$, where c is a context, w is a possible world, i is an information state comprising a set of worlds, a is an assignment function, i' is the set of worlds not excluded by what is known by the extension of α at $\langle c, w, i, a \rangle$ at w and the time denoted by τ at $\langle c, w, i, a \rangle$, and w' is some world in i' (op. cit.). MacFarlane writes: '[A] speaker considering ⌜FAK$_{now}^{I}$: Φ⌝ and ⌜Might: Φ⌝ from a particular context c should hold that an occurrence of either at c would have the same truth value. This vindicates the intuition that it is correct to say "It is possible that p" just when what one knows does not exclude p' (2011: 167).

Przyjemski (2017) endorses a conception of epistemic possibility according to which it satisfies the condition of being 'strong' , i.e. a 'proposition p is epistemically possible only if it is supported by (non-overridden) evidence' (190). 'Strong epistemic possibility' contrasts to 'weak epistemic possibility' according to which a 'proposition p is epistemically possible only if p is compatible with the relevant body of evidence' (op. cit.). Weak epistemic possibility is the limiting case of strong epistemic possibility (op. cit.). I

---

[7]'We normally say that is *epistemically possible* for a subject that p, when it might be that p for all the subject knows' (60).

examine the relation between epistemic possibility and evidence in chapter **6**.[8]

A fourth approach to epistemic possibility defines the notion in relation to logical reasoning (Jago, 2009; Bjerring, 2012). Bjerring writes: '[W]e can now spell out deep epistemic necessity and possibility by appeal to provability in n steps of logical reasoning using the rules in R. To that end, let a proof of A in n steps of logical reasoning be a derivation of A from a set Γ of sentences – potentially the empty set – consisting of at most n applications of the rules in R. Let a disproof of A in n steps of logical reasoning be a derivation of ¬A from A – or from the set Γ of sentences such that A∈Γ – consisting of at most n applications of the rules in R. Similarly, let a set Γ of sentences be disprovable in n steps of logical reasoning whenever there is a derivation of A and ¬A from Γ consisting of at most n applications of the rules in R. For simplicity, I will assume that agents can rule out sets of sentence [sic] that contain {A, ¬A} non-inferentially. Finally, let '$\Box_n$' and '$\Diamond_n$' be metalinguistic operators, where '$\Diamond_n$' is defined as ¬$\Box_n$¬. Read '$\Box_n$' as 'A is provable in n steps of logical reasoning using the rules in R', and read '$\Diamond_n$' as 'A is not disprovable in n steps of logical reasoning using the rules in R'. We can then define:

(Deep-Nec$_n$) A sentence A is deeply$_n$ epistemically necessary iff $\Box_n$.

(Deep-Pos$_n$) A sentence A is deeply$_n$ epistemically possible iff $\Diamond_n$' (op. cit.).

In Chapter **8**, I mention the identification of epistemic possibility with consistent logical reasoning. I examine, however, the bearing of hyperintensional apriority, where hyperintensional apriority is interpreted as an epistemically necessary truthmaker, on absolute decidability (see §**2.2.1**). In the remaining chapters of this book, I define epistemic possibility in this distinct, fifth manner. The fifth way to understand epistemic possibility is via apriority, such that ϕ is epistemically possible iff ϕ is primary conceivable, where primary conceivability ($\Diamond$) is the dual of apriority (¬$\Box$¬, i.e. not apriori ruled out).[9] Chalmers (2002) distinguishes between primary and secondary conceivability. Secondary conceivability is counterfactual, so rejecting the

---

[8] See Fine (2023), for further discussion of the relation between evidence and truthmaker semantics for static epistemic logic.

[9] The understanding of epistemic possibility tied to knowledge, as in epistemic logic, is examined in chapter **5**. Kripke (1980: 34-39) discusses the definition of apriority as knowledge independent of experience, and argues that '[i]t's not trivial' whether apriori truths are of necessary truths known independently of experience (39).

metaphysical necessity of the identity between Hesperus and Phosphorus is not secondary conceivable. Primary conceivability targets epistemically possible worlds considered as actual rather than counterfactual worlds. Chalmers also distinguishes between positive and negative conceivability and prima facie and ideal conceivability. A scenario is positively conceivable when it can be imagined with perceptual detail. A scenario is negatively conceivable when nothing rules it out apriori, as above. A scenario is prima facie conceivable when it is conceivable 'on first appearances'. E.g. a formula might be prima facie conceivable if it does not lead to contradiction after a finite amount of reasoning. A scenario is ideally conceivable if it is prima facie conceivable with a justification that cannot be defeated by subsequent reasoning (op. cit.).

Chalmers distinguishes between deep and strict epistemic possibilities. He writes: '[W]e might say that the notion of *strict epistemic possibility* – ways things might be, for all we know – is undergirded by a notion of *deep epistemic possibility* – ways things might be, prior to what anyone knows. Unlike strict epistemic possibility, deep epistemic possibility does not depend on a particular state of knowledge, and is not obviously relative to a subject' (2011: 62). About deep epistemic necessity, he writes: 'For example, a sentence s is deeply epistemically possible when the thought that s expresses cannot be ruled out a priori / This idealized notion of apriority abstracts away from contingent limitations' (66). All references to epistemic possibility in what follows will be to Chalmers' notion of deep epistemic possibility.

Chalmers' conception of deep epistemic necessity is similar to the proposal in Edgington (2004: 6) according to which 'a priori knowledge is independent of the state of information of the subject'. While being states of information, epistemic states are yet parts of deeply epistemically possible worlds, because they are not relativized to the contingent knowledge bases of particular epistemic agents.

Rossi and Özgün (2023) countenance epistemic possibility hyperintensionally, by defining it as a strict epistemic possibility operator, i.e. a for all one knows operator, interpreted analogously to positive instead of negative conceivability (2, 6-7). Hyperintensionality is secured via topic-sensitivity, and supposed to entail a condition on agent non-ideality which they refer to as satisfying the property of 'epistemic reach'. They write: 'the boundaries of S's epistemic reach are determined by their cognitive, computational, or conceptual limitations' (6). The significance of the hyperintensionality condition is, similarly to this book, that it is supposed to circumvent the problem of omniscience based on intensionalism about propositions. Hyperintensional knowledge is defined as being satisfied by a 'model-theoretic condition' (3), MOD, truth in all worlds , and a 'hyperintensionality condition', HYPE, i.e. 'grasping ϕ's topic', i.e. 'a total function defined from the object language of the underlying logic to the set $\{0, 1\}$' (4). Thus,

KNOW(ϕ) = 1 iff MOD(ϕ) =1 and HYPE(ϕ) = 1 (4).

Hyperintensional positive epistemic possibility is defined thus,

POSS(ϕ) = 1 iff MOD$(\neg\phi)$ = 0 and HYPE(ϕ) = 1 (8).

Rossi and Özgün apply the foregoing to Stalnaker (2006)'s conception of strong or full belief as subjective certainty, where 'believing implies believing that one knows': B$\phi \to$ BKϕ; or, B$\phi \iff \langle K\rangle K\phi$ and B$\phi \iff \neg K\neg K\phi$ (Rossi and Özgün, op. cit.: 10; Stalnaker, 2006: 179). $\neg K\neg$ and $\langle K \rangle$ are not, however, equivalent, because $\neg K\neg$ is interpreted as negative epistemic possibility and $\langle K \rangle$ is interpreted as positive epistemic possibility (Rossi and Özgün, op. cit.: 8). B$\phi \iff \neg K\neg K\phi$ entails: BEL$_{Stal}(\phi)$ = 1 iff MOD$(\neg K\phi)$ = 0 and HYPE$(\neg K\phi)$ = 0 (11). Construed as a positive operator because a condition on the operator is grasp of topic (8, 17), when B$\phi \iff \langle K\rangle K\phi$: BEL*$_{Stal}(\phi)$ = 1 [i.e., POSS(Kϕ) = 1] iff MOD$(\neg K\phi)$ = 0 and HYPE(Kϕ) = 1(Rossi and Özgün, op. cit.: 11-12). 'B$\overline{\phi}$, [...] K$\overline{\phi}$ and $\langle K\rangle\overline{\phi}$ express that "the agent has grasped the topic of ϕ"'(15).*$\in$\{K, B\} (16). $\Box$ is an analyticity or apriority modality (14). Strong negative introspection, $\vdash \neg B\phi \to K\neg B\phi$, is invalid in Rossi and Özgün's logic.

An axiom of Rossi and Özgün's logic is a restricted closure axiom, $[\Box(\phi \to \psi) \land *\phi \land *\overline{\psi}] \to *\psi$, interpreted as 'the agent knows/believes a priori consequences of what they know/believe as long as they grasp the topics of these consequences' (op. cit.). One issue with Rossi and Özgün's restricted closure axiom is that grasping the topic of the antecedent might, too, be necessary, for grasping the topic of the consequent, instead of only grasp of the topic of the consequent, $\overline{\psi}$. One might want one's account of logic and semantics to satisfy what Berto (2022: 25) refers to as 'Yablo's Thesis' or 'Parry Implication'. (See Parry, 1968, 1989; Yablo, 2014). Yablo's Thesis states that 'B is part of A iff the inference from A to B is (i) truth-preserving – A implies B (ii) aboutness-preserving – A's subject matter includes that of B' (Yablo, 2014: 15). Intuitively: 'Content-inclusion is implication plus subject-matter inclusion' (op. cit.). The left-to-right direction of Yablo's Thesis is referred to as Weak Yablo's Thesis (Berto, 2022: 25). The biconditional is referred to as Full Yablo's Thesis, and 'gives an account of same-saying as two-way containment: ϕ and ψ say the same ... just in case they are both mutually entailing and topic-equivalent' (25-26). Yablo refers to a type of closure according to which '[s]ome conclusions are such that you should *already* know them, to know the premise' as 'immanent closure' or 'topical closure' (116-117). 'Transeunt' closure occurs when 'you are assured of knowing the conclusion only if you engage in some reasoning' (116) and 'knowledge of conclusions [is] drawn from premises that do not contain them' (127). Immanent closure states that: 'If S knows that P, and Q is part of P, then S knows that Q' (117).

Berto addresses the problem of possible mathematical and logical omniscience entailed by Full Yablo's Thesis and immanent closure by separating topic-inclusion from its metalinguistic and epistemic dimensions. Thus, the topic of (i) '16 + 16 = 32', $\{16, +, =, 32\}$, is part of the topic of (ii) '16 + 32 = 48', $\{16, +, 32, =, 48\}$, yet Full Yablo's Thesis and immanent closure would entail that metalinguistic knowledge of (i) entails metalinguistic knowledge of (ii), which is false (Berto, 2022: 55). Berto argues that the maneuver does not entail, however, that all mathematical knowledge is metalinguistic knowledge (56).

Another maneuver would be to reject immanent closure, because of an objection from Alexandru Baltag (56): from $\phi \to \psi$, one can infer $[\phi \wedge (\psi \vee \neg \psi)] \to \psi$, yet, taking ϕ to be a mathematical theory and ψ to be a theorem of the theory, knowledge that a theorem in the consequent is true and grasping its topic ought not to be entailed by the disjunction in the antecedent with regard to whether the theorem is true (57).

Another maneuver might be to add logically impossible worlds to one's ontology rather than taking topic-inclusion to have the property of semantic necessity. Berto suggests that denying that the axioms of a mathematical theory entail a theorem in the theory might require incorporating logically impossible worlds into one's ontology, 'where entailment laws are violated' (58).

One issue with Rossi and Özgün's restricted closure axiom is that it satisfies Full Yablo's Thesis and immanent closure, without addressing the separability of the metalinguistic and epistemic dimensions from the topic-inclusion in Full Yablo's Thesis and immanent closure, or Baltag's objection and the possible addition of logically impossible worlds to one's ontology in order to countenance the invalidity of a disjunct being entailed by a disjunction without an instance of either disjunct.

In order for grasp of the topic of the antecedent to entail grasp of the topic of the consequent, a third type of topic-sensitive closure, distinct from immanent closure and transeunt closure, might too be required.[10]

Chalmers defines epistemic possibility as (i) not being apriori ruled out (2011: 63, 66),[11] i.e. as the dual of epistemic necessity i.e. apriority (65),[12] and as (ii) being true at an epistemic scenario i.e. epistemically possible world (62, 64). He accepts a Plenitude principle according to which: 'A thought T is epistemically possible iff there exists a scenario S such that S verifies T' (64). Chalmers advances both epistemic and metaphysical constructions of epistemic scenarios. Chalmers (private correspondence) writes: '[T]he definition of epistemic possibility does not depend on either construction'. In the metaphysical construction of epistemic scenarios, epistemic scenarios are centered metaphysically possible worlds (69). Canonical descriptions of epistemic scenarios i.e. epistemically possible worlds on the metaphysical construction are required to be specified using only 'semantically neutral' vocabulary, which is 'non-twin-earthable' by having the same extensions when worlds are considered as actual or counterfactual (Chalmers, 2006: §3.5). In the epistemic construction of epistemic scenarios, an epistemic scenario consists in a set of sentence types comprising an infinitary ideal language, M, with vocabulary restricted to epistemically invariant expressions (Chalmers, 2011: 75). He defines epistemically invariant expressions thus: '[W]hen s is epistemically invariant, then if some possible competent utterance of s is epistemically necessary, all possible competent utterances of s are epistemically necessary' (op. cit.). The sentence types in the infinitary language must also be epistemically complete. A sentence s is epistemically complete if s is epistemically possible and there is no distinct sentence t such that both $s \wedge t$ and $s \wedge \neg t$ are epistemically possible (76). The epistemic construction of epistemic scenarios transforms the Plenitude principle into an Epistemic Plenitude principle according to which: 'For all sentence tokens s, if s is epistemically possible, then some epistemically complete sentence of [M] implies s' (op. cit.).[13] The value of a sentence in epistemically constructed epistemic scenarios determines the

---

[10] See Hawke and Özgün (2023) and Fine (2023), for preliminary discussion of this point.

[11] 'One might also adopt a conception on which every proposition that is not logically contradictory is deeply epistemically possible, or on which every proposition that is not ruled out a priori is deeply epistemically possible. In this paper, I will mainly work with the latter understanding' (63).

[12] 'We can say that s is deeply epistemically necessary when s is a priori: that is when s expresses actual or potential a priori knowledge' (65).

[13] Chalmers (private correspondence) writes: '[E]pistemic possibility isn't defined in terms of epistemic completeness (the epistemic plenitude principle doesn't define epistemic possibility) but rather in terms of apriority, so circularity is avoided'.

value of the sentence in metaphysically possible worlds when a super-rigidity condition is satisfied (Chalmers, 2012: 239, 468, 474). Chalmers writes: 'I accept Apriority/Necessity and Super-Rigid Scrutability. (Relatives of these theses play crucial roles in "The Two-Dimensional Argument against Materialism")' (241). The Apriority/Necessity Thesis is defined as the 'thesis that if a sentence S contains only super-rigid expressions, S is a priori iff S is [metaphysically] necessary' (468), and Super-Rigid Scrutability is defined as the 'thesis that all truths are scrutable from super-rigid truths and indexical truths' (474).

I will assume the epistemic construction of epistemic scenarios in this book. I concur, as well, that epistemic possibility is the dual of epistemic necessity i.e. apriority, but argue for an epistemic two-dimensional truthmaker semantics which avails of hyperintensional epistemic states, i.e. epistemic truthmakers or verifiers for a proposition, which comprise a state space (see chapter **4**). Epistemic states are parts of epistemically possible worlds, rather than whole worlds themselves. Apriority is thus redefined in the hyperintensional semantics (see chapter **2**). The epistemic two-dimensional truthmaker semantics is motivated by its capacity (i) to model conceivability arguments involving hyperintensional metaphysics (see chapters **6-7**), and (ii) to avoid the problem of mathematical omniscience entrained by intensionalism about propositions (see chapters **2, 8-11**).

1.1.3 The Literature

In the contemporary literature, there is a paucity of works devoted to the nature of epistemic modality and its relation to other modalities. Recent books and edited volumes which examine aspects of epistemic modality include Gendler and Hawthorne (2002); Yablo (2008); Gendler (2010); Egan and Weatherson (2011); Chalmers (2012); and Berto (2022). The present work is focused on the foundations and philosophical significance of the epistemic interpretation of modal logic and semantics. For the sake of completeness, a critical summary of the relevant literature is thus included below.

The Gendler and Hawthorne volume includes seminal contributions to the theory of the relationship between epistemic and metaphysical modality. By contrast, this book provides foundations for the nature of epistemic modality, when the modality concerns apriority and conceivability, as well as the logic of knowledge and belief; makes contributions to our understanding of the ontology of consciousness, by regimenting the ontology of consciousness using hyperintensional grounding operators; examines the nature and philosophical extensions of epistemic logic; and examines the relations between epistemic modality and the variety of other modalities (e.g., metaphysical and mathematical modalities and the types of intention in the setting of evidential decision theory).

The papers on modal epistemology in Yablo (2008) predominantly concern the relation between epistemic and metaphysical modalities, and, in particular, non-trivial conditions on error theory in modal epistemology.[14] Issues for the epistemic interpretation of two-dimensional intensional semantics are examined; e.g., the conditions on ascertaining when an epistemic possibility is actual, and a dissociation in the case of recognitional concepts, e.g. concepts of geometric shapes, between the conceptual necessity of their definitions and the apriority of their definitions. The discussion is similar, in scope, to the discussions in the Gendler and Hawthorne volume. This book

---

[14]Kripke (1980) accepts an error theory for modal epistemology. He writes: 'What, then, does the intuition that the table might have turned out to have been made of ice or of anything else, that it might even have turned out not to be made of molecules, amount to? I think that it means simply that there might have been *a table* looking and feeling just like this one and placed in this very position in the room, which was in fact made of ice. In other words, I (or some conscious being) could have been *qualitatively in the same epistemic situation* that in fact obtains, I could have the same sensory evidence that I in fact have, about *a table* which was made of ice. The situation is thus akin to the one which inspired the counterpart theorists; when I speak of the possibility of the table turning out to be made of various things, I am speaking loosely. *This* table itself could not have had an origin different from the one it in fact had, but in a situation qualitatively identical to this one with respect to all the evidence I had in advance, the room could have contained *a table made of ice* in place of this one. Something like counterpart theory is thus applicable to the situation, but it applies only because we are *not* interested in what might have been true of *this particular* table, but in what might or might not be true of *a table* given certain evidence. It is precisely because it is not true that this table might have been made of ice from the Thames that we must turn here to qualitative descriptions and counterparts' (142). 'Since we are concerned with how things might have turned out otherwise, our general paradigm is to redescribe both the prior evidence and the statement qualitatively and claim that they are only contingently related. In the case of identities, using two rigid designators, such as the Hesperus-Phosphorus case above, there is a simpler paradigm which is often usable to at least approximately the same effect. Let "R_1" and "R_2" be the two rigid designators which flank the identity sign. Then "$R_1 = R_2$" is necessary if true. The references of "R_1" and "R_2", respectively, may well be fixed by nonrigid designators "D_1" and "D_2" [...] in the Hesperus and Phosphorus cases these have the form "the heavenly body in such-and-such position in the sky in the evening (morning)". Then although "$R_1 = R_2$" is necessary, "$D_1 = D_2$" may well be contingent, and this is often what leads to the erroneous view that "$R_1 = R_2$" might have turned out otherwise' (143-144).

aims to redress the limits mentioned in the foregoing, and to proffer the positive proposals delineated above. Yablo (2014) countenances thick or directed propositions, which combine a possible worlds semantics with subject matters (p. 21, fn. 28; p. 49). Yablo writes: 'The subject matter of [a sentence,] $S =$ the [similarity, i.e. reflexive and symmetric] relation m such that worlds are m-dissimilar iff S is differently true at them' (41; 36). 'Where the identity of a set is given by its members, the identity of a subject matter m is given by the pattern of trans-world changes where m is concerned: (4) $m_1 = m_2$ iff worlds differing where the first is concerned differ also with respect to the second, and vice versa. Shouldn't we know, to grasp a subject matter m, the proposition m(w) that specifies how matters stand in w where m is concerned? / But, subject matters as just explained *do* tell us what w is like where m is concerned. The proposition we're looking for is meant to be true in all and only worlds in the same m-condition as w; on an intensional view of propositions, it *is* the set of worlds in the same m-condition as w. That proposition is already in our possession. To be in the same m-condition as w is to be m-[similar] to w, and the set of worlds m-[similar] to w is just w's cell in the partition. A world's m-cell is thus the proposition saying how matters stand in it m-wise' (27). 'A subject matter, whatever else it may do, determines a function from worlds w to propositions stating how matters stand in w where it – the subject matter in question – is concerned. **The number of stars**, to use Lewis's example, maps worlds with equally many stars to the proposition that there are that many stars. Thinking of propositions as sets of worlds, we're talking about a function from worlds w to collections of sets of worlds – where, since the ways a world is **m**-wise are propositions true in that world, each set in the collection has w as a member. **m** will in the simplest case be a partition of logical space, with each world being mapped to its cell in the partition' (Yablo, 2015). 'Each specification function m(. . .) has associated with it a set of propositions, expressing between them the various ways matters can stand where m is concerned. (A proposition goes into the set if it is m(w) for some world w.) The operation is again reversible: to find m(w), look for the proposition to which w belongs' (Yablo, 2014: 28). Berto (2022: ch. 2) refers to directed propositions as two-component (2C) propositional contents and develops a semantics for 2C contents which combines possible worlds with atomic topics. The semantics developed in this book makes propositions multi-hyperintensional, by combining atomic topic-, i.e. subject matter, sensitivity, with a second aspect of their subject matter being captured via epistemic and metaphysical truthmakers (see chs.

2, **4**; and Fine, 2017a,b, for truthmaker semantics), a third aspect of their hyperintensionality captured via subject matter similarity relations between truthmakers, and a fourth aspect of their hyperintensionality captured by structured hyperintensions.

The Egan and Weatherson volume is comprised of papers which predominantly analyze epistemic modals in the setting of natural language semantics. Four papers in the volume target epistemic possibilities as imaginable or conceptual possibilities; those by Chalmers ("The Nature of Epistemic Space"), Jackson ("Possibilities for Representation and Credence"), MacFarlane ("Epistemic Modals are Assessment-Sensitive"), and Yalcin ("Nonfactualism about Epistemic Modality").[15]

Chalmers' paper examines some principles governing epistemic space and its interaction with metaphysical modality, as well as Kaplan's paradox. This book endeavors to account for the distinct conditions on formal and informal domains in epistemic space (see chapter **6**); examines the interaction between epistemic modality and metaphysical modality, as well as various other types of modality; and examines the role that epistemic modality plays in resolving the alethic paradoxes, as well as undecidable sentences in the philosophy of mathematics.

Jackson's paper argues that conceptual possibilities and metaphysical possibilities ought to be defined within a single space, in order both to avoid cases in which a sentence is conceptually possible although metaphysically impossible and to secure the representational adequacy of conceptually possible terms. Chapter **4** adopts, by contrast, the modal dualist proposal to the effect that epistemic modality and metaphysical modality, as well as epistemic and metaphysical states, occupy distinct spaces.

Yalcin's paper argues that epistemic modal sentences in natural language semantics mirror the structure of the beliefs of speakers. Epistemic mental states are taken, then, to be expressive rather than representational, because the communication of epistemic modal and interrogative updates on an informational background shared by speakers is not truth-conditional. The present approach contrasts to the foregoing, by not taking the values of expressions in natural language semantics to be a guide to the nature of mental states (see Evans, 1982). We here take epistemic possibility to concern conceivability and epistemic necessity to concern apriority, as well as

---

[15]The semantics for the 'for all I know' interpretation of epistemic possibility in MacFarlane's paper was discussed above.

taking the box operator to be interpreted so as to concern knowledge and belief as in epistemic and doxastic logic, as well. Chapter **2** discusses Hawke and Steinert-Threlkeld (2021)'s expressivist semantics for epistemic modals because it converges with the metaphysical expressivism about epistemic modality there adumbrated.

Yalcin (2011; 2016) countenances the notion of epistemic possibility as strict, that is, a for all one knows operator, and adds a condition of sensitivity to subject matter. Subject matters partition logical space into the propositions which answer or resolve an interrogative concerning the subject matter (2011: Section 6).[16] The propositions which resolve the interrogative are said to be 'visible' (318). Doxastic states are partial functions from interrogatives to resolutions (op. cit.). A 'view' is the set of possibilities in the resolution of logical space (op. cit.). The true propositions in the resolution are said to comprise the agent's 'commitments concerning the subject matter' (op. cit.). A proposition is compatible with an agent's view only if it is true at one of the worlds in the view (319). Then, 'to believe that a proposition is possible, or might be, is for the proposition to be compatible with one's view, and moreover for it to be an answer to a question one is sensitive to' (320). Semantic values for epistemic modals are defined relative to worlds and parameters for 'non-factual' states of information (324, 329).

Bueno (2017) argues that the epistemology of modality is similar to the epistemology of mathematics, and endorses a modalist conception of possibility and necessity. Modalism rejects the existence of possible worlds [see Prior (1968); Plantinga (1976); and Bueno and Shalkowski, 2015]. Bueno writes: 'If ... the goal is to understand, to figure out the modal constraints on various philosophical claims about the world, then the exploration of the relevant possibilities (and impossibilities) is clearly relevant, albeit no categorical modal knowledge of the extraordinary possibilities in question is forthcoming. On this modalist account, we may have a lot of ordinary modal knowledge (that is, knowledge of ordinary possibilities and necessities), but less of extraordinary categorical modal knowledge (that is, knowledge of philosophical possibilities and necessities independently of any assumptions), although we may have some conditional extraordinary modal knowledge, by taking notice of the relevant philosophical assumptions, determining their consequences,

---

[16]The view that subject matters, broadly construed, have the form of an interrogative update on a set of worlds is anticipated by Hamblin (1958, 1973); Lewis (1988/1998); and further defended by Yalcin (2008, 2016) and Yablo (2014). For further discussion of subject matters, see chapter **4**.

and resisting the temptation to discharge the assumptions. Interestingly, this conditional modal knowledge can be obtained without much commitment, since only the logical relations between assumptions and results are highlighted rather than any claim about the correctness of the assumptions in question' (Bueno, 2017: 81).

Another development which is worth mentioning is Holliday and Mandelkern (2024)'s orthologic and possibility semantics for epistemic modals, which is non-classical by rejecting the laws of distributivity, disjunctive syllogism, and orthomodularity, while negation is defined as orthocomplementation rather than psuedocomplentation such that the inference from 'p $\wedge$ $\Diamond \neg$p $\vdash \bot$' to '$\Diamond \neg$p $\vdash \neg$p' does not hold. Possibility semantics rejects a primeness condition according to which a world x makes disjunction true iff it makes the disjuncts true. Rather, in possibility semantics, x makes a disjunction true just in case for every refinement x' $\sqsubseteq$ x, there is a further refinement x" $\sqsubseteq$ x' which makes one of the disjuncts true (see Holliday, 2021, for further discussion).

Chalmers (2012) provides a book-length examination of the scrutability of truth, and the apriori entailment relations between different types of truths. The rigidity of intensions is availed of, in order to explain the relation between epistemic modality and metaphysical modality. The relation between epistemic modality and metaphysical modality is examined in Part II of this book, but I aim to examine novel philosophical extensions of epistemic two-dimensional semantics and the role of epistemic modality and hyperintensionality in the philosophy of mathematics and logic.

Gendler (2010) is a rare, empirically informed study of the limits of representational capacities, when they target counterfactual assignments of values to variables in thought experiments – e.g., the conditions under which there might be resistance attending the states of imagining that fictional characters have variant value-theoretic properties – and when implicit biases and unconscious sub-doxastic states affect the veridicality conditions of one's beliefs. One crucial distinction between Gendler's approach and the one pursued in this chapter, however, is that the former does not examine the interaction between epistemic modal and hyperintensional semantics and epistemic logic.

In the literature on modal epistemology, Hale (2013a) argues that modal knowledge ought to be pursued via the epistemology of essential definitions which specify conditions on sortal membership. Apriori knowledge of essence is explained in virtue of knowledge of the purely general terms – embedding no singular terms – which figure in the definitions. Thus – by being purely general – the essential properties of objects, and thus the objects, have necessary being. Aposteriori knowledge of essential definitions can be pursued via theoretical identity statements, yet, because the terms figuring therein are not purely general, both the essential properties and the objects which satisfy the properties have contingent being. As mentioned, the book redefines the extant proposals in the ontology of consciousness using hyperintensional grounding operators. The ground-theoretic interpretation of the ontology of consciousness, and an examination of the bearing of the latter for the relation between (hyperintensional) conceivability and metaphysical possibility and hyperintensionality is, as noted, examined in chapter **7**. Hale's higher-order and first-order Necessitist proposals are examined in further detail, in chapter **9**.

Nichols (2006) features three essays on modal epistemology. Nichols' "Imaginative Blocks and Impossibility" examines introspection-based tasks in developmental psychology, in order to account for the interaction between imaginative exercises and counterfactual judgments. Hill's "Modality, Modal Epistemology, and the Metaphysics of Consciousness" examines the interaction between conceptual and metaphysical possibility, where conceptual possibilities are construed as Fregean thoughts, and the relation between conceivability and metaphysical possibility is then analyzed as the relation between Fregean thoughts (augmented by satisfaction-conditions such as conceptual coherence) and empirical propositions. Sorensen's paper, "Meta-conceivability and Thought Experiments", argues that meta-conceivable thought experiments are distinct from both conscious perceptual states and conceivable possibilities. Sorensen (1999) argues that (thought) experiments track the consequences of reassignments of values to variables.[17] My approach differs from Hill's by arguing in favor of both a possible worlds semantics as well as a hyperintensional, epistemic two-dimensional truthmaker semantics for thoughts, which is able to recover the virtues attending the Fregean model, as well as in accounting for the relations between epistemic states and various other interpretations of states and their hyperintensionality, including the mathematical interpretation (see chapters **8-12** for further discussion). My approach is similar in methodology to Nichols', although I endeavor to account to for the relation between (hyperintensional) conceivability and metaphysical possibility and hyperintensionality by availing of the epistemic interpretation of two-dimensional semantics. Finally, my approach is simi-

---

[17]See Gendler (2000), for further discussion.

lar to Sorensen's, in targeting both a formal semantic analysis of epistemic and related modality and hyperintensionality, as well as of the operators of knowledge and belief in the setting of epistemic logic.

Waxman (ms$_1$) endeavors to account for the interaction between the imagination and mathematics. Whereas I avail of conceivability as defined in epistemic two-dimensional semantics in chapters **8-12** – which I refer to in the mathematical setting as epistemic mathematical modality and hyperintensionality – in order to account for how the epistemic possibility and hyperintensionality of abstraction principles and large cardinal axioms relates to their metaphysical possibility and hyperintensionality, Waxman's aim is to account for how imagining a model of a mathematical theory entrains justification to believe its consistency (op. cit.). Unlike Waxman, epistemic mathematical modality and hyperintensionality are ideal, whereas imagination is, on his account, non-ideal (Waxman, op. cit.: 18; Chalmers, 2002), where ideal (hyperintensional) conceivability means true at the limit of apriori reflection unconstrained by finite limitations. Unlike Waxman, I believe, further, that imaginative contents are sensitive to hyperintensional subject-matters or topics (see chapters **2, 4**; Berto, 2018; Canavotto, Berto, and Giordani, 2022).

Finally, a class of views in the epistemology of modality can be characterized as being broadly empiricist. Stalnaker (2003) and Williamson (2007; 2013) refrain from countenancing the notion of epistemically possible worlds; and argue instead either that the imagination is identifiable with cognitive processes taking the form of counterfactual presupposition (Williamson, 2007); that one's choice of the axioms governing modal logic should satisfy abductive criteria on theory choice (Williamson, 2013a); or that metaphysical modalities are properties of the actual world (Stalnaker, op. cit.). Vetter (2013) argues for a reduction of modal notions to actual dispositional properties, and Roca-Royes (2016) and Schoonen (2020) pursue a corresponding modal empiricist approach, according to which knowledge of the *de re* possibilities of objects consists in the extrapolation of properties from acquaintance with objects in one's surround to formally similar objects, related by reflexivity and symmetry. Generally, according to the foregoing approaches, the method of modal epistemology proceeds by discerning the modal truths – captured, e.g., by abductively preferred theorems in modal logic; conditional propositions; and dispositional and counterfactual properties – and then working backward to the exigent incompleteness of an individual's epistemic states concerning such truths. By contrast, the approach advanced in this work both retains and provides explanatory foundations for epistemic modal and hyperintensional space, and augments the examination by empirical research and an abductive methodology.

The foregoing texts either examine epistemic modality via natural language semantics; restrict their examination to the interaction between conceivable possibilities and metaphysical possibilities; eschew epistemic possibilities; provide a naturalistic approach to the analysis of epistemic modality, without drawing on formal methods; or provide a formal analysis of epistemic modality, without drawing on empirical results.

The book endeavors, by contrast, to examine the interaction between epistemic modality and hyperintensionality and the computational theory of mind; metaphysical modality and hyperintensionality; the types of mathematical modality and hyperintensionality; the hyperintensional semantics and modal logic of rational intuition; and the types of intention, when the latter are interpreted as hyperintensional mental states.

The models developed here are of interest in their own right. However, this work is principally concerned with, and examines, their philosophical significance, as witnessed by the new distinctions and properties that they induce. Beyond conditions on theoretical creativity, both formal regimentation and empirical confirmation are the best methods available for truth-apt philosophical inquiry into both the space of epistemic modality and hyperintensionality and the multiple points of convergence between epistemically possible and hyperintensionally verified truth and the most general, fundamental structure of metaphysically possible worlds and states.

1.2 Chapter Summary

In chapter **2**, I provide a mathematically tractable background against which to model both modal and hyperintensional cognitivism and modal and hyperintensional expressivism. I argue that epistemic modal algebras, endowed with a hyperintensional, topic-sensitive epistemic two-dimensional truthmaker semantics, comprise a materially adequate fragment of the language of thought. I demonstrate, then, how modal expressivism can be regimented by modal coalgebraic automata, to which the above epistemic modal algebras are categorically dual. I examine five methods for modeling the dynamics of conceptual engineering for intensions and hyperintensions. I develop a novel topic-sensitive truthmaker semantics for dynamic epistemic logic, and develop a novel dynamic epistemic two-dimensional hyperintensional semantics. My novel dynamic two-dimensional semantics is grounded in two-dimensional hyperintensional Turing machines. I examine then the virtues unique to the modal expressivist approach here proffered in the setting of the foundations of mathematics, by contrast to competing approaches based upon both the inferentialist approach to concept-individuation and the codification of speech acts via intensional semantics.

In chapter **3**, I aim to vindicate the thesis that cognitive computational properties are abstract objects implemented in physical systems. I avail of Voevodsky's Univalence Axiom and function type equivalence in Homotopy Type Theory, in order to specify an abstraction principle for two-dimensional (hyper-)intensions. The homotopic abstraction principle for two-dimensional (hyper-)intensions provides an epistemic conduit for our knowledge of (hyper-)intensions as abstract objects. Higher observational type theory might be one way to make first-order abstraction principles defined via inference rules, although not higher-order abstraction principles, computable. The truth of my first-order abstraction principle for hyperintensions is grounded in its being possibly recursively enumerable i.e. Turing computable and the Turing machine being physically implementable. Epistemic modality and hyperintensionality can thus be shown to be both a compelling and a materially adequate candidate for the fundamental structure of mental representational states, comprising a fragment of the language of thought.

In chapter **4**, I endeavor to establish foundations for the interaction between hyperintensional semantics and two-dimensional indexing. I examine the significance of the semantics, by developing three, novel interpretations of the framework. The first interpretation provides a characterization of the distinction between fundamental and derivative truths. The second interpretation demonstrates how the elements of decision theory are definable within the semantics, and provides a novel account of the interaction between probability measures and hyperintensional grounds. The third interpretation concerns the contents of the types of intentional action, and the semantics is shown to resolve a puzzle concerning the role of intention in action. Two-dimensional truthmaker semantics can be interpreted epistemically and metasemantically, as well, and epistemic two-dimensional truthmaker semantics is examined in the chapter, as well as appealed to in chapters **7-12**.

In chapter **5**, I provide a novel account of iterated epistemic states. The essay argues that states of epistemic determinacy might be secured by countenancing iterated epistemic states on the model of fixed points in the modal

μ-calculus. Despite the epistemic indeterminacy witnessed by the invalidation of modal axiom 4 in the sorites paradox – i.e. the KK principle: $\Box\phi \to \Box\Box\phi$ – an epistemic hyperintensional μ-automaton permits fixed points to entrain a principled means by which to iterate epistemic states and account thereby for necessary conditions on self-knowledge. The epistemic hyperintensional μ-calculus is applied to the iteration of the epistemic states of a single agent instead of the common knowledge of a group of agents, and is thus a novel contribution to the literature.

In chapter **6**, I aim to redress the contention that epistemic possibility cannot be a guide to the principles of modal metaphysics. I introduce a novel epistemic two-dimensional truthmaker semantics. I argue that the interaction between the two-dimensional framework and the mereological parthood relation, which is super-rigid, enables epistemic possibilities and truthmakers with regard to parthood to be a guide to its metaphysical profile. I specify, further, a two-dimensional formula encoding the relation between the epistemic possibility and verification of essential properties obtaining and their metaphysical possibility or verification. I then generalize the approach to haecceitistic properties. I also examine the Julius Caesar problem as a test case. I conclude by addressing objections from the indeterminacy of ontological principles relative to the space of epistemic possibilities, and from the consistency of epistemic modal space.

In chapter **7**, I argue that Chalmers (1996; 2010)'s two-dimensional conceivability argument against the derivation of phenomenal truths from physical truths risks being obviated by a hyperintensional regimentation of the ontology of consciousness. The regimentation demonstrates how ontological dependencies between truths about consciousness and about physics cannot be witnessed by epistemic constraints, when the latter are recorded by the conceivability – i.e., the epistemic possibility – thereof. Generalizations and other aspects of the philosophical significance of the hyperintensional regimentation are further examined.

Chapters **8-12** provide cases demonstrating how the two-dimensional hyperintensions of epistemic two-dimensional semantics solve the access problem in the epistemology of mathematics. In his (1973), Benacerraf inquires into how the semantics for mathematics might interact with the theory of knowledge for mathematics. He raises the inquiry concerning how knowledge of acausal abstract objects such as those of mathematics (numbers, functions, and sets) is possible, assuming that the best theory of knowledge is that deployed in the empirical sciences and thus presupposes a condition of causal interaction. This is known in the literature in philosophy of mathematics as the access problem. Field (1989) generalizes Benacerraf's problem by no longer presupposing the condition of causal interaction, and inquiring into what might explain the reliability of mathematical beliefs. Clarke-Doane (2016) has argued that the Benacerraf-Field problem might no longer be thought to be pressing in light of mathematical beliefs satisfying conditions of safety and sensitivity. A belief is safe if it could not easily have been different. A belief is sensitive if, had the contents of the belief been false, we would not believe them (see Nozick, 1981: 172-176). Mathematical beliefs are thus sensitive, because mathematical truths are metaphysically necessary, true at all worlds. Clarke-Doane quotes David Lewis, who writes: '[I]f it is a necessary truth that so-and-so, then believing that so-and-so is an infallible method of being right. If what I believe is a necessary truth, then there is no possibility of being wrong. That is so whatever the subject matter [...] and no matter how it came to be believed' (1986: 114-115). Mathematical beliefs are safe, because mathematical truths hold at all nearby worlds, indeed at all of them, and 'there are reasons to think that we could not have easily had different mathematical beliefs. Our "core" mathematical beliefs might be thought to be evolutionarily inevitable. Given that our mathematical theories best systematize those beliefs, there is a "bootstrapping" argument for the safety of our belief in those theories' (24).

Two-dimensional intensions provide a conduit from conceivability to metaphysical possibility, and can thus explain the connection between the conceivability of mathematical formulas and their metaphysical possibility. By bridging the epistemic and metaphysical universes, the two-dimensional intensions of epistemic two-dimensional semantics can explain how our epistemic states about mathematical formulas can be a guide to their metaphysical profiles. An abstraction principle for (hyper-)intensions modeled on Voevodsky's Univalence Axiom and function type equivalence is countenanced. The truth of the first-order abstraction principle for two-dimensional hyperintensions is supposed to secure the existence of two-dimensional hyperintensions, and its truth is grounded in its possibly being recursively enumerable i.e. Turing machine computable, owing to results in higher observational type theory, and the machine being physically implementable.

When hyperintensional resources are availed of, the topics of truthmakers for mathematical truths will be relevant to capturing their distinctively mathematical subject matter. Topic-sensitive two-dimensional hyperintensions are similarly such that epistemic states can be a guide to metaphysical states for mathematical truths, given the satisfaction of a number of other conditions specified below.

In chapter **8**, I aim to contribute to the analysis of the nature of mathematical modality and hyperintensionality, and to the applications of the latter to absolute decidability. Rather than countenancing the interpretational type of mathematical modality as a primitive, I argue that the interpretational type of mathematical modality is a species of epistemic modality. I argue, then, that the framework of two-dimensional semantics ought to be applied to the mathematical setting. The framework permits of a formally precise account of the priority and relation between epistemic mathematical modality and metaphysical mathematical modality. The discrepancy between the modal systems governing the parameters in the two-dimensional intensional setting provides an explanation of the difference between the metaphysical possibility of absolute decidability and our knowledge thereof. The topic-sensitive epistemic two-dimensional truthmaker semantics from chapters **4** and **2** is advanced, if hyperintensional approaches are to be preferred to possible worlds semantics. I examine the relation between two-dimensional hyperintensional states and epistemic set theory, providing two-dimensional hyperintensional formalizations of epistemic set theory, large cardinal axioms, and the modal axioms governing Ω-logic.

In chapter **9**, I aim to provide hyperintensional foundations for mathematical platonism. I examine Hale and Wright (2009)'s objections to the merits and need, in the defense of mathematical platonism and its epistemology, of the thesis of Necessitism. In response to Hale and Wright's objections to the role of epistemic and metaphysical modalities in providing justification for both the truth of abstraction principles and the success of mathematical predicate reference, I examine the Necessitist commitments of the abundant conception of properties endorsed by Hale and Wright and examined in Hale (2013a,b); examine cardinality issues which arise depending on whether Necessitism is accepted at first- and higher-order; and demonstrate how a two-dimensional semantic approach to the epistemology of mathematics is consistent with Hale and Wright's notion of there being epistemic entitlement rationally to trust that abstraction principles are true. A choice point that I flag is that between availing of intensional or hyperintensional semantics. The hyperintensional semantics approach that I favor is an epistemic two-dimensional truthmaker semantics, for which I define a model. I countenance a hyperintensional semantics for novel epistemic abstractionist modalities. Epistemic and metaphysical states and possibilities may thus be shown to play a constitutive role in vindicating the reality of mathematical objects and truth, and in explaining our possible knowledge thereof.

In chapter **10**, I examine the philosophical significance of Ω-logic in Zermelo-Fraenkel set theory with choice (ZFC). I argue that the philosophical significance of the foregoing is two-fold. First, because the epistemic and modal and hyperintensional profiles of Ω-logical validity correspond to those of second-order logical consequence, Ω-logical validity is genuinely logical. Second, the foregoing provides a hyperintensional account of the interpretation of mathematical and metamathematical vocabulary.

In chapter **11**, I aim to provide a modal logic and hyperintensional semantics for Charles Parsons (1980)'s treatment of rational intuition as a mathematical modality. Similarly to treatments of the property of knowledge in epistemic logic, I argue that rational intuition can be codified by a modal operator governed by the modal μ-calculus. Via correspondence results between fixed point modal propositional logic and the bisimulation-invariant fragment of monadic second-order logic, a precise translation can then be provided between the notion of 'intuition-of', i.e., the cognitive phenomenal properties of thoughts, and the modal operators regimenting the notion of 'intuition-that'. I argue that intuition-that can further be shown to entrain conceptual elucidation, by way of figuring as a dynamic-interpretational modality which induces the reinterpretation of both domains of quantification and the intensions and hyperintensions of mathematical concepts that are formalizable in monadic first- and second-order formal languages. Hyperintensionality is countenanced via a topic-sensitive epistemic two-dimensional truthmaker semantics. I relate the modal logic and hyperintensional semantics for the modality to self-knowledge, because of Gödel's footnotes and fragment in which (I) Gödel relates rational intuition to (i) the realistic aspect of idealism,[18] and to (ii) self-knowledge,[19] and (II) I countenance fixed points in automata for the epistemic modal μ-calculus (iii) to characterize the iteration of epistemic states,[20] and (iv) to specify how Gödel's claim to Wang that intuition might be computational might be characterized.[21]

In chapter **12**, I endeavor to define the concept of indefinite extensibility in the setting of category theory. I argue that the generative property of indefinite extensibility for set-theoretic truths in category theory is identifi-

---

[18]See chapter **11**, fns. 11, 12.
[19]See chapter **5**.
[20]See chapter **5**.
[21]See chapter **11**.

able with the Grothendieck Universe Axiom and the elementary embeddings in Vopenka's principle. The interaction between the interpretational and objective modalities of indefinite extensibility is defined via the epistemic interpretation of two-dimensional semantics. The semantics can be defined intensionally or hyperintensionally. By characterizing the modal profile of Ω-logical validity, and thus the absoluteness i.e. the generic invariance of mathematical truth, modal coalgebras are further capable of capturing the notion of definiteness for set-theoretic truths, in order to yield a non-circular definition of indefinite extensibility.

In chapter **13**, I target a series of potential issues for the discussion of, and modal resolution to, the alethic paradoxes advanced by Scharp (2013). I proffer four novel extensions of the theory, and detail six issues that the theory faces. I provide a counter-example to epistemic closure for reductio proofs.

In chapter **14**, I argue that the types of intention can be modeled both as modal operators and via a multi-hyperintensional semantics. I delineate the semantic profiles of the types of intention, and provide a precise account of how the types of intention are unified in virtue of both their operations in a single, encompassing, epistemic space, and their role in practical reasoning. I endeavor to provide reasons adducing against the proposal that the types of intention are reducible to the mental states of belief and desire, where the former state is codified by subjective probability measures and the latter is codified by a utility function. I argue, instead, that each of the types of intention – i.e., intention-in-action, intention-as-explanation, and intention-for-the-future – has as its aim the value of an outcome of the agent's action, as derived by her partial beliefs and assignments of utility, and as codified by the value of expected utility in evidential decision theory.

Part I: A Framework for Epistemic Modality and Hyperintensionality

Chapter 2

Modal and Hyperintensional Cognitivism and Modal and Hyperintensional Expressivism

2.1 Introduction

This essay endeavors to reconcile two approaches to the modal foundations of thought: modal and hyperintensional cognitivism and modal and hyperintensional expressivism. The novel contribution of the essay is its argument for a reconciliation between the two positions, by providing a hybrid account in which both internal cognitive architecture, on the model of epistemic possibilities, as well as modal automata, are accommodated, while retaining what is supposed to be their unique and inconsistent roles.

The notions of cognitivism and expressivism here targeted concern the role of internal – rather than external – factors in countenancing the nature of thought and information (see Fodor, 1975; Haugeland, 1978). Possible worlds or hyperintensional semantics is taken then to provide the most descriptively adequate means of countenancing the structure of the foregoing.[1] Whereas the type of modal and hyperintensional cognitivism examined here assumes that thoughts and information take exclusively the form of internal

---

[1] Delineating cognitivism and expressivism by whether the positions avail of internal representations is thus orthogonal to the eponymous dispute between realists and antirealists with regard to whether mental states are truth-apt, i.e., have a representational function, rather than being non-representational and non-factive, even if real (see Dummett, 1959; Blackburn, 1984; Price, 2013).

representations, the target modal and hyperintensional expressivist proposals assume that information states are exhaustively individuated by both linguistic behavior and conditions external to the cognitive architecture of agents.

Modal and hyperintensional cognitivism is thus the proposal that the internal representations comprising the language of thought can be modeled via either a possible world or hyperintensional semantics.[2] Modal expres-

---

[2]See Fodor (1975). I endorse (i) 'the representational theory of thought'. (Rescorla, 2024: 1.1) writes: 'Fodor (1981: 177–203; 1987: 16–26) proposes a theory of propositional attitudes that assigns a central role to mental representations. A mental representation is a mental item with semantic properties (such as a denotation, or a meaning, or a truth-condition, etc.). To believe that p, or hope that p, or intend that p, is to bear an appropriate relation to a mental representation whose meaning is that p. For example, there is a relation belief* between thinkers and mental representations, where the following biconditional is true no matter what English sentence one substitutes for "p":
'X believes that p iff there is a mental representation S such that X believes* S and S means that p.
'More generally:
'(1) Each propositional attitude A corresponds to a unique psychological relation A*, where the following biconditional is true no matter what sentence one substitutes for "p": X As that p iff there is a mental representation S such that X bears A* to S and S means that p.
'On this analysis, mental representations are the most direct objects of propositional attitudes. A propositional attitude inherits its semantic properties, including its truth-condition, from the mental representation that is its object.
'Proponents of (1) typically invoke functionalism to analyze A*. Each psychological relation A* is associated with a distinctive functional role: a role that S plays within your mental activity just in case you bear A* to S. When specifying what it is to believe* S, for example, we might mention how S serves as a basis for inferential reasoning, how it interacts with desires to produce actions, and so on. Precise functional roles are to be discovered by scientific psychology. Following Schiffer (1981), it is common to use the term "belief-box" as a placeholder for the functional role corresponding to belief*: to believe* S is to place S in your belief box. Similarly for "desire-box", etc.
'According to Fodor (1987: 17), thinking consists in chains of mental events that instantiate mental representations:
'(2) Thought processes are causal sequences of tokenings of mental representations. A paradigm example is deductive inference: I transition from believing* the premises to believing* the conclusion. The first mental event (my belief* in the premises) causes the second (my belief* in the conclusion).
'(1) and (2) fit together naturally as a package that one might call the representational theory of thought (RTT). RTT postulates mental representations that serve as the objects of propositional attitudes and that constitute the domain of thought processes'; (ii) 'the compositionality of mental representations: Compositionality of mental representations

sivism has, in turn, been delineated in two ways. On the first approach, the presuppositions shared by a community of speakers have been modeled as possibilities (see Kratzer, 1979; Stalnaker, 1978, 1984). Speech acts have in turn been modeled as modal operators which update the common ground of possibilities, the semantic values of which are then defined relative to an array of intensional parameters (Stalnaker, op. cit.; Veltman, 1996; Yalcin, 2007). On the second approach, the content of concepts is supposed to be individuated via the ability to draw inferences. Modally expressive normative inferences are taken then to have the same subjunctive form as that belonging to the alethic modal profile of descriptive theoretical concepts (Brandom, 2014: 211-212).[3] Both the modal approach to shared information and the speech acts which serve to update the latter, and the inferential approach to concept-individuation, are consistent with mental states having semantic values or truth-conditional characterizations. Hyperintensional expressivism is countenanced by Hawke (2024: 1120, 1127-1129) and is defined

---

(COMP): Mental representations have a compositional semantics: complex representations are composed of simple constituents, and the meaning of a complex representation depends upon the meanings of its constituents together with the constituency structure into which those constituents are arranged' (Rescorla, 2024: 1.2); (iii) that mental representations are logically structured: 'Logically structured mental representations (LOGIC): Some mental representations have logical structure. The compositional semantics for these mental representations resembles the compositional semantics for logically structured natural language expressions' (Rescorla, 2024: 1.3); (iv) 'the classical computational theory of mind (CCTM). According to CCTM, the mind is a computational system similar in important respects to a Turing machine, and certain core mental processes are computations similar in important respects to computations executed by a Turing machine' (Rescorla, 2024: 3); and reject (v) 'the formal-syntactic conception of computation (FSC). According to FSC, computation manipulates symbols in virtue of their formal syntactic properties but not their semantic properties' [op. cit.; see **§3.3**, **§10.3.2**, and Rescorla (2015)]. Chalmers (2023) endorses the representational language of thought hypothesis without the classical computational language of thought hypothesis. Chalmers endorses the representational language of thought hypothesis with 'subsymbolic versions of nonclassical computational LOT'. See Chalmers (1990); Kleyko et al. (2022); Piantadosi (2021).

[3]Brandom writes, e.g.: 'For modal *expressivism* tells us that modal vocabulary makes explicit normatively significant relations of subjunctively robust material consequence and incompatibility among claimable (hence propositional) contents in virtue of which ordinary empirical descriptive vocabulary *describes* and does not merely *label, discriminate*, or *classify*. And modal *realism* tells us that there are modal facts, concerning the subjunctively robust relations of material consequence and incompatibility in virtue of which ordinary empirical descriptive properties and facts are determinate. Together, these two claims give a definite sense to the possibility of the correspondence of modal claimings with modal facts' (op. cit.: 2012).

by way of combining a topic-sensitive epistemic truthmaker semantics and a two-component assertability semantics.[4]

So defined, the modal and hyperintensional cognitivist and modal and hyperintensional expressivist approaches have been assumed to be in constitutive opposition. While the cognitivist proposal avails of modal resources in order to model the internal representations comprising an abstract language of thought, the expressivist proposal targets informational properties which extend beyond the remit of internal cognitive architecture: both the form and the parameters relevant to determining the semantic values of linguistic

---

[4]'A formal assert[a]bility semantics models the *assert[a]bility relation* $\Vdash$, holding between a unified body of information **s** (an *information state*) and a meaningful declarative ϕ, exactly when: were an agent's knowledge state to contain exactly information **s**, she would be correct to assert ϕ, from a purely semantic and epistemic perspective [...] We assume that an information state can be identified with a proposition and use **I** to denote the set of all information states. Call a subset of **I** a *cognitive feature* [...]

'**Definition 1 (Expressed Feature)** Relative to a model and an account of $\Vdash$, the cognitive feature expressed by ϕ is: $[\![\phi]\!] := \{s \in \mathbf{I} : s \Vdash \phi\}$.

'So, ϕ expresses the type of information state that renders ϕ assert[a]ble.

'A *TF frame* has five components: W, T, @, knowledge function **K**, and belief function **B**. W and @ are as before. T is a set of possible *topics*; call a subset of T a *subject matter* (denoted **m**). We now model a proposition, or information state, as a pair $\langle \mathbf{i}, \mathbf{m} \rangle$: an intension **i** plus a subject matter **m**. The first component gives the verification/truth conditions of a proposition; the second fixes what it is about. A proposition is *veridical at w* iff its intension includes w, and veridical iff it is *veridical* at @.

'Per fragmentation, an acceptance state is now modeled as a set of propositions, called fragments. Thus, **K** and **B** map a world to a set of propositions: $\mathbf{K}(w)$ is Smith's *total knowledge state* at w and $\mathbf{B}(w)$ is Smith's *total belief state* at w. We stipulate that every proposition in $\mathbf{K}(w)$ is veridical at w and that every knowledge fragment is a type of belief fragment: $\mathbf{K}(w) \subseteq \mathbf{B}(w)$, for all w.

'**Definition 7 (FaTE)** ['Fragmented and Topic-sensitive Expressivism' (27)]
'For arbitrary **s**, p, ϕ, and ψ, relative to TF model T:
$\mathbf{s} \Vdash p$ iff $\mathbf{t}(p) \subseteq \mathbf{s}$ and $\mathbf{s} \subseteq \mathbf{v}(p)$
$\mathbf{s} \dashv p$ iff $\mathbf{t}(p) \subseteq \mathbf{s}$ and $\mathbf{s} \cap \mathbf{v}(p) = \varnothing$
$\mathbf{s} \Vdash \neg\phi$ iff $\mathbf{s} \dashv \phi$
$\mathbf{s} \dashv \neg\phi$ iff $\mathbf{s} \Vdash \phi$
$\mathbf{s} \Vdash \phi \wedge \psi$ iff $\mathbf{s} \Vdash \phi$ and $\mathbf{s} \Vdash \psi$
$\mathbf{s} \dashv \phi \wedge \psi$ iff there are **u** and **v** s.t. $\mathbf{s} = \mathbf{u} \cup \mathbf{v}$ and $\mathbf{u} \dashv \phi$ and $\mathbf{v} \dashv \psi$
$\mathbf{s} \Vdash \Diamond\phi$ iff $\mathbf{t}(\phi) \subseteq \mathbf{s}$ and $\mathbf{s} \not\dashv \phi$
$\mathbf{s} \dashv \Diamond\phi$ iff $\mathbf{s} \dashv \phi$
$\mathbf{s} \Vdash K\phi$ iff $\mathbf{t}(K\phi) \subseteq \mathbf{s}$ and $\forall w \in \mathbf{s}$: $\exists k \in \mathbf{K}(w)$: $k \Vdash \phi$
$\mathbf{s} \dashv K\phi$ iff $\mathbf{t}(K\phi) \subseteq \mathbf{s}$ and $\forall w \in \mathbf{s}$: $\forall k \in \mathbf{K}(w)$: $k \not\Vdash \phi$
$\mathbf{s} \Vdash B\phi$ iff $\mathbf{t}(B\phi) \subseteq \mathbf{s}$ and $\forall w \in \mathbf{s}$: $\exists b \in \mathbf{B}(w)$: $b \Vdash \phi$
$\mathbf{s} \dashv B\phi$ iff $\mathbf{t}(B\phi) \subseteq \mathbf{s}$ and $\forall w \in \mathbf{s}$: $\forall b \in \mathbf{B}(w)$: $b \not\Vdash \phi$'.

utterances, where the informational common ground is taken to be reducible to possibilities; and the individuation of the contents of concepts on the basis of inferential behavior.

In this essay, I provide a background mathematical theory, in order to account for the reconciliation of the cognitivist and expressivist proposals. I avail, in particular, of the duality between Boolean-valued models of epistemic modal algebras and coalgebras; i.e., labeled transition systems defined in the setting of category theory.[5] The mappings of coalgebras permit of flexible interpretations, such that they are able to characterize both modal logics as well as discrete-state automata. I argue that the correspondence between epistemic modal algebras and modal coalgebraic automata is sufficient then for the provision of a mathematically tractable, modal foundation for thought and action.

In Section 2, I provide the background mathematical theory, in order to account for the reconciliation of the cognitivist and expressivist proposals.

In Section 3, I provide reasons adducing in favor of modal and hyperintensional cognitivism, and argue for the material adequacy of epistemic modal algebras as a fragment of the language of thought.

In Section 4, I compare my approach with those advanced in the historical and contemporary literature.

In Section 5, I provide new models for the dynamics of conceptual engineering of intensions and hyperintensions. The first method is via announcements in dynamic epistemic logic. The second method is via dynamic interpretational modalities which redefine intensions and hyperintensions which reassign topics to atomic formulas. The third method is via dynamic hyperintensional belief revision. The fourth method is via rendering epistemic two-dimensional semantics dynamic, such that updates to the epistemic space for the first parameter of a formula will determine an update to the metaphysical space for the second parameter of the formula. The fifth method models updates to two-dimensional intensions via the Logic of Epistemic Dependency in the parameter for epistemic space which then constrains interventions to structural equation models in the parameter for metaphysical space.[6]

---

[5]For an algebraic characterization of dynamic-epistemic logic, see Kurz and Palmigiano (2013). Baltag (2003) develops a coalgebraic semantics for dynamic-epistemic logic, where coalgebraic mappings are intended to record the informational dynamics of single- and multi-agent systems.

[6]For the origins of two-dimensional intensional semantics, see Kamp (1967); Vlach (1973); and Segerberg (1973). Crusius (1745), Kant (1763), Lessing (1777), and Jacobi

In Section **6**, I countenance a hyperintensional construal of the Epistemic Church-Turing Thesis, to ground my dynamic two-dimensional semantics.

In Section **7**, I examine reasons adducing in favor of an expressivist natural language semantics for epistemic modals, to complement the metaphysical expressivism for epistemic modality examined in the chapter.

In Section **8**, modal coalgebraic automata are argued, finally, to be preferred as models of modal expressivism, by contrast to the speech-act and inferentialist approaches, in virtue of the advantages accruing to the model in the philosophy of mathematics. The interest in modal coalgebraic automata consists, in particular, in the range of mathematical properties that

---

(1787) anticipate the question of how epistemic possibilities and states are related to metaphysical possibilities and states, by distinguishing between logical and real connection. Beiser (1987: 96) writes of Mendelssohn that 'Having seen Lessing's ditch, Mendelssohn still attempts to hurdle it. He thinks that reason can cross it, though only at definite points, namely, those where a concept is self-validating, or where it would be absurd to deny its referent. We are told that there are only two such concepts. The first is the concept of a thinking being; and the second is the concept of the most perfect being, God. Here Mendelssohn has in mind Descartes's *cogito* and Anselm's ontological argument. Like Wolff, he adheres to modified versions of both these arguments'. [See Mendelssohn (1764/1997: §§2-3).] Kant (1787/1998) anticipates two-dimensional semantics by inquiring into the objective validity of the categories in the Transcendental Deduction in the *Critique of Pure Reason*. See Book I of the 'Transcendental Analytic', the 'Analytic of Concepts', which includes the Metaphysical Deduction (A66–83, B92–116) and the Transcendental Deduction (A84–130, B116–169). The distinction between epistemic and metaphysical possibilities, as they pertain to the values of mathematical formulas, is further anticipated by Gödel (1951/1995: §§11-12)'s distinction between mathematics in its subjective and objective senses, where the former targets all 'demonstrable mathematical propositions', and the latter includes 'all true mathematical propositions'. Gödel (op. cit.: p. 312, fn. 17) writes of mathematical ontology that: 'The true situation, on the contrary, is that if the objectivity of mathematics is assumed, it follows at once that its objects must be totally different from sensual objects because

'1. Mathematical propositions, if properly analyzed, turn out to assert nothing about the actualities of the space-time world. This is particularly clear in applied propositions such as: Either it has or it has not rained yesterday. The existence of purely conceptual knowledge (besides mathematics) satisfying these requirements is not excluded by this remark.

'2. The mathematical objects are known precisely, and general laws can be recognized with certainty, that is, by deductive, not inductive, inference.

'3. *They can be known (in principle) without using the senses (that is, by means of reason alone) for this very reason, that they don't concern actualities about which the senses (the inner sense included) inform us, but possibilities and impossibilities.*' [My emphasis added - D.E.] See van Atten (2015: ch. 7), for further discussion.

can be recovered on the basis thereof.[7] By contrast to the above competing approaches to modal expressivism, the mappings of modal coalgebraic automata are able both to model and explain elementary embeddings; the intensions of mathematical terms; as well as the modal profile of Ω-logical consequence.

Section **9** provides concluding remarks.

2.2 The Hybrid Proposal

2.2.1 Epistemic Modal Algebra

An epistemic modal algebra is defined as $U = \langle A, 0, 1, \neg, \cap, \cup, \mathbf{l}, \mathbf{m} \rangle$, with A a set containing 0 and 1 (Bull and Segerberg, 2001: 28).[8]
$\mathbf{l}1 = 1$,
$\mathbf{l}(a \cap b) = \mathbf{l}a \cap \mathbf{l}b$
$\mathbf{m}a = \neg \mathbf{l} \neg a$,
$\mathbf{m}0 = 0$,
$\mathbf{m}(a \cup b) = \mathbf{m}a \cup \mathbf{m}b$, and
$\mathbf{l}a = \neg \mathbf{m} \neg a$ (op. cit.).

A valuation v on U is a function from propositional formulas to elements of the algebra, which satisfies the following conditions:
$v(\neg A) = \neg v(A)$,
$v(A \wedge B) = v(A) \cap v(B)$,
$v(A \vee B) = v(A) \cup v(B)$,
$v(\Box A) = \mathbf{l}v(A)$, and
$v(\Diamond A) = \mathbf{m}v(A)$ (op. cit.).

A frame $F = \langle W, R \rangle$ consists of a set W and a binary relation R on W (op. cit.). R[w] denotes the set $\{v \in W \mid (w, v) \in R\}$. A valuation V on F is a function such that $V(A, x) \in \{1, 0\}$ for each propositional formula A and $x \in W$, satisfying the following conditions:
$V(\neg A, x) = 1$ iff $V(A, x) = 0$,
$V(A \wedge B, x) = 1$ iff $V(A, x) = 1$ and $V(B, x) = 1$,
$V(A \vee B, x) = 1$ iff $V(A, x) = 1$ or $V(B, x) = 1$ (op. cit.)

---

[7]See Wittgenstein (2001: IV, 4-6, 11, 30-31), for a prescient expressivist approach to the modal profile of mathematical formulas.
[8]Boolean algebras with operators were introduced by Jónsson and Tarski (1951, 1952).

Epistemic Two-dimensional Truthmaker Semantics

Chalmers endorses a principle of plenitude according to which 'For all sentences s, s is epistemically possible iff there exists a scenario [i.e. epistemically possible world - D.E.] such that w verifies s' (2011: 64), where '[w]hen w verifies s, we can say that s is true at w' (63). In this essay, I accept, instead, a hyperintensional truthmaker approach to epistemic possibility, defined by the notion of exact verification in a state space, where states are parts of whole worlds (Fine 2017a,b; Hawke and Özgün, 2023). According to truthmaker semantics for epistemic logic, a modalized state space model is a tuple ⟨S, P, ≤, v⟩, where S is a non-empty set of states, i.e. parts of the elements in A in the foregoing epistemic modal algebra U, P is the subspace of possible states where states s and t comprise a fusion when s ⊔ t∈P, ≤ is a partial order, and v: Prop → (2^S x 2^S) assigns a bilateral proposition ⟨p^+, p^-⟩ to each atom p∈Prop with p^+ and p^- incompatible (Hawke and Özgün, 2023). Exact verification (⊢) and exact falsification (⊣) are recursively defined as follows (Fine, 2017a: 19; Hawke and Özgün, 2023):

s ⊢ p if s∈⟦p⟧$^+$
(s verifies p, if s is a truthmaker for p i.e. if s is in p's extension);
s ⊣ p if s∈⟦p⟧$^-$
(s falsifies p, if s is a falsifier for p i.e. if s is in p's anti-extension);
s ⊢ ¬p if s ⊣ p
(s verifies not p, if s falsifies p);
s ⊣ ¬p if s ⊢ p
(s falsifies not p, if s verifies p);
s ⊢ p ∧ q if ∃v,u, v ⊢ p, u ⊢ q, and s = v ⊔ u
(s verifies p and q, if s is the fusion of states, v and u, v verifies p, and u verifies q);
s ⊣ p ∧ q if s ⊣ p or s ⊣ q
(s falsifies p and q, if s falsifies p or s falsifies q);
s ⊢ p ∨ q if s ⊢ p or s ⊢ q
(s verifies p or q, if s verifies p or s verifies q);
s ⊣ p ∨ q if ∃v,u, v ⊣ p, u ⊣ q, and s = v ⊔ u
(s falsifies p or q, if s is the fusion of the states v and u, v falsifies p, and u falsifies q);
s ⊢ ∀xϕ(x) if ∃s_1, ..., s_n, with s_1 ⊢ ϕ(a_1), ..., s_n ⊢ ϕ(a_n), and s = s_1 ⊔ ... ⊔ s_n

[s verifies $\forall x\phi(x)$ "if it is the fusion of verifiers of its instances $\phi(a_1)$, ..., $\phi(a_n)$" (Fine, 2017c)];

s ⊣ $\forall x\phi(x)$ if s ⊣ $\phi(a)$ for some individual a in a domain of individuals (op. cit.)

[s falsifies $\forall x\phi(x)$ "if it falsifies one of its instances" (op. cit.)];

s ⊢ $\exists x\phi(x)$ if s ⊢ $\phi(a)$ for some individual a in a domain of individuals (op. cit.)

[s verifies $\exists x\phi(x)$ "if it verifies one of its instances $\phi(a_1)$, ..., $\phi(a_n)$" (op. cit.)];

s ⊣ $\exists x\phi(x)$ if $\exists s_1$, ..., s_n, with s_1 ⊣ $\phi(a_1)$, ..., s_n ⊣ $\phi(a_n)$, and s = s_1 ⊔ ... ⊔ s_n (op. cit.)

[s falsifies $\exists x\phi(x)$ "if it is the fusion of falsifiers of its instances" (op. cit.)];

s exactly verifies p if and only if s ⊢ p if s∈⟦p⟧;

s inexactly verifies p if and only if s ▷ p if $\exists s' \leq S$, s' ⊢ p; and

s loosely verifies p if and only if, $\forall v$, s.t. s ⊔ v ⊢ p, where ⊔ is the relation of compatibility (35-36);

s ⊢ Aϕ if and only if for all u∈P there is a u'∈P such that u' ⊔ u∈P and u' ⊢ ϕ, where Aϕ denotes the apriority of ϕ;[9] and

s ⊣ Aϕ if and only if there is a v∈P such that for all u∈P either v ⊔ u∉P or u ⊣ ϕ;

s ⊢ A(Aϕ) if and only if for all u∈P there is a u'∈P such that u' ⊔ u ∈P and u' ⊢ ϕ and there is a u"∈P such that u' ⊔ u"∈P and u" ⊢ ϕ;

s ⊢ A($\forall x\phi(x)$) if and only if for all u∈P there is a u'∈P such that u ⊢ [u' ⊢ $\exists s_1$, ..., s_n, with s_1 ⊢ $\phi(a_1)$, ..., s_n ⊢ $\phi(a_n)$, and u' = s_1 ⊔ ... ⊔ s_n];

---

[9] In epistemic two-dimensional semantics, epistemic possibility is defined as the dual of apriority or epistemic necessity, i.e. as not being ruled-out apriori (¬□¬), and follows Chalmers (2011: 66). Apriority receives, however, different operators depending on whether it is defined in truthmaker semantics or possible worlds semantics. Both operators are admissible, and the definition in terms of truthmakers is here taken to be more fundamental. The definition of apriority here differs from that of DeRose (1991: 593-594) – who defines the epistemic possibility of P as being true iff "(1) no member of the relevant community knows that P is false and (2) there is no relevant way by which members of the relevant community can come to know that P is false" – by defining epistemic possibility in terms of apriority rather than knowledge. It differs from that of Huemer (2007: 129) – who defines the epistemic possibility of P as it not being the case that P is epistemically impossible, where P is epistemically impossible iff P is false, the subject has justification for ¬P "adequate for dismissing P", and the justification is "Gettier-proof" – by not availing of impossibilities, and rather availing of the duality between apriority as epistemic necessity and epistemic possibility.

$s \vDash A(\exists x\phi(x))$ if and only if or all u∈P there is a u'∈P such that $u \vDash [u' \vDash \phi(a)]$ for some individual a in a domain of individuals (op. cit.).

Epistemic (primary), subjunctive (secondary), and 2D hyperintensions can be defined as follows, where hyperintensions are functions from states to extensions, and intensions are functions from worlds to extensions:[10]

- Epistemic Hyperintension:

    $\text{pri}(x) = \lambda s.\llbracket x \rrbracket^{s,s}$, with s a state in the state space defined over the foregoing epistemic modal algebra, U;

- Subjunctive Hyperintension:

    $\sec_{v_@}(x) = \lambda w.\llbracket x \rrbracket^{v_@,w}$, with w a state in metaphysical state space W;

In epistemic two-dimensional semantics, the value of a formula or term relative to a first parameter ranging over epistemic scenarios determines the value of the formula or term relative to a second parameter ranging over metaphysically possible worlds. The dependence is recorded by 2D-intensions. Chalmers (2006: 102) provides a conditional analysis of 2D-intensions to characterize the dependence. See chapter **1**, fn. 4, for further discussion.

- 2D-Hyperintension:

    $2D(x) = \lambda s \lambda w \llbracket x \rrbracket^{s,w} = 1$.

An abstraction principle for two-dimensional hyperintensions can be defined as follows:

For all types, A,B, there is a homotopy[11]:

---

[10] The notation for intensions follows the presentation in Chalmers and Rabern (2014: 211-212) and von Fintel and Heim (2011).

[11] '[A] *homotopy* between a pair of continuous maps $f: X \to Y$ and $g: X \to Y$ is a continuous map $H: X \times [0, 1] \to Y$ satisfying $H(x, 0) = f(x)$ and $H(x, 1) = g(x)$' (Awodey et al., 2013: 1164). '[T]he logical notion of identity $a = b$ of two objects $a, b: A$ of the same type A can be understood as the existence of a path $p: a \leadsto b$ from point a to point b in the space A. This also means that two functions $f, g: A \to B$ are identical just in case they are homotopic, since a homotopy is just a family of paths $p_x: f(x) \leadsto g(x)$ in B, one for each $x:A$. In type theory, for every type A there is a (formerly somewhat mysterious) type Id_A of identities between objects of A; in homotopy type theory, this is just the *path space* A^I of all continuous maps $I \to A$ from the unit interval' (op. cit.: 1165).

H := [(f ~ g) :≡ $\prod_{x:A}$(f(x) = g(x))], where
$\prod_{f:A \to B}$[(f ~ f) ∧ (f ~ g → g ~ f) ∧ (f ~ g → g ~ h → f ~ h)],
such that, via Voevodsky (2006)'s Univalence Axiom, for all type families A,B:U, there is a function:
`idtoeqv` : (A =$_U$ B) → (A ≃ B),
which is itself an equivalence relation:
(A =$_U$ B) ≃ (A ≃ B).

Abstraction principles for two-dimensional hyperintensions take, then, the form of a function type equivalence:

- ∀x[#f(x) = #g(x)] ≃ [f(x) ≃ g(x)].[12]

2.2.2 Modal Coalgebraic Automata

Modal coalgebraic automata can be thus characterized. Let a category C be comprised of a class Ob(C) of objects and a family of arrows for each pair of objects C(A,B) (Venema, 2007: 421). A functor from a category C to a category D, **E**: C → D, is an operation mapping objects and arrows of C to objects and arrows of D (422). An endofunctor on C is a functor, **E**: C → C (op. cit.).

---

[12]**§2.4** argues that two-dimensional hyperintensions are computable functions. See Herbrand's and Gödel's correspondence, in Gödel (2003) for Herbrand's definition of computable function. See, too, Gödel (1934/1986: 26) and (193?/1995: 11-15) and Skolem (1923). Gödel (1934/1986: 26) writes of Herbrand's definition: 'If ϕ denotes an unknown function, and ψ_1, ... , ψ_k are known functions, and if the ψ's and ϕ are substituted in one another in the most general fashions and certain pairs of the resulting expressions are equated, then, if the resulting set of functional equations has one and only one solution for ϕ, ϕ is a recursive function'. Grassmann (1844/1995) anticipates Herbrand's definition of computable function; Cantù (2022: 2.3.1) writes of Grassmann's general theory of forms: 'Forms are determined by their generating law, and are therefore equal if the same law from the the same initial law generates them ... Given that forms are not given objects but the results of an act of thought that generates them according to a certain law, only the characteristics that depend on the specific way in which forms have been generated will be taken into account in the comparison'. See chapter **1**, fn. 3. Abstraction principles modeled on propositional equality and definitional equality in Homotopy Type Theory are discussed by Sambin and Valentini (1998), Rodin (2014), Shulman (2017), and Catren (2023). See Awodey (2019), for a discussion of the relation between senses and equivalence types. See chapter **3**, for further discussion.

A **E**-coalgebra is a pair $\mathbb{A} = (A, \mu)$, with A an object of C referred to as the carrier of $\mathbb{A}$, and μ: A → **E**(A) is an arrow in C, referred to as the transition map of $\mathbb{A}$ (390).

As, further, a coalgebraic model of modal logic, $\mathbb{A}$ can be defined as follows (407):

For a set of formulas, Φ, let $\nabla \Phi := \Box \bigvee \Phi \wedge \bigwedge \Diamond \Phi$, where $\Diamond \Phi$ denotes the set $\{\Diamond \phi \mid \phi \in \Phi\}$ (op. cit.). Then,

$\Diamond \phi \equiv \nabla \{\phi, T\}$,
$\Box \phi \equiv \nabla \varnothing \vee \nabla \phi$ (op. cit.).
$[\![\nabla \Phi]\!] = \{w \in W \mid R[w] \subseteq \bigcup \{[\![\phi]\!] \mid \phi \in \Phi\}$ and $\forall \phi \in \Phi, [\![\phi]\!] \cap R[w] \neq \varnothing\}$ (Fontaine, 2010: 17).

Let an **E**-coalgebraic modal model, $\mathbb{A} = \langle S, \lambda, R[.]\rangle$, where $\lambda(s)$ is 'the collection of proposition letters true at s in S, and R[s] is the successor set of s in S', such that $\mathbb{S}, s \Vdash \nabla \Phi$ if and only if, for all (some) successors σ of $s \in S$, $[\Phi, \sigma(s) \in \mathbf{E}(\Vdash_{\mathbb{A}})]$ (Venema, 2007: 407), with $\mathbf{E}(\Vdash_{\mathbb{A}})$ a relation lifting of the satisfaction relation $\Vdash_{\mathbb{A}} \subseteq S \times \Phi$. Let a functor, **K**, be such that there is a relation $\overline{\mathbf{K}} \subseteq \mathbf{K}(A) \times \mathbf{K}(A')$ (Venema, 2012: 17)). Let Z be a binary relation s.t. $Z \subseteq A \times A'$ and $\wp \overline{Z} \subseteq \wp(A) \times \wp(A')$, with

$\wp \overline{Z} := \{(X, X') \mid \forall x \in X \exists x' \in X'$ with $(x, x') \in Z \wedge \forall x' \in X' \exists x \in X$ with $(x, x') \in Z\}$ (op. cit.). Then, we can define the relation lifting, $\overline{\mathbf{K}}$, as follows:

$\overline{\mathbf{K}} := \{[(\pi, X), (\pi', X')] \mid \pi = \pi'$ and $(X, X') \in \wp \overline{Z}\}$ (op. cit.), with π a projection mapping of $\overline{\mathbf{K}}$.[13]

The relation lifting, $\overline{\mathbf{K}}$, associated with the functor, **K**, satisfies the following properties (Enqvist et al, 2019: 586):

- $\overline{\mathbf{K}}$ extends **K**. Thus $\overline{\mathbf{K}}f = \mathbf{K}f$ for all functions $f: X_1 \to X_2$;

- $\overline{\mathbf{K}}$ preserves the diagonal. Thus $\overline{\mathbf{K}}\mathrm{Id}_X = \mathrm{Id}_{KX}$ for any set X and functor, Id, where Id_C maps a set S to the product S x C (583, 586);

- $\overline{\mathbf{K}}$ is monotone. $R \subseteq Q$ implies $\overline{\mathbf{K}}R \subseteq \overline{\mathbf{K}}Q$ for all relations $R, Q \subseteq X_1 \times X_2$;

- $\overline{\mathbf{K}}$ commutes with taking converse. $\overline{\mathbf{K}}R° = (\overline{\mathbf{K}}R)°$ for all relations $R \subseteq X_1 \times X_2$;

---

[13] The projections of a relation R, with R a relation between two sets X and Y such that $R \subseteq X \times Y$, are
$X \xleftarrow{(\pi_1)} R \xrightarrow{(\pi_2)} Y$ such that $\pi_1((x, y)) = x$, and $\pi_2((x, y)) = y$. See Rutten (2019: 240).

- $\overline{K}$ distributes over relation composition. $\overline{K}(R \mathbin{;} Q) = \overline{K}R \mathbin{;} \overline{K}Q$, for all relations $R \subseteq X_1 \times X_2$ and $Q \subseteq X_2 \times X_3$, provided that the functor K preserves weak pullbacks (op. cit.). Venema and Vosmaer (2014: §4.2.2) define a weak pullback as follows: 'A weak pullback of two morphisms $f : X \to Z$ and $g : Y \to Z$ with a shared codomain Z is a pair of morphisms $p_X : P \to X$ and $p_Y : P \to Y$ with a shared domain P, such that (1) $f \circ p_X = g \circ p_Y$, and (2) for any other pair of morphisms $q_X : Q \to X$ and $q_Y : Q \to Y$ with $f \circ q_X = g \circ q_Y$, there is a morphism $q : Q \to P$ such that $p_X \circ q = q_X$ and $p_Y \circ q = q_Y$. This pullback is "weak" because we are not requiring q to be unique. Saying that [a set functor] $T : \mathbf{Set} \to \mathbf{Set}$ preserves weak pullbacks means that if $p_X : P \to X$ and $p_Y : P \to Y$ form a weak pullback of $f : X \to Z$ and $g : Y \to Z$, then $Tp_X : TP \to TX$ and $Tp_Y : TP \to TY$ form a weak pullback of $Tf : TX \to TZ$ and $Tg : TY \to TZ$'.

A coalgebraic model of deterministic automata can finally be thus defined (Venema, 2007: 391). An automaton is a tuple, $\mathbb{A} = \langle A, a_I, C, \Xi, F \rangle$, such that A is the state space of the automaton $\mathbb{A}$; $a_I \in A$ is the automaton's initial state; C is the coding for the automaton's alphabet, mapping numerals to the natural numbers; $\Xi: A \times C \to A$ is a transition function, and $F \subseteq A$ is the collection of admissible states, where F maps A to $\{1, 0\}$, such that $F: A \to 1$ if $a \in F$ and $A \to 0$ if $a \notin F$ (op. cit.).

Modal automata are defined over a modal one-step language (Venema, 2020: 7.2). With A being a set of propositional variables the set, $\mathtt{Latt}(X)$, of lattice terms over X has the following grammar:

$$\phi ::= \bot \mid \top \mid x \mid \phi \wedge \phi \mid \phi \vee \phi,$$

with $x \in X$ and $\phi \in \mathtt{Latt}(A)$ (op. cit.).
The set, $\mathtt{1ML}(A)$, of modal one-step formulas over A has the following grammar:

$$\alpha \in A ::= \bot \mid \top \mid \Diamond\phi \mid \Box\phi \mid \alpha \wedge \alpha \mid \alpha \vee \alpha \text{ (op. cit.).}$$

A modal P-automaton $\mathbb{A}$ is a triple, (A, Θ, a_I), with A a non-empty finite set of states, $a_I \in A$ an initial state, and the transition map
$\Theta: A \times \wp P \to \mathtt{1ML}(A)$
maps states to modal one-step formulas (op. cit.: 7.3).

The crux of the reconciliation between algebraic models of cognitivism and the formal foundations of modal expressivism is based on the duality between categories of algebras and coalgebras: $\mathbb{A} = \langle A, \alpha{:}A \rightarrow \mathbf{E}(A)\rangle$ is dual to the category of algebras over the functor α (417-418). For a category C, object A, and endofunctor $\mathbf{E}$, define a new arrow, α, s.t. $\alpha{:}\mathbf{E}A \rightarrow A$. A homomorphism, f, can further be defined between algebras $\langle A, \alpha\rangle$, and $\langle B, \beta\rangle$. Then, for the category of algebras, the following commutative square can be defined: (i) $\mathbf{E}A \rightarrow \mathbf{E}B$ ($\mathbf{E}f$); (ii) $\mathbf{E}A \rightarrow A$ (α); (iii) $\mathbf{E}B \rightarrow B$ (β); and (iv) $A \rightarrow B$ (f) (see Hughes, 2001: 7-8). The same commutative square holds for the category of coalgebras, such that the latter are defined by inverting the direction of the morphisms in both (ii) [$A \rightarrow \mathbf{E}A$ (α)], and (iii) [$B \rightarrow \mathbf{E}B$ (β)] (op. cit.)

The significance of the foregoing is twofold. First and foremost, the above demonstrates how a formal correspondence can be effected between algebraic models of cognition and coalgebraic models which provide a natural setting for modal logics and automata. The second aspect of the philosophical significance of modal coalgebraic automata is that – as a model of modal expressivism – the proposal is able to countenance fundamental properties in the foundations of mathematics, and circumvent the issues accruing to the attempt so to do by the competing expressivist approaches.

2.3 Material Adequacy

The material adequacy of epistemic modal algebras as a fragment of the representational theory of mind is witnessed by the prevalence of possible worlds and hyperintensional semantics – the model theory for which is algebraic (see Blackburn et al., 2001: ch. 5) – in cognitive psychology and artificial intelligence.

In artificial intelligence, the subfield of knowledge representation draws on epistemic logic, where belief and knowledge are interpreted as necessity operators (Meyer and van der Hoek, 1995; Fagin et al., 1995). Possibility and necessity may receive other interpretations in mental terms, such as that of conceivability and apriority (i.e. truth in all epistemic possibilities, or inconceivability that not ϕ). The language of thought hypothesis maintains that thinking occurs in a mental language with a computational syntax and a semantics. The philosophical significance of cognitivism about epistemic modality and hyperintensionality is that it construes epistemic intensions and hyperintensions as abstract, computational functions in the mind, and thus provides an explanation of the relation that human beings bear to epistemic possibilities. Intensions and hyperintensions are semantically imbued abstract functions comprising the computational syntax of the language of thought. The functions are semantically imbued because they are defined relative to a parameter ranging over either epistemically possible worlds or epistemic states in a state space, and extensions or semantic values are defined for the functions relative to that parameter. Cognitivism about epistemic modality or hyperintensionality argues that thoughts are composed of epistemic intensions or hyperintensions. Cognitivism about epistemic modality provides a metaphysical explanation or account of the ground of thoughts, arguing that they are grounded in epistemic possibilities and either intensions or hyperintensions which are themselves internal representations comprising the syntax and semantics for a mental language. This is consistent with belief and knowledge being countenanced in an epistemic logic for artificial intelligence, as well. Epistemic possibilities are constitutively related to thoughts, and figure furthermore in the analysis of notions such as apriority and conceivability, as well as belief and knowledge in epistemic logic for artificial intelligence.

My claim is only that epistemic intensions and hyperintensions – i.e. functions from epistemically possible worlds or epistemic states to extensions – are computable functions comprising a fragment of the language of thought, leaving it open whether the mind is more generally a Turing machine. I thus hope to avoid taking a position here on whether human cognition is generally computational in light of Gödel (1931/1986)'s incompleteness theorems. See §**6.4**, for proofs of the incompleteness theorems. A theory is recursively enumerable if the valid strings in the theory can be enumerated by a Turing machine. A theory is recursive if the Turing machine halts on every input. Gödel's Disjunction claims that either (I) the mind surpasses the computability via the recursive enumerability of sentences in a Turing machine, and currently undecidable sentences are provable i.e. decidable owing to (i) mathematical intuition instead of computable mechanism,[14] and

---

[14]See Gödel (1953/9-III/1995: §29; 1953/9-V/1995: §§3, 4), for proofs of admissible rules / consistency using mathematical intuition: 'On the grounds of these results it can be said that *the scheme of the syntactical program to replace mathematical intuition by rules for the use of symbols fails because this replacing destroys any reason for expecting consistency, which is vital for both pure and applied mathematics, and because for the consistency proof one either needs a mathematical intuition of the same power as for*

(ii) Gödel's acceptance of rational optimism; or (II) the mind is a Turing machine and thus there are sentences which are undecidable, i.e. not provable, because (i) formal theories are recursively enumerable, and (ii) the first incompleteness theorem entails that, in *consistent* formal systems, the provability via the recursive enumerability of sentences is distinct from the truth of Gödel sentences (1931/1986: 195). For further discussion, see Gödel (1951/1995); Lucas (1961); Penrose (1989; 1994); the essays in Horsten and Welch (2016); and Koellner (2018a,b). See **§8.3**, fn. 14, and **§10.3.2**, fn. 14, for further discussion. I account for the convergence between modal and hyperintensional computational automata and rational intuition in chapter 11.

2.4 Precedent

The proposal that possible worlds semantics comprises the model for thoughts and propositions is anticipated by Wittgenstein (1921/1974: 2.15-2.151, 3-3.02); Chalmers (2011); and Jackson (2011). Their approaches depart, however, from the one here examined in the following respects.

Wittgenstein (op. cit.: 1-1.1) has been interpreted as endorsing an identity theory of propositions, which does not distinguish between internal thoughts and external propositions (see McDowell, 1994: 27; and Hornsby, 1997: 1-3). How the identity theory of propositions is able to accommodate Wittgenstein's suggestion that a typed hierarchy of propositions can be generated – only if the class of propositions has a general form and the sense of propositions over which operations range is invariant by being individuated by the possibilities figuring as their truth and falsity conditions (see Wittgenstein, 1979: 21/11/16, 23/11/16, 7/11/17; and Potter, 2009: 283-285 for detailed discussion) – is an open question. Wittgenstein (1921/1974: 5.5561) writes that 'Hierarchies are and must be independent of reality', although provides no account of how the independence can be effected.

Jackson (2008: 48-50) distinguishes between personal and subpersonal theories by the role of neural science in individuating representational states (see Shea, 2013, for further discussion), and argues in favor of a 'personal-level implicit theory' for the possible worlds semantics of mental representations.

---

discerning the truth of the mathematical axioms or a knowledge of empirical facts involving an equivalent mathematical content' (1953/9-III/1995: §29).

Chalmers' approach comes closest to the one here proffered, because he argues for a hybrid cognitivist-expressivist approach as well, according to which epistemic intensions – i.e. functions from epistemically possible worlds to extensions – are individuated by their inferential roles (2012: 462-463). Chalmers endorses what he refers to as 'anchored inferentialism', and in particular 'acquaintance inferentialism' for intensions, according to which 'there is a limited set of primitive concepts, and all other concepts are grounded in their inferential role with respect to these concepts', where 'the primitive concepts are acquaintance concepts' (463, 466) and '[a]cquaintance concepts may include phenomenal concepts and observational concepts: primitive concepts of phenomenal properties, spatiotemporal properties, and secondary qualities' (2010b: 11). According to Chalmers, 'anchored inferential role determines a primary intension. The relevant role can be seen as an internal (narrow or short-armed) role, so that the content is a narrow content' (5). The inferences in question are taken to be 'suppositional' inferences, from a base class of truths, $PQTI$ – i.e. truths about physics, consciousness, and indexicality, and a that's all truth – determining canonical specifications of epistemically possible worlds, to other truths (3). With regard to how suppositional inference, i.e. 'scrutability', plays a role in the definitions of intensions, Chalmers writes that '[t]he primary intension of [a sentence] S is true at a scenario [i.e. epistemically possible world] w iff [A] epistemically necessitates S, where [A] is a canonical specification of w', where '[A] epistemically necessitates S iff a [material] conditional of the form '[A] → S' is apriori' and the apriori material entailment is the relation of scrutability (2006).[15]

---

[15]'We can define a priori scrutability in parallel to definitional entailment: a sentence S is a priori scrutable from (or a priori entailed by) a class of sentences C if S can be logically derived from some members of C along with some a priori truths. Given weak assumptions, the right-hand side is equivalent to the claim that there is a conjunction D of sentences in C such that the material conditional "If D, then S" (which is equivalent to "¬(D ∧ ¬S)" is a priori', thus a strict conditional with an epistemic necessity i.e. apriority operator (Chalmers, 2012: 7). See chapter **1**, fn. 4. Chalmers (private correspondence) writes: '[I] use strict implication (a priori material implication) [i.e. $\Box_1(p \to q)$; see Chalmers, 2006], not material implication, so avoid the paradoxes of the latter, and accept the paradoxes of the former'. Mares (2024) writes: '[T]he strict implication (p $\Box\to$ q) is true whenever it is not possible that p is true and q is false — i.e., ¬◊(p ∧ ¬q). Among the paradoxes of strict implication are the following:
'(p ∧ ¬p) → q,
'p → (q → q),
'p → (q ∨ ¬q).
'The first asserts that a contradiction strictly implies every proposition; the second and

Chalmers (2012: 245) is explicit about this: 'The intension of a sentence S (in a context) is true at a scenario w iff S is a priori scrutable from [A] (in that context), where [A] is a canonical specification of w (that is, one of the epistemically complete sentences in the equivalence class of w) ... A Priori Scrutability entails that this sentence S is a priori scrutable (for me) from a canonical specification [A] of my actual scenario, where [A] is something along the lines of $PQTI$'. 'The secondary intension of S is true at a world w iff [A] metaphysically necessitates S', where '[A] metaphysically necessitates S when a subjunctive conditional of the form 'if [A] had been the case, S would have been the case' is true' (op. cit.). Thus, suppositional inference, i.e. scrutability, determines the intensions of two-dimensional semantics.

On the approach advanced here, intensions and hyperintensions are countenanced as semantically imbued functions. Intensions and hyperintensions as functions comprise the computational syntax for the language of thought, but they are semantically imbued because they are functions from epistemic possibilities to extensions.

An anticipation of this proposal is Tichy (1969), who defines intensions as Turing machines. Adriaans (2020) provides an example of intensions modeled using a Turing machine, as well.[16] The expression

$$U_j(\overline{T_i}x) = y$$

has the following components. 'The universal Turing machine U_j is a **context** in which the computation takes place. It can be interpreted as a **possible computational world** in a modal interpretation of computational semantics. / The sequences of symbols $\overline{T_i}x$ and y are **well-formed data**. / The sequence $\overline{T_i}$ is a self-delimiting description of a program and it can be interpreted as a piece of well-formed **instructional data**. / The sequence $\overline{T_i}x$ is an **intension**. The sequence y is the corresponding **extension**. / The expression $U_j(\overline{T_i}x) = y$ states the result of the program $\overline{T_i}x$ in world U_j is y. It is a **true sentence**'.

I will avail, in this book, of Adriaans (2020)'s definition of intensions as Turing machines. The variable, x, in the (hyper-)intension, $\overline{T_i}x$, ranges over epistemically possible worlds or states and metaphysically possible worlds or states, and $\overline{T_i}x$ is a function from epistemic states verifying sentences, where

---

third imply that every proposition strictly implies a tautology'.

[16] Approaches to conceiving of intensions as computable functions have been pursued, as well, by Muskens (2005), Moschovakis (2006), and Lappin (2014). The computational complexity of algorithms for intensions has been investigated by Mostowski and Wojtyniak (2004), Mostowski and Szymanik (2012), and Kalocinski and Godziszewski (2018).

the epistemic states are taken as actual, to the value of the sentences verified by metaphysical states, to the sentences' extensions.

This is consistent with the inferences of scrutability playing a role in the individuation of intensions and hyperintensions, but whereas Chalmers grounds inferences in dispositions (2010: 10; 2021), I claim that the inferences drawn from the canonical specifications of epistemic possibilities to arbitrary truths are apriori computations between mental representations.[17]

Schroeder (2008) provides a protracted examination of variations on the expression relation. Schroeder argues that expressivists ought to opt for an assertability account of the expression relation, such that the propositions expressed by sentences are governed by assertability conditions for the sentences rather than their truth conditions, and the expression thus doesn't concern the conveyance of information but rather norms on correct assertion of the sentence. He writes: 'Every sentence in the language is associated with conditions in which it is semantically correct to use that sentence assertorically ... Assertability conditions, so conceived, are a device of the semantic theorist. They are not a kind of information that speakers intend to convey. So there is no sense in which a community of speakers could get by, managing to communicate information to each other about the world, by means of assertability conditions alone. It is only because some assertability conditions mention beliefs, and beliefs have contents about the world, that speakers can manage to convey information about the world' (op. cit.: 108, 110). The present account is not committed to Schroeder's proposed assertability expressivism. However, I note in Section **2.6** that Hawke and Steinert-Threlkeld

---

[17]See Wittgenstein (1975, 1976: pp. 105-106), for an anticipation – by Turing, in discussion with Wittgenstein – of Kripke (1982)'s infinity argument against dispositional theories of meaning:

'[Wittgenstein:] That I should take this procedure as the standard procedure means a whole lot: that it is the right procedure and at the same time removed from possible tests [...]

'Turing: The difficulty is that there is not a finite number of multiplications. You can only put a finite number of multiplications in your archives; and when I do a multiplication that is not in your archives, what then?

'Wittgenstein: Well, what then?–That is like counting to a number which has not been counted to. / [...] Couldn't there be in the archives rules for using these rules one used? Couldn't this go on forever? / But this has nothing to do with the fact that the number of multiplications is infinite. In fact, that it has no connexion with it is an important point. The idea that it is connected with it comes from the idea that the examples, being infinite, are too numerous to go in to the archives'.

See Shagrir (2014), for further discussion.

(2021)'s assertability semantics for epistemic modals is consistent with the model-theoretic account of expressivism here advanced. The present account might also converge with a view which Schroeder attributes to Gibbard (1990, 2003), which he refers to as indicator expressivism, according to which mental states do not express propositional contents, but rather express ur-contents owing to an agent's intentions (§4.1). Ur-contents differ from propositional contents, by the differences in their roles in expressing normative and non-normative contents. Schroeder objects to the appeal to ur-contents, arguing that they play a role too similar to that of propositional contents because they convey descriptive information, while Gibbard simultaneously rejects the similarity (107). I think that because ur-contents express normative contents rather than non-normative ones, they are sufficiently distinct from propositional contents, and that it is innocuous for them to be descriptive in part. The present model-theoretic account of expressivism might thus be thought to be consistent with indicator expressivism.

2.5 Conceptual Engineering of Intensions and Hyperintensions

How can intensions and hyperintensions be revised, given that they are here countenanced as computable functions comprising the syntax of the language of thought? Note that the epistemically possible worlds or hyperintensional truthmakers, and the topics to which they are sensitive, which figure as input to intensions and hyperintensions, can be externally individuated. If so, then they are susceptible to updates by external sources. One might want further to engage in the project of the conceptual engineering one's intensions and hyperintensions, perhaps in order to engage in an ameliorative project relevant to using more socially just concepts (see Haslanger, 2012, 2020 for further discussion). Conceptual engineering of intensions and hyperintensions can then be effected by five methods. The first is via announcements in dynamic epistemic logic. The second method is via dynamic interpretational modalities which concern the possible reassignment of topics to atomic formulas. The third method is via dynamic hyperintensional belief revision. We here propose a novel truthmaker semantics for the first and second methods.

The language of public announcement logic has the following grammar (see Baltag and Renne, 2016):

$$\phi := p \mid \phi \wedge \phi \mid \neg\phi \mid [a]\phi \mid [\phi!]\psi$$

$[a]\phi$ is interpreted as the 'the agent knows ϕ'. $[\phi!]\psi$ is an announcement formula, and is intuitively interpreted as 'whenever ϕ is true, ψ is true after we eliminate all not-ϕ possibilities (and all arrows to and from these possibilities)'.

Semantics for public announcement logic is as follows:

$\langle M, w \rangle \Vdash \phi$ if and only if $w \in V(\phi)$

$\langle M, w \rangle \Vdash \phi \wedge \psi$ if and only if $M, w \Vdash \phi$ and $M, w \Vdash \psi$

$\langle M, w \rangle \Vdash \neg\phi$ if and only if $M, w \nVdash \phi$

$\langle M, w \rangle \Vdash [a]\phi$ if and only if $M, w \Vdash \phi$ for each v satisfying $wR_a v$

$\langle M, w \rangle \Vdash [\phi!]\psi$ if and only if $M, w \nVdash \phi$ or $M[\phi!], w \Vdash \psi$,

where $M[\phi!] = (W[\phi!], R[\phi!], V[\phi!])$ is defined by

$W[\phi!] := (v \in W \mid M, v \Vdash \phi)$ (intuitively, 'retain only the worlds where ϕ is true' (op. cit.),

$xR[\phi!]_a y$ if and only if $xR_a y$ (intuitively, 'leave arrows between remaining words unchanged'), and

$v \in V[\phi!](p)$ if and only if $v \in V(p)$ (intuitively, 'leave the valuation the same at remaining worlds').

Fine (2006) and Uzquiano (2015) countenance interpretational modalities. Fine (2005b)'s modality is simultaneously postulational, dynamic, and prescriptive. The dynamic modality is interpreted so as to concern the execution of computer programs which entrain e.g. the introduction of objects into a domain which conform to a certain property. Fine (2006) advances a postulational interpretational modality which concerns the possible reinterpretation of quantifier domains in accounting for indefinite extensibility. Uzquiano's modality is interpretational and also relevant to capturing the property of indefinite extensibility. The modality is mathematical, and concerns the possible reinterpretations of the intensions of non-logical vocabulary such as the membership relation, $\in$.

In this chapter, I propose to render Fine's and Uzquiano's interpretational modalities dynamic. The dynamic interpretational modalities are interpreted as program executions which entrain reinterpretations of intensions as well as reinterpretations of hyperintensions which reassign topics to atomic formulas.

My proposal is that both announcement formulas, $[\phi!]\psi$, and Fine and Uzquiano's modalities ought to be rendered hyperintensional, such that the box operators are further interpreted as necessary truthmakers as specified in the clause for $A(\phi)$ above. The dynamic interpretational modalities can just take the clause for $A(\phi)$. For announcement formulas, $[\phi!]\psi$ if and only if either (i) for all $t \in P$ there is no $t' \in P$ such that $t' \sqcup t \in P$ and $t' \vdash \phi$ or (ii) $M[\phi!], s \vdash \psi$,

where $M[\phi!] = \langle S[\phi!], \leqslant[\phi!], v[\phi!]\rangle$ is defined by

$S[\phi!] := s' \in S \mid M, s' \vdash \phi$ (intuitively, retain only states which verify ϕ),

$\leqslant[\phi!]$ if and only if $s \leqslant s'$ (intuitively, leave relations between remaining states unchanged), and

$v[\phi!]$ if and only if v: Prop $\rightarrow (2^S \times 2^S)$ which assigns a bilateral proposition $\langle \phi^+, \phi^-\rangle$ to $\phi \in$ Prop (intuitively, leave the valuation the same at remaining states).

This would suffice for what Chalmers (2020) refers to as conceptual re-engineering, rather than 'de novo' conceptual engineering, of intensions and hyperintensions. Conceptual re-engineering concerns the refinement or replacement of extant concepts, while de novo engineering concerns the introduction of new concepts. The third method for conceptual re-engineering contents would be via Berto and Özgün (2021)'s logic for dynamic hyperintensional belief revision, which includes a topic-sensitive upgrade operator. On this method, the worlds and topics for formulas are both updated in cases of belief revision.

A fourth novel method can be countenanced, namely making epistemic two-dimensional semantics dynamic. On this approach, an epistemic action such as an announcement which updates the first, epistemic parameter for a formula would entrain an update to a second parameter ranging over metaphysically possible worlds or states in a state space. Using two-dimensional (hyper-)intensions, such that the value of a formula relative to a first parameter ranging over epistemic states determines the value of the formula relative to a second parameter ranging over metaphysical states, an update (announcement, epistemic action) to the epistemic space over which the first parameter of a formula ranges induces an update to the metaphysical space over which a second parameter for a formula ranges. With M* a model including a class of epistemic states, S, and a class of metaphysical states, W, two-dimensional updates have the form:

$M^*, w \Vdash [\phi!]\psi$ if and only if $M^*, w \nVdash \phi$ or $M^*[\phi!], w \Vdash \psi$,

where $M^*[\phi!] = (S[\phi!], W[\phi!]^{S[\phi!]}, R[\phi!], V[\phi!])$. $W[\phi!]^{S[\phi!]}$ records the dynamic two-dimensional update of metaphysical states, W, conditional on the update of epistemic states, S, and the rest is defined as above.

A fifth method for modeling updates might be via the interventions of structural equation models which reassign values to exogenous variables which then determines the values of endogenous variables (see e.g. Pearl, 2009).[18] Using two-dimensional (hyper-)intensions, the updates to the epistemic parameter of a formula might be modeled using Baltag (2016)'s Logic of Epistemic Dependency. As Baltag writes: 'An *epistemic dependency formula* $K_a^{x_1,...,x_n} y$ says that an agent knows the value of some variable y *conditional on being given the values of the variables* $x_1, \ldots, x_n \ldots$ if we use the abbreviation $(w(\vec{x})) = (v(\vec{x}))$ for the conjunction $(w(x_1)) = (v(x_1)) \wedge (w(x_n)) = (v(x_n))$, then we put

$w \Vdash K_a^{x_1,...,x_n} y$ iff $\forall v \sim_a w\ (w(\vec{x})) = (v(\vec{x})) \Rightarrow v(y) = w(y)$.

In words: an agent knows y given $x_1, \ldots, x_n$ if the value of y is the same in all the epistemic alternatives that agree with the actual world on the values of $x_1, \ldots, x_n$. This operator has connections with Dependence Logic and allows us to "pre-encode" the dynamics of the value-announcement operator $[!x]\phi$' (136).

Epistemic updates via announcements would then, via two-dimensional intensions and hyperintensions, induce an intervention in the metaphysical space in the parameter defining the second dimension of a formula, by reassigning values of exogenous variables so as to constrain the values of endogenous variables in structural equations.

2.6 Two-dimensional Hyperintensionality and the Epistemic Church-Turing Thesis

The Epistemic Church-Turing Thesis can receive a similar two-dimensional hyperintensional formalization. Carlson (2016: 132) presents the schema for the Epistemic Church-Turing Thesis as follows:

With □ interpreted as a knowledge operator, '□∀x∃y□ϕ → ∃e□∀x∃y[E(e, x, y) ∧ ϕ],

'where e does not occur free in ϕ and E is a fixed formula of L_{PA} [i.e. the language of Peano Arithmetic] with free variables v_0, v_1, v_2 such that, letting N be the standard model of arithmetic,

'N ⊩ E(e, x, y)[e, x, y | a, m, n]

'iff on input m, the ath Turing machine halts and outputs n. For convenience, we will write $\{t_1\}\{t_2\} \simeq t_3$ for E(t_1, t_2, t_3) when t_1, t_2, t_3 are terms'.

---

[18] Thanks here to Hannes Leitgeb for mentioning interventions in structural equation models with regard to a possible example of updates in metaphysical space.

Carlson defines $(x_1, \ldots, x_n) \mid (y_1, \ldots, y_1)$ as denoting the 'function which maps x_i to y_i for each i = 1, ..., n' (op. cit.: 130). Hyperintensionally reformalized, the Epistemic Church-Turing Thesis is then:
$A\forall x \exists y A \phi \rightarrow \exists e A \forall x \exists y [E(e, x, y) \wedge \phi]$.

The two-dimensional hyperintensional profile of the Epistemic Church-Turing Thesis can be countenanced by adding a topic-sensitive truthmaker from a metaphysical state space and making its value dependent on the value of the epistemically necessary truthmaker $A(\phi)$. Thus:
$A^{(w \cap t)} \forall x \exists y A^{(w \cap t)} \phi \rightarrow \exists e A^{(w \cap t)} \forall x \exists y [E(e, x, y) \wedge \phi]$.

An application of the two-dimensional Epistemic Church-Turing Thesis is to the above dynamic epistemic two-dimensional semantics. Two-dimensional Turing machines can be availed of in order to provide mechanistic, constructive definitions of the epistemic actions and metaphysical interventions and their dependence in the two-dimensional semantics. Aside from defining epistemic hyperintensions as computable functions, where the functions comprise a fragment of the computable syntax of the language of thought, I record here my preference for non-mechanistic approaches to epistemic modality, such as the interpretation thereof as informal provability or as an inference package.

In the remainder of the essay, I outline an expressivist semantics for epistemic modality. I endeavor, then, to demonstrate the advantages accruing to the present approach to countenancing modal expressivism via modal coalgebraic automata, via a comparison of the theoretical strength of the proposal when applied to characterizing the fundamental properties of the foundations of mathematics, by contrast to the competing approaches to modal expressivism and the limits of their applications thereto.

2.7 Expressivist Semantics for Epistemic Possibility

I assume a dissociation between the natural language semantics for epistemic modals and an account of mental states as epistemic possibilities or hyperintensional epistemic states. However, my expressivism about epistemic modality might be thought to adduce in favor of expressivism about epistemic modals.

Let expressivism about a domain of discourse be the claim that an utterance from that domain expresses a mental state, rather than states a fact

(Hawke and Steinert-Threlkeld, 2021). Hawke and Steinert-Threlkeld (op. cit., 480) distinguish between semantic expressivism and pragmatic expressivism. Expressivism about epistemic possibility takes the property expressed by $\Diamond \phi$ to be $\{\mathbf{s} \subseteq W: \mathbf{s} \not\Vdash \neg p\}$, where $\mathbf{s}$ is a state of information, W is a set of possible worlds, and $\mathbf{s} \Vdash \phi$ if and only if ϕ is assertible relative to $\mathbf{s}$, if and only if the state of information is compatible with ϕ (op. cit.). Semantic expressivism incorporates a 'psychologistic semantics' according to which the value of ϕ is a partial function from information states to truth-values, such that 'the mental type expressed by ϕ is characterized in terms of the assert[a]bility relation $\Vdash$' and 'the definition of $\Vdash$ is an essential part of that of $[\![\ ]\!]$' (481). Pragmatic expressivism rejects the psychologistic semantics condition, and 'allows for a *gap* between the compositional semantic theory and $\Vdash$' (op. cit.).

Hawke and Steinert-Threlkeld (op. cit.) argue that satisfying the following conditions is a desideratum of any expressivist account about epistemic possibility (§3.5):

(Weak) Wide-scope Free Choice (**WFC** (§3.1)):
$\Diamond p \vee \Diamond \neg p \Vdash \Diamond p \wedge \Diamond \neg p$
Disjunctive Inheritence (**DIN** (§3.2)):
$(\Diamond p \wedge q) \vee r \Vdash [\Diamond(p \wedge q) \wedge q] \vee r$
Disjunctive Syllogism and Schroeder's Constraints (§3.4):
DSF $\{\Diamond \neg q, p \vee \Box q \not\Vdash p\}$
SCH $\{\Diamond \neg p, p \vee \Box q \not\Vdash \Box q\}$

DSF and **SCH** record the failure of disjunctive syllogism in the presence of epistemic contradictions.

WFC is vindicated by the contention that when someone asserts $p \vee \neg p$, they neither believe p nor believe $\neg p$, and so are in a position to assert both $\Diamond p$ and $\Diamond \neg p$.

DIN is vindicated by the equivalence of the content of the utterances, e.g.,
(1) David is at home and might be watching a film.
(2) David is at home and might be watching a film at home (§3.2).

Hawke and Steinert-Threlkeld's modal propositional assert[a]bility semantics is then as follows (§5.1).

Reading $t \subseteq s$: $[\![\phi]\!]^t \neq 1$ as 's refutes ϕ':

- if p is an atom: $[\![p]\!]^s = 1$ iff $s \subseteq V(p)$
  if p is an atom $[\![p]\!]^s = 0$ iff s refutes p
- $[\![\neg\phi]\!]^s = 1$ iff $[\![\phi]\!]^s = 0$
  $[\![\neg\phi]\!]^s = 0$ iff $[\![\phi]\!]^s = 1$
- $[\![\phi \wedge \psi]\!]^s = 1$ iff $[\![\phi]\!]^s = 1$ and $[\![\psi]\!]^s = 1$
  $[\![\phi \wedge \psi]\!]^s = 0$ iff s refutes $\phi \wedge \psi$
- $[\![\phi \vee \psi]\!]^s = 1$ iff there exists s_1, s_2 such that $s = s_1 \cup s_2$, $[\![\phi]\!]^{s_1} = 1$ and $[\![\psi]\!]^{s_2} = 1$
  $[\![\phi \vee \psi]\!]^s = 0$ iff s refutes $\phi \vee \psi$
- $[\![\Diamond\phi]\!]^s = 1$ iff $[\![\phi]\!]^s \neq 0$
  $[\![\Diamond\phi]\!]^s = 0$ iff s refutes $\Diamond\phi$
- $\Box\phi := \neg\Diamond\neg\phi$
- $\Diamond\phi := \neg\Box\neg\phi$.[19]

Unlike Yalcin (2007)'s domain semantics (4.1), Veltman (1996)'s update semantics (4.2), and Moss (2015; 2018)'s probabilistic semantic expressivism (6.2), Hawke and Steinert-Threlkeld's assert[a]bility semantics satisfies **WFC**, **DIN**, **DSF**, and **SCH** (Hawke and Steinert-Threlkeld, 2020: 507). As a preliminary, suppose

Proposition 1 If ϕ is $\Diamond$-free, then $s \Vdash \Diamond\phi$ holds iff there exists $w \in s$ such that: $\{w\} \Vdash \phi$ (op. cit.).

Proof: $s \Vdash \Diamond\phi$ holds iff $[\![\phi]\!]^s \neq 0$. $[\![\phi]\!]^s = 0$ iff $[\![\phi]\!]^{\{w\}} = 0$ for every $w \in s$. So, $[\![\phi]\!]^s \neq 0$ iff $[\![\phi]\!]^w \neq 0$ for some $w \in s$ iff $\{w\} \Vdash \phi$ for some $w \in s$ (op. cit.).

For **WFC**, suppose that $s \Vdash \Diamond p \vee \Diamond \neg p$. So, there exists s_1, s_2 that cover s and $s_1 \Vdash \Diamond p$ and $s_2 \Vdash \Diamond \neg p$. By Proposition 1, there exist u,v∈s such that $\{u\} \Vdash p$ and $\{v\} \Vdash \neg p$. Thus, $s \Vdash \Diamond p$ and $s \Vdash \Diamond \neg p$ (op. cit.).

---

[19] I have revised the previous clause, and further added this clause to Hawke and Steinert-Threlkeld's model. The clause states that epistemic possibility is defined as the dual of apriority or epistemic necessity, i.e. as not being ruled-out apriori ($\neg\Box\neg$), and follows Chalmers (2011: 66).

For **DIN**, suppose that $s \Vdash (\Diamond p \wedge q) \vee r$. So, there exists s_1, s_2, such that $s = s_1 \cup s_2$ with $s_1 \Vdash \Diamond p$, $s_1 \Vdash q$, and $s_2 \Vdash r$. For every $w \in s_1$, $\{w\} \Vdash q$. There also exists $u \in s_1$ such that $\{u\} \Vdash p$. Hence, $\{u\} \Vdash p \wedge q$ and – by Proposition 1 – $s_1 \Vdash \Diamond(p \wedge q)$. Thus $s \Vdash [\Diamond(p \wedge q) \wedge q] \vee r$ (op. cit.).

For **DSF** and **SCH**, suppose that there is an s such that every world in s is either a $p \wedge \neg q$ world or a $\neg p \wedge q$ world. Suppose that there exists at least one $p \wedge \neg q$ world in s and at least one $\neg p \wedge q$ world in s (op. cit.).

2.8 Modal Expressivism and the Philosophy of Mathematics

When modal expressivism is modeled via speech acts on a common ground of presuppositions, the application thereof to the foundations of mathematics is limited by the manner in which necessary propositions are characterized.

Because for example a proposition is taken, according to the proposal, to be identical to a set of possible worlds, all necessarily true mathematical formulas can only express a single proposition; namely, the set of all possible worlds (see Stalnaker, 1978; 2003: 51). Thus, although distinct set-forming operations will be codified by distinct axioms of a language of set theory, the axioms will be assumed to express the same proposition: The axiom of Pairing in set theory – which states that a unique set can be formed by combining an element from each of two extant sets: $\forall x,y.\exists z.\forall w. w \in z \iff w = x \vee w = y$ – will be supposed to express the same proposition as the Power Set axiom – which states that a set can be formed by taking the set of all subsets of an extant set: $\forall x.\exists y.\forall z. z \in y \iff z \subseteq x$. However, that distinct operations – i.e., the formation of a set by selecting elements from two extant sets, by contrast to forming a set by collecting all of the subsets of a single extant set – are characterized by the different axioms is readily apparent. As Williamson (2016a: 244) writes: '...if one follows Robert Stalnaker in treating a proposition as the set of (metaphysically) possible worlds at which it is true, then all true mathematical formulas literally express the same proposition, the set of all possible worlds, since all true mathematical formulas literally express necessary truths. It is therefore trivial that if one true mathematical proposition is absolutely provable, they all are. Indeed, if you already know one true mathematical proposition (that $2 + 2 = 4$, for example), you thereby already know them all. Stalnaker suggests that what mathematicians really learn are in effect new contingent truths about which mathematical formulas we use to express the one necessary truth, but his view faces grave internal problems, and the conception of the content of mathematical knowledge as contingent and metalinguistic is in any case grossly implausible.'

Thomasson (2007) argues for a version of modal expressivism which she refers to as 'modal normativism', according to which alethic modalities are to be replaced by deontic modalities taking the form of object-language, modal indicative conditionals (op. cit.: 136, 138, 141). The modal indicative conditionals serve to express constitutive rules pertaining, e.g., to ontological dependencies which state that: 'Necessarily, if an entity satisfying a property exists then a distinct entity satisfying a property exists' (143-144), and generalizes to other expressions, such as analytic conditionals which state, e.g., that: 'Necessarily, if an entity satisfies a property, such as being a bachelor, then the entity satisfies a distinct yet co-extensive property, such as being unmarried' (148). A virtue of Thomasson's interpretation of modal indicative conditionals as expressing both analytic and ontological dependencies is that it would appear to converge with the 'If-thenist' proposal in the philosophy of mathematics. 'If-thenism' is an approach according to which, if an axiomatized mathematical language is consistent, then (i) one can either bear epistemic attitudes, such as fictive acceptance, toward the target system (see Leng, 2010: 180) or (ii) the system (possibly) exists [see Russell (op. cit.: §1)]; Hilbert (1899/1980: 39); Menger (1930/1979: 57); Putnam (1967); Shapiro (2000: 95); Chihara (2004: Ch. 10); and Awodey (2004: 60-61)].[20]

According, finally, to Brandom (op. cit.)'s modal expressivist approach, terms are individuated by their rules of inference, where the rules are taken to have a modal profile translatable into the counterfactual forms taken by the transition functions of automata (see Brandom, 2008: 142). In order

---

[20]See Leng (2009), for further discussion. Field (1980/2016: 11-21; 1989: 54-65, 240-241) argues in favor of the stronger notion of conservativeness, according to which consistent mathematical theories must be satisfiable by internally consistent theories of physics. More generally, for a class of assertions, A, comprising a theory of fundamental physics, and a class of sentences comprising a mathematical language, M, any sentences derivable from A + M ought to be derivable from A alone. See Gödel (1953/9-III: §34), for an anticipation. Another variation on the 'If-thenist' proposal is witnessed in Field (2001: 333-338), who argues that the existence of consistent forcing extensions of set-theoretic ground models adduces in favor of there being a set-theoretic pluriverse, and thus entrains indeterminacy in the truth-values of undecidable sentences. For a similar proposal, which emphasizes the epistemic role of examining how instances of undecidable sentences obtain and fail so to do relative to forcing extensions in the set-theoretic pluriverse, see Hamkins (2012: §7).

to countenance the metasemantic truth-conditions for the object-level, pragmatic abilities captured by the automata's counterfactual transition states, Brandom augments a first-order language comprised of a stock of atomic formulas with an incompatibility function (141). An incompatibility function, I, is defined as the incoherence of the union of two sentences, where incoherence is a generalization of the notion of inconsistency to nonlogical vocabulary.

$x \cup y \in Inc \iff x \in I(y)$ (141-142).

Incompatibility is supposed to be a modal notion, such that the union of the two sentences is incompossible (126). A sentence, β is an incompatibility-consequence, $\Vdash_I$, of a sentence, α, iff there is no sequence of sentences, $<\gamma_1, \ldots, \gamma_n>$, such that it can be the case that $\alpha \Vdash_I <\gamma_1, \ldots, \gamma_n>$, yet not be the case that $\beta \Vdash_I <\gamma_1, \ldots, \gamma_n>$ (125). To be incompatible with a necessary formula is to be compatible with everything that does not entail the formula (129-130). Dually, to be incompatible with a possible formula is to be incompatible with everything compatible with something compatible with the formula (op. cit.).

There are at least two, general issues for the application of Brandom's modal expressivism to the foundations of mathematics.

The first issue is that the mathematical vocabulary – e.g., the set-membership relation, $\in$ – is axiomatically defined. I.e., the membership relation is defined by, inter alia, the Pairing and Power Set axioms of set-theoretic languages. Thus, mathematical terms have their extensions individuated by the axioms of the language, rather than via a set of inference rules that can be specified in the absence of the mention of truth values. Even, furthermore, if one were to avail of modal notions in order to countenance the intensions of the mathematical vocabulary at issue – i.e., functions from terms or sentences in worlds to their extensions – the modal profile of the intensions is orthogonal to the properties encoded by the incompatibility function. Fine (2006) avails, e.g., of postulational interpretational modalities in order to countenance the possibility of reinterpreting quantifier domains, and of thus accounting for variance in the range of the domains of quantifier expressions. The interpretational possibilities are specified as operational conditions on tracking increases in the size of the cardinality of the universe. Uzquiano (2015b) argues that it is always possible to reinterpret the intensions of non-logical vocabulary, as one augments one's language with stronger axioms of infinity and climbs thereby farther up the cumulative hierarchy of sets. The reinterpretations of, e.g., the concept of set are effected by the addition of new large cardinal axioms, which stipulate the existence of larger inaccessible cardinals. However, it is unclear how the incompatibility function – i.e., a modal operator defined via Boolean negation and a generalized condition on inconsistency – might similarly be able to model the intensions pertaining to the ontological expansion of the cumulative hierarchy.

The second issue is that Brandom's inferential expressivist semantics is not compositional (Brandom, 2008: 135-136). While the formulas of the semantics are recursively formed – because the decomposition of complex formulas into atomic formulas is decidable[21] – formulas in the language are not compositional, because they fail to satisfy the subformula property to the effect that the value of a logically complex formula is calculated as a function of the values of the component logical connectives applied to subformulas therein (op. cit.).[22]

By contrast to the limits of Brandom's approach to modal expressivism, modal coalgebraic automata can circumvent both of the issues mentioned in the foregoing. In response to the first issue, concerning the axiomatic individuation and intensional profiles of mathematical terms, mappings of modal coalgebraic automata can be interpreted in order to provide a precise delineation of the (hyper-)intensions of the target vocabulary. In response, finally, to the second of the above issues, the values taken by modal coalgebraic automata are both decidable and computationally feasible, while the duality of coalgebras to Boolean-valued models of modal algebras ensures that the formulas therein retain their compositionality. The decidability of coalgebraic automata can further be witnessed by the role of modal coalgebras in countenancing the modal profile of Ω-logical consequence, where – given a proper class of Woodin cardinals – the values of mathematical formulas can remain invariant throughout extensions of the ground models comprising the set-theoretic universe (see Woodin, 2010; and chapter **10**). The individuation of large cardinals can further be characterized by the functors of modal coalgebras, when the latter are interpreted so as to countenance the elementary

---

[21]Let a decision problem be a propositional function which is feasibly decidable, if it is a member of the polynomial time complexity class; i.e., if it can be calculated as a polynomial function of the size of the formula's input [see Dean (2021) for further discussion].

[22]Note that Incurvati and Schlöder (2020) advance a multilateral inferential expressivist semantics for epistemic modality which satisfies the subformula property. (Thanks here to Luca Incurvati.) Incurvati and Schlöder (2021) extend the semantics to normative vocabulary, but it is an open question whether the semantics is adequate for mathematical vocabulary as well.

embeddings constitutive of large cardinal axioms in category theory.

2.9 Concluding Remarks

In this essay, I have endeavored to account for a mathematically tractable background against which to model both modal and hyperintensional cognitivism and modal and hyperintensional expressivism. I availed, to that end, of the duality between epistemic modal and hyperintensional algebras and modal and hyperintensional coalgebraic automata. Epistemic modal and hyperintensional algebras were shown to comprise a materially adequate fragment of the language of thought, given that models thereof figure in both cognitive psychology and artificial intelligence. With regard to conceptual engineering of intensions and hyperintensions, I introduced a novel topic-sensitive truthmaker semantics for dynamic epistemic logic as well as a novel dynamic epistemic two-dimensional hyperintensional semantics. It was then shown how the approach to modal and hyperintensional expressivism here proffered, as regimented by the modal and hyperintensional coalgebraic automata to which the epistemic modal and hyperintensional algebras are dual, avoids the pitfalls attending to the competing modal and hyperintensional expressivist approaches based upon both the inferentialist approach to concept-individuation and the approach to codifying the speech acts in natural language via intensional semantics. The present modal and hyperintensional expressivist approach was shown, e.g., to avoid the limits of the foregoing in the philosophy of mathematics, as they concerned the status of necessary propositions; the inapplicability of inferentialist-individuation to mathematical vocabulary; and failures of compositionality.

Chapter 3

Cognitivism about Epistemic Modality and Hyperintensionality

3.1 Introduction

This essay aims to vindicate the thesis that cognitive computational properties are abstract objects implemented in physical systems.[1] A recent approach to the foundations of mathematics is Homotopy Type Theory.[2] In Homotopy Type Theory, homotopies can be defined as equivalence relations on (hyper-)intensional functions. In this essay, I argue that homotopies can thereby figure in abstraction principles for two-dimensional (hyper-)intensions, i.e. functions from epistemically possible worlds or states to extensions.[3] Homotopies for two-dimensional hyperintensions thus comprise identity criteria for some cognitive mechanisms. The philosophical significance of the foregoing is twofold. First, the proposal demonstrates how epistemic modality and hyperintensionality are viable candidates for frag-

---

[1] See Turing (1950); Putnam (1967b); Newell (1973); Fodor (1975); and Pylyshyn (1978).

[2] See The Univalent Foundations Program (2013).

[3] For the first proposal to the effect that abstraction principles can be used to define abstracta such as cardinal number, see Frege (1884/1980: 68; 1893/2013: 20). For the locus classicus of the contemporary abstractionist program, see Hale and Wright (2001).

ments of the language of thought.⁴ Second, the proposal serves to delineate one conduit for our epistemic access to two-dimensional hyperintensions as abstract objects. The truth of my first-order abstraction principle for hyperintensions is grounded in its being possibly recursively enumerable i.e. Turing computable and the Turing machine being physically implementable.⁵

Bealer (1982) proffers a *non-modal* algebraic logic for intensional entities – i.e., properties, relations, and propositions – which avails of a λ-definable variable-binding abstraction operator (op. cit.: 46-48, 209-210). Bealer reduces modal notions to logically necessary conditions-cum-properties, as defined in his non-modal algebraic logic (207-209). The present approach differs from the foregoing by: (i) countenancing a *modal* algebra, on an epistemic interpretation thereof; (ii) availing of Voevodsky's Univalence Axiom in Homotopy Type Theory – which collapses identity and isomorphism – in order to provide an equivalence relation for the relevant abstraction principle; and (iii) demonstrating how the model is availed of in various branches of the cognitive sciences, such that Epistemic Modal Algebra may be considered a viable candidate for the language of thought.

Katz (1998) proffers a view of the epistemology of abstracta, according to which the syntax and the semantics for the propositions are innate (35). Katz suggests that the proposal is consistent with both a Fregean approach to propositions, according to which they are thoughts formed by the composition of senses, and a Russellian approach, according to which they are structured tuples of non-conceptual entities (36). He endorses an account of senses according to which they are correlated to natural language sentence types (114-115). One difference between Katz's proposal and the one here presented is that Katz rejects modal approaches to propositions, because the latter cannot distinguish between distinct contradictions (p. 38, fn. 6). Following, Lewis (1973: I.6), the present approach does not avail

---

⁴Given a metalanguage, a precedent to the current approach – which models thoughts and internal representations via possible worlds semantics – can be found in Wittgenstein (1921/1974: 2.15-2.151, 3-3.02).

⁵The proposal that epistemic intensions might be sui generis abstract objects, not reducible to sets, is proffered by Chalmers (2011: 101) who writes: 'It is even possible to introduce a special sort of abstract object corresponding to these intensions. Of course these abstract objects cannot be sets of ordered pairs. But we might think of an intension formally as an abstract object which when combined with an arbitrary scenario yields a truth value (or an extension)'. Chalmers (2014c) defines senses, i.e. cognitive modes of presentation, as constructions from intensions, and mentions axiomatic definitions of intensions (Zalta, 2001) and intensional functions in Boolean algebras (Bealer, 1993).

of impossible worlds which distinguish between distinct contradictions. For approaches to epistemic space and conceivability which do admit of impossible worlds, see Rantala (1982); Jago (2009; 2014); Berto (2014); Berto and Schoonen (2018); and Priest (2019). However, chapter 4 advances an epistemic two-dimensional truthmaker semantics, such that impossible states can be constructed which distinguish between distinct contradictory states (see Fine, 2021, for further discussion). A second difference is that, on Katz's approach, the necessity of mathematical truths is argued to consist in reductio proofs, such that the relevant formulas will be true on all interpretations, and thus true of logical necessity (39). Section **13.4.7** argues that modal axiom K, i.e. epistemic closure, is invalid for reductio proofs. However, the endeavor to reduce the necessity of mathematical truths to the necessity of logical consequence would result in the preclusion, both of cases of informal proofs in mathematics, which can, e.g., involve diagrams (see Azzouni, 2004; Giaquinto, 2008: 1.2), and of mathematical truths which obtain in axiomatizable, yet non-logical mathematical languages such as Euclidean geometry. Finally, Katz rejects abstraction principles, and thus implicit definitions for abstract objects (105-106).

In Section **2**, I provide an abstraction principle for two-dimensional hyperintensions, by availing of Voevodsky's Univalence axiom and function type equivalence relations countenanced in Homotopy Type Theory. In Section **3**, I describe how models of Epistemic Modal Algebra are availed of when perceptual representational states are modeled in Bayesian perceptual psychology; when speech acts are modeled in natural language semantics; and when knowledge, belief, intentional action, and rational intuition are modeled in philosophical approaches to the nature of propositional attitudes. This provides abductive support for the claim that Epistemic Modal Algebra is both a compelling and materially adequate candidate for a fragment of the language of thought. In Section **4**, I argue that the proposal resolves objections to the relevant abstraction principles advanced by both Dean (2016) and Linnebo and Pettigrew (2014). Section **5** provides concluding remarks.

3.2 An Abstraction Principle for Two-dimensional (Hyper-)Intensions

In this section, I specify a homotopic abstraction principle for two-dimensional (hyper-)intensions. Intensional isomorphism, as a jointly necessary and sufficient condition for the identity of intensions, is first proposed in Carnap (1947: §14). The isomorphism of two intensional structures is argued to consist in their logical, or L-, equivalence, where logical equivalence is co-extensive with the notions of both analyticity (§2) and synonymy (§15). Carnap writes that: '[A]n expression in S is L-equivalent to an expression in S' if and only if the semantical rules of S and S' together, without the use of any knowledge about (extra-linguistic) facts, suffice to show that the two have the same extension' (p. 56), where semantical rules specify the intended interpretation of the constants and predicates of the languages (4).[6] The current approach differs from Carnap's by defining the equivalence relation necessary for an abstraction principle for two-dimensional (hyper-)intensions on Voevodsky (2006)'s Univalence Axiom, which collapses identity with isomorphism in the setting of intensional type theory.[7]

Topological Semantics

In the topological semantics for modal logic, a frame is comprised of a set of points in topological space, a domain of propositions, and an accessibility relation:
$F = \langle X, R \rangle$;

---

[6] For criticism of Carnap's account of intensional isomorphism, based on Carnap (1937: 17)'s 'Principle of Tolerance' to the effect that pragmatic desiderata are a permissible constraint on one's choice of logic, see Church (1954: 66-67).

[7] Note further that, by contrast to Carnap's approach, epistemic intensions are here distinguished from contextual linguistic intensions (see chapter **6**, for further discussion of the difference between epistemic and contextual intensions), and the current work examines the philosophical significance of the convergence between epistemic intensions and formal, rather than natural, languages. For a translation from type theory to set theory – which is of interest to, inter alia, the definability of epistemic hyperintensions in the setting of set theory (see chapter **8**, below) – see Linnebo and Rayo (2012). For topological Boolean-valued models of Epistemic Set Theory – i.e., a variant of ZF with the axioms augmented by epistemic modal operators interpreted as informal provability and having a background logic satisfying S4 – see Scedrov (1985), Flagg (1985a), and Goodman (1990). For Epistemic Type Theory, see Flagg (1985b).

$X = (X_x)_{x \in X}$; and
$R = (Rxy)_{x,y \in X}$ iff $R_x \subseteq X_x \times X_x$, s.t. if Rxy, then $\exists o \subseteq X$, with x∈o s.t. $\forall y \in o(Rxy)$,
where the set of points accessible from a privileged node in the space is said to be open.[8] A model defined over the frame is a tuple, $M = \langle F, V \rangle$, with V a valuation function from subsets of points in F to propositional variables taking the values 0 or 1. Necessity is interpreted as an interiority operator on the space:
M,x ⊩ $\Box \phi$ iff $\exists o \subseteq X$, with x∈o, such that $\forall y \in o$ M,y ⊩ ϕ.

Homotopy Theory

Homotopy Theory countenances the following identity, inversion, and concatenation morphisms, which are identified as continuous paths in the topology. The formal clauses, in the remainder of this section, evince how homotopic morphisms satisfy the properties of an equivalence relation.[9]

Reflexivity

$\forall x,y:A \forall p(p : x =_A y) : \tau(x,y,p)$, with A and τ designating types, 'x:A' interpreted as 'x is a token of type A', p • q is the concatenation of p and q, refl_x: x $=_A$ x for any x:A is a reflexivity element, $\prod_{x:A} B(x)$ is a dependent function type, and e:$\prod_{x:A} \tau(a, a, \text{refl}_a)$ is a dependent function[10]:
$\forall \alpha : A \exists e(\alpha) : \tau(\alpha, \alpha, \text{refl}_\alpha)$;
p,q : (x $=_A$ y)
$\exists r \in e : p =_{(x=_A y)} q$
$\exists \mu : r = _{(p=_{(x=_A y)} q)}$ s.

Symmetry

$\forall A \forall x,y:A \exists H_\Sigma(x = y \rightarrow y = x)$

---

[8] In order to ensure that the Kripke semantics matches the topological semantics, X must further be Alexandrov; i.e., closed under arbitrary unions and intersections. Thanks here to Peter Milne.

[9] The definitions and proofs at issue can be found in the Univalent Foundations Program (2013: 2.0-2.1).

[10] A dependent function is a function type 'whose codomain type can vary depending on the element of the domain to which the function is applied' (Univalent Foundations Program (op. cit.: §1.4).

$H_\Sigma := p \mapsto p^{-1}$, such that
$\forall x{:}A(\texttt{refl}_x \equiv \texttt{refl}_x{}^{-1})$.

Transitivity

$\forall A \forall x,y{:}A \exists H_T(x = y \to y = z \to x = z)$
$H_T := p \mapsto q \mapsto p \bullet q$, such that
$\forall x{:}A[\texttt{refl}_x \bullet \texttt{refl}_x \equiv \texttt{refl}_x]$.

Homotopic Abstraction

For all type families A,B, there is a homotopy:

$H := [(f \sim g) :\equiv \prod_{x:A}(f(x) = g(x)]$, where
$\prod_{f:A \to B}[(f \sim f) \wedge (f \sim g \to g \sim f) \wedge (f \sim g \to g \sim h \to f \sim h)]$,
such that, via Voevodsky (op. cit.)'s Univalence Axiom, for all type families A,B:U, there is a function:
$\texttt{idtoeqv} : (A =_U B) \to (A \simeq B)$,
which is itself an equivalence relation:
$(A =_U B) \simeq (A \simeq B)$.

Topic-sensitive two-dimensional hyperintensions take the form,
$2D(x) = \lambda s \lambda w [\![x]\!]^{s,w} = 1$,
with s a topic-sensitive epistemic state and w a topic-sensitive metaphysical state.

Abstraction principles for two-dimensional hyperintensions take, then, the form of a function type equivalence:

- $\forall x[\#f(x) = \#g(x)] \simeq [f(x) \simeq g(x)]$.

Observational type theory countenances 'structure identity principles' which are type equivalences between identification types, and the theory is said to be observational because the type formation rules satisfy structure-preserving definitional equality. Higher observational type theory holds for propositional equality. 'The idea of higher observational type theory is to make these and analogous structural characterizations of identification types be part of their definitional inference rules, thus building the structure identity principle right into the rewrite rules of the type theory' (2023:

https://ncatlab.org/nlab/show/higher+observational+type+theory). Shulman (2022) argues that higher observational type theory is one way to make the Univalence Axiom computable. Tennant (1984; 1987; 2009; 2022; 2024) and Wright (2012a: 120) define Hume's Principle via inference rules, and higher observational type theory might be one way to make first-order abstraction principles defined via inference rules, although not higher-order abstraction principles, computable.[11]

The truth of my first-order abstraction principle for hyperintensions is grounded in its being possibly recursively enumerable i.e. Turing computable and the Turing machine being physically implementable.

I avoid the Burali-Forti paradox in my abstraction principle for two-

---

[11] Wright's inference rules for Hume's Principle are the following:
Introduction Rule:
'Γ ⇒ ∃R(F1-1G)
'Γ ⇒ #F = #G'.
Elimination Rule:
'Γ ⇒ #F = #G
'Γ ⇒ ∃R(F1-1G)'. See **§6.3.2**, for the definition of Hume's Principle.
Shulman (2022) defines proof-rules for identity and one-to-one correspondence thus:

$$\frac{A:U \qquad a:A \qquad b:A}{Id_A(A, B):U}$$

Ferrari and Adam-Day (2017) write: 'We say that a type A is contractible if there is a a:A, called the center of contraction, such that a = x for all x:A.
'isContr(A) $\overset{def}{=} \sum_{a:A} \prod_{x:A} (a = x)$'.
'If R : A → B → U is 1-1, the centers of contraction yield f : A → B and g : B → A, which form an equivalence' (Shulman: 2022).
Then,

$$\frac{R : \text{1-1-Corr}(A, B)}{R\!\uparrow\; :\; Id_U(A, B)}$$

$$\frac{A2 : Id_U(A_0, A_1)}{R\!\downarrow\; :\; \text{1-1-Corr}(A_0, A_1)}$$

$$\frac{p : \text{1-1-Corr}(A, B)}{p\!\uparrow\!\downarrow\; \equiv\; p}$$

dimensional hyperintensions because the definition is not augmented to second-order logic, like in the abstractionist foundations of mathematics, is instead taken in isolation, and the definition defines functions from classes of epistemic states taken as actual to classes of metaphysical states to extensions.[12]

3.3 Examples in Philosophy and Cognitive Science

The material adequacy of epistemic modal algebras as a fragment of the language of thought is witnessed by the prevalence of possible worlds semantics – the model theory for which is algebraic (see Blackburn et al., 2001: ch. 5) – in cognitive psychology. Possible worlds model theory is availed of in the computational theory of mind, Bayesian perceptual psychology, and natural language semantics.

Marcus (2001) writes that: 'A multilayer perceptron consists of a set of *input nodes*, one or more sets of *hidden nodes*, and a set of *output nodes* ... These nodes are attached to each other through *weighted connections*; the weights of these connections are generally adjusted by some sort of *learning algorithm* ... *Nodes* are *units* that have activation [real] values ... *Input* and *output nodes* also have *meanings* or *labels* that are assigned by an external programmer ... The *meanings* of nodes (their labels) play no direct role in the computation: a network's computations depend only on the activation values of nodes and not on the labels of those nodes' (7-8). Both a single and multiple nodes can serve to represent the variables for a target domain. A target domain for variables is universally quantified over and the function is one-one, mapping a number of inputs to an equivalent number of outputs (35-36). Models of the above algebraic rules can be defined in both classical and weighted, connectionist systems (42-45). Temporal synchrony or dynamic variable-bindings are stored in short-term memory (56-57), while information relevant to long-term variable-bindings are stored in 'binary registers' i.e. 'bits' (41, 54-56). Marcus writes of bits that: 'Operations are defined in parallel over these sets of binary bits. When a programmer issues a command to copy the contents of variable **x** into variable **y**, the computer copies in parallel each of the bits that represents variable **x** into the corresponding bits that represent variable **y**' (41). Examples of the foregoing algebraic rules on

---

[12]See Burali-Forti (1897/1967); Hodes (1984); and Hazen (1985).

variable-binding include both the syntactic concatenation of morphemes and noun phrase reduplication in linguistics (37-39, 70-72), as well as learning algorithms (45-48). Conditions on variable-binding are further examined, including treating the binding relation between variables and values as tensor products – i.e., an application of a multiplicative axiom for variables and their values treated as vectors (53-54, 105-106). In order to account for recursively formed, complex representations, which he refers to as structured propositions, Marcus argues instead that the syntax and semantics of such representations can be modeled via an ordered set of registers, which he refers to as 'treelets' (108).

A strengthened version of the algebraic rules on variable-binding can be accommodated in models of epistemic modal algebras, when the latter are augmented by cylindrifications, i.e., operators on the algebra simulating the treatment of quantification, and diagonal elements.[13] By contrast to Boolean Algebras with Operators, which are propositional, cylindric algebras define first-order logics. Intuitively, valuation assignments for first-order variables are, in cylindric modal logics, treated as possible worlds of the model, while existential and universal quantifiers are replaced by, respectively, possibility and necessity operators ($\Diamond$ and $\Box$) (Venema, 2013: 249). For first-order variables, $\{v_i \mid i < \alpha\}$ with α an arbitrary, fixed ordinal, $v_i = v_j$ is replaced by a modal constant $\mathbf{a}_{i,j}$ (op. cit: 250). The following clauses are valid, then, for a model, M, of cylindric modal logic, with $E_{i,j}$ a monadic predicate and T_i for i,j $< \alpha$ a dyadic predicate:

M,w $\Vdash$ p $\iff$ w$\in$V(p);
M,w $\Vdash$ $\mathbf{a}_{i,j}$ $\iff$ w$\in$E$_{i,j}$;
M,w $\Vdash$ $\Diamond_i \psi$ $\iff$ there is a v with wT$_i$v and M,v $\Vdash$ ψ (252).

Cylindric frames need further to satisfy the following axioms (op. cit.: 254):

1. p $\rightarrow$ $\Diamond_i$p
2. p $\rightarrow$ $\Box_i \Diamond_i$p
3. $\Diamond_i \Diamond_i$p $\rightarrow$ $\Diamond_i$p
4. $\Diamond_i \Diamond_j$p $\rightarrow$ $\Diamond_j \Diamond_i$p
5. $\mathbf{a}_{i,i}$
6. $\Diamond_i(\mathbf{a}_{i,j} \wedge p) \rightarrow \Box_i(\mathbf{a}_{i,j} \rightarrow p)$

[Translating the diagonal element and cylindric (modal) operator into,

---

[13]See Henkin et al (op. cit.: 162-163) for the introduction of cylindric algebras, and for the axioms governing the cylindrification operators.

respectively, monadic and dyadic predicates and universal quantification:
$\forall xyz[(T_i xy \wedge E_{i,j} y \wedge T_i xz \wedge E_{i,j} z) \rightarrow y = z]$ (op. cit.)]

7. $\mathbf{a}_{i,j} \iff \Diamond_k(\mathbf{a}_{i,k} \wedge \mathbf{a}_{k,j})$.

Finally, a cylindric modal algebra of dimension α is an algebra, $\mathbb{A} = \langle A, +, \bullet, -, 0, 1, \Diamond_i, \mathbf{a}_{ij} \rangle_{i,j<\alpha}$, where $\Diamond_i$ is a unary operator which is normal ($\Diamond_i 0 = 0$) and additive [$\Diamond_i(x + y) = \Diamond_i x + \Diamond_i y$] (257).

The philosophical interest of cylindric modal algebras to Marcus' cognitive models of algebraic variable-binding is that the valuation assignments to variables in the Epistemic Modal Algebra are epistemically possible worlds, while universal quantification is interpreted as epistemic necessitation. The interest of translating universal generalization into operations of epistemic necessitation is, finally, that – by identifying epistemic necessity with apriority – both the algebraic rules for variable-binding and the recursive formation of structured propositions can be seen as operations, the implicit knowledge of which is apriori.

In Bayesian perceptual psychology, the problem of underdetermination is resolved by availing of a gradational possible worlds model. The visual system is presented with a set of possibilities with regard, e.g., to the direction of a light source. So, for example, the direction of light might be originating from above, or it might be originating from below. The visual system computes the constancy, i.e. the likelihood that one of the possibilities is actual.[14] The computation of the perceptual constancy is an unconscious statistical inference, as anticipated by Helmholtz (1878)'s conjecture.[15] The constancy places, then, a condition on the accuracy of the attribution of properties – such as boundedness and volume – to distal particulars.[16]

In the program of natural language semantics in empirical and philosophical linguistics, the common ground or 'context set' is the set of possibilities presupposed by a community of speakers.[17] Kratzer (1979: 121) refers to cases in which the above possibilities are epistemic as an 'epistemic conversa-

---

[14] See Mamassian et al. (2002).

[15] For the history of the integration of algorithms and computational modeling into contemporary visual psychology, see Johnson-Laird (2004).

[16] See Burge (2010), and Rescorla (2013), for further discussion. A distinction ought to be drawn between unconscious perceptual representational states – as targeted in Burge (op. cit.) – and the inquiry into whether the properties of phenomenal consciousness have accuracy-conditions – where phenomenal properties are broadly construed, so as to include, e.g., color-phenomenal properties, as well as the property of being aware of one's perceptual states.

[17] See Stalnaker (1978).

tional background', where the epistemic possibilities are a subset of objective or circumstantial possibilities (op. cit.). Modal operators are then defined on the space, encoding the effects of various speech acts in entraining updates on the context set.[18] So, e.g., assertion is argued to provide a truth-conditional update on the context set, whereas there are operator updates, the effects of which are not straightforwardly truth-conditional and whose semantic values must then be defined relative to an array of intensional parameters (including a context – agent, time, location, et al. – and a tuple of indices).

Finally, Epistemic Modal Algebra, as a fragment of the language of thought, is able to delineate the fundamental structure of the propositional attitudes targeted in 20th century philosophy; notably knowledge, belief, intentional action, and rational intuition. In chapter **14**, I argue, e.g., that the types of intention – acting intentionally; referring to an intention as an explanation for one's course of action; and intending to pursue a course of action in the future – can be modeled as modal operators, whose semantic values are defined relative to an array of intensional parameters. E.g., an agent can be said to act intentionally iff her 'intention-in-action' receives a positive semantic value, where a necessary condition on the latter is that there is at least one world in her epistemic modal space at which – relative to a context of a particular time and location, which constrains the admissibility of her possible actions as defined at a first index, and which subsequently constrains the outcome thereof as defined at a second index – the intention is realized:

$[\![\text{Intenton-in-Action}(\phi)]\!]_w = 1$ only if $\exists w' [\![\phi]\!]^{w',c(=t,l),a,o} = 1$.

The agent's intention to pursue a course of action at a future time – i.e., her 'intention-for-the-future' – can receive a positive value only if there is a possible world and a future time, relative to which the possibility that a state, ϕ, is realized can be defined. Thus:

$[\![\text{Intention-for-the-future}(\phi)]\!]_w = 1$ only if $\exists w' \forall t \exists t' [t < t' \land [\![\phi]\!]^{w',t'} = 1]$.

In the setting of epistemic logic, epistemic necessity can further be modeled in a relational semantics encoding the properties of knowledge and belief (see Hintikka, 1962; Fagin et al., 1995; Meyer and van der Hoek, 1995; Williamson, 2009; chapters **5, 13**). In chapter **11**, I treat Gödel (1953/9-III; 1953/9-V/1995)'s conception of rational intuition as a modal operator in the setting of dynamic logic, and demonstrate how – via correspondence theory – the notion of 'intuition-of', i.e. a property of awareness of one's cognitive

---

[18]See Kratzer (op. cit.); Stalnaker (op. cit.); Lewis (1980); Heim (1992); Veltman (1996); von Fintel and Heim (2011); and Yalcin (2012).

states, can be shown to be formally equivalent to the notion of 'intuition-that', i.e. a modal operator concerning the value of the propositional state at issue. The correspondence results between (fixed point) modal propositional and bisimulation-invariant first-order logic and monadic second-order logic are advanced in van Benthem (1983; 1984/2003) and Janin and Walukiewicz (1996). Availing of correspondence theory in order to account for the relationship between the notions of 'intuition-of' and 'intuition-that' resolves the inquiry about the foregoing posed by Parsons (1993: 233). As a dynamic interpretational modality, rational intuition can further serve as a guide to possible reinterpretations both of quantifier domains (see Fine, 2005b) and of the intensions of mathematical vocabulary such as the membership-relation (see Uzquiano, 2015a). This provides an account of Gödel (op. cit.; 1961)'s suggestion that rational intuition can serve as a guide to conceptual elucidation.

3.4 Objections and Replies

Dean (2016) raises two issues for a proposal similar to the foregoing, namely that algorithms – broadly construed – can be defined via abstraction principles which specify equivalence relations between implementations of computational properties in isomorphic machines.[19] Dean's candidate abstraction principle for algorithms as abstracts is: that the algorithm implemented by M_1 = the algorithm implemented by M_2 iff $M_1 \simeq M_2$.[20] Both issues target the uniqueness of the algorithm purported to be identified by the abstraction principle.

The first issue generalizes Benacerraf (1965)'s contention that, in the reduction of number theory to set theory, there must be, and is not, a principled reason for which to prefer the identification of natural numbers with von Neumann ordinals (e.g., $2 = \{\varnothing, \{\varnothing\}\}$), rather than with Zermelo ordinals

---

[19]Fodor (2000: 105, n.4) and Piccinini (2004) note that the identification of mental states with their functional roles ought to be distinguished from identifying those functional roles with abstract computations. Conversely, a computational theory of mind need not be committed to the identification of abstract, computational operations with the functional organization of a machine. See chapter **11**, for further discussion of the computational profile of rational intuition, with rational intuition being contrasted to mechanism. Identifying abstract computational properties with the functional organization of a creature's mental states is thus a choice point, in theories of the nature of mental representation.

[20]See Moschovakis (1998).

(e.g., $2 = \{\{\emptyset\}\}$). The issue is evinced by the choice of whether to define algorithms as isomorphic *iterations* of state transition functions (see Gurevich, 1999), or to define them as isomorphic *recursions* of functions which assign values to a partially ordered set of elements (see Moschovakis, op. cit.). Linnebo and Pettigrew (2014: 10) argue similarly that, for two 'non-rigid' structures which admit of non-trivial automorphisms, one can define a graph which belies their isomorphism. E.g., let an abstraction principle be defined for the isomorphism between S and S*, such that

$\forall S, S^* [\mathbf{AS} = \mathbf{AS}^*$ iff $\langle S, R_1 \ldots R_n \rangle \simeq \langle S^*, R^*_1 \ldots R^*_n \rangle]$.

However, if there is a graph, G, such that:
$S = \{v_1, v_2\}$, and $R = \{\langle v_1, v_2 \rangle, \langle v_2, v_1 \rangle\}$,

then one can define an automorphism, $f : G \simeq G$, such that $f(v_1) = v_2$ and $f(v_2) = v_1$, such that $S^* = \{v_1\}$ while $R^* = \{\langle v^*_1, v^*_1 \rangle\}$. Then S* has one element via the automorphism, while S has two. So, S and S* are not, after all, isomorphic.

The second issue is that complexity is crucial to the identity criteria of algorithms. Two algorithms might be isomorphic, while the decidability of one algorithm is proportional to a deterministic *polynomial* function of the size of its input – with k a member of the natural numbers, N, and TIME referring to the relevant complexity class: $\bigcup_{k \in N} \text{TIME}(n^k)$ – and the decidability of the second algorithm will be proportional to a deterministic *exponential* function of the size of its input – $\bigcup_{k \in N} \text{TIME}(2^{n^k})$. The deterministic polynomial time complexity class is a subclass of the deterministic exponential time complexity class. However, there are problems decidable by algorithms only in polynomial time (e.g., the problem of primality testing, such that, for any two natural numbers, the numbers possess a greatest common divisor equal to 1), and only in exponential time (familiarly from logic, e.g., the problem of satisfiability – i.e., whether, for a given formula, there exists a model which can satisfy it – and the problem of validity – i.e. whether a satisfiable formula is valid).[21]

Both issues can be treated by noting that Dean's discussion targets abstraction principles for the very notion of a computable function, rather than for abstraction principles for cognitive computational properties. It is a virtue of homotopic abstraction principles for cognitive intensional functions that both the temporal complexity class to which the functions belong, and the applications of the model, are subject to variation. Variance in the cognitive

---

[21] For further discussion, see Dean (2015).

roles, for which Epistemic Modal Algebra provides a model, will crucially bear on the nature of the representational properties unique to the interpretation of the intensional functions at issue. Thus, e.g., when the internal representations in the language of thought – as modeled by Epistemic Modal Algebra – subserve perceptual representational states, then their contents will be individuated by both the computational constancies at issue and the external, environmental properties – e.g., the properties of lightness and distance – of the perceiver.[22]

The examples of instances of Epistemic Modal Algebra – witnessed by the possible worlds models in Bayesian perceptual psychology, linguistics, and philosophy of mind – provide abductive support for the existence of the intensional functions specified in homotopic abstraction principles. The philosophical significance of independent, abductive support for the existence of epistemic modalities in the philosophy of mind and cognitive science is that the latter permits a circumvention of the objections to the abstractionist foundations of number theory that have accrued since its contemporary founding (see Wright, 1983). Eklund (2006) suggests, e.g., that the existence of the abstract objects which are the referents of numerical term-forming operators might need to be secured, prior to assuming that the abstraction principle for cardinal number is true. While Hale and Wright (2009) maintain, in response, that the truth of the relevant principles will be prior to the inquiry into whether the terms defined therein refer, they provide a preliminary endorsement of an 'abundant' conception of properties, according to which identifying the sense of a predicate will be sufficient for predicate reference.[23] One aspect of the significance of empirical and philosophical instances of models of Epistemic Modal Algebra is thus that, by providing independent, abductive support for the truth of the homotopic abstraction principles for epistemic hyperintensions, the proposal remains neutral on the status of 'sparse' versus 'abundant' conceptions of properties.[24] The truth

---

[22]The computational properties at issue can also be defined over non-propositional information states, such as cognitive maps possessed of geometric rather than logical structure. See, e.g., O'Keefe and Nadel (1978); Camp (2007); and Rescorla (2009).

[23]For identity conditions on abundant properties – where the domain of properties, in the semantics of second-order logic, is a subset of the domain of objects, and the properties are definable in a metalanguage by predicates whose satisfaction-conditions have been fixed – see Hale (2013b). For a generalization of the abundant conception, such that the domain of properties is isomorphic to the power set of the domain of objects, see Cook (2014b).

[24]Finding abductive support for abstraction principles is suggested by Rayo (2003). Hale and Wright (2009) and Wright (2012a, 2014, 2016) argue that there is prima facie,

of my first-order abstraction principle for hyperintensions is, as mentioned, grounded in its being possibly recursively enumerable i.e. Turing computable and the Turing machine being physically implementable. Another aspect of the philosophical significance of possible worlds semantics being availed of in Bayesian vision science and empirical linguistics is that it belies the purportedly naturalistic grounds for Quine (1963/1976)'s scepticism of *de re* modality.[25]

3.5 Concluding Remarks

In this essay, Voevodsky's Univalence Axiom and function type equivalence in Homotopy Type Theory were availed of, in order to specify an abstraction principle for hyperintensional, computational properties. The homotopic abstraction principle for two-dimensional hyperintensions provides an epistemic conduit for our knowledge of hyperintensions as abstract objects. The truth of my first-order abstraction principle for hyperintensions is grounded in its being possibly recursively enumerable i.e. Turing computable and the Turing machine being physically implementable. Because hyperintensions in Epistemic Modal Algebra are deployed as core models in the philosophy of mind, Bayesian visual psychology, and natural language semantics, there is independent abductive support for the truth of homotopic abstraction. Epistemic modality and hyperintensionality may thereby be recognized as both compelling and materially adequate candidates for the fundamental structure of mental representational states, and as thus comprising a fragment of the language of thought.

---

default non-evidential entitlement to accept that abstraction principles are true.

[25]See Barcan Marcus (1993: 66-67), for a defense of Aristotelian essentialism, according to which essentialist modalities are temporal and 'causal and physical modalities'. Barcan Marcus writes, too, that 'What has gone wrong in recent discussions of essentialism is the assumption of surface synonymy between "is essentially" and *de re* occurrences of "is necessarily"' (60), and examines the distinction in various systems of quantified modal logic (Ch. 4, §III).

Chapter 4

Topic-Sensitive Two-Dimensional Truthmaker Semantics

4.1 Introduction

Philosophical applications of two-dimensional semantics have demonstrated that an account of representation which is sensitive to an array of parameters can play a crucial role in explaining the values of linguistic expressions (Kamp, 1967; Kaplan, 1979); the role of speech acts in affecting shared contexts of information (Stalnaker, 1978; Lewis, 1980a/1998; MacFarlane, 2005); the relationship between conceivability and metaphysical possibility (Chalmers, 1996); and the viability of modal realism (Russell, 2010).

In order to circumvent issues for the modal analysis of counterfactuals (2012a,b), and to account for the general notions of aboutness and subject matters (2015), a hyperintensional, 'truthmaker' semantics has recently been developed by Fine (2017a,b). In this essay, I examine the status of two-dimensional indexing in truthmaker semantics, and specify the two-dimensional profile of the grounds for the truth of a formula (**§4.2-4.3**). I proceed, then, to outline three novel interpretations of the two-dimensional, hyperintensional framework, beyond the interpretations of multiply indexed intensional semantics that are noted above (**§4.4**). The first interpretation provides a formal setting in which to define the distinction between fundamental and derivative truths. The second interpretation concerns the interaction between the two-dimensional profile of the verifiers for a proposition, subjective probability, and decision theory. Finally, a third interpretation of the two-dimensional hyperintensional framework concerns the types of intentional action. I demonstrate, in particular, how multiply indexed truthmaker semantics is able to resolve a puzzle concerning the role of intention in action. Section 5 provides concluding remarks.

4.2 Two-dimensional Truthmaker Semantics

4.2.1 Intensional Semantics

In his (1979/1985), Evans endeavors to account for the phenomenon of the contingent apriori by distinguishing between two types of modality. Suppose that the descriptive name, 'Plotinus', is introduced via the reference fixer, 'the author of the *The Enneads*'. Then the sentence, 'if anyone uniquely is the author of *The Enneads*, then Plotinus is the author of the *The Enneads*' is 'epistemically equivalent' to the sentence, 'if anyone uniquely is the author of *The Enneads*, then the author of the *The Enneads* is the author of the *The Enneads*' (see Hawthorne, 2002). Informative identity statements – such as that (i) 'Plotinus = the author of *The Enneads*' – are thus taken to be epistemically equivalent to vacuously true identity statements – e.g., (ii) 'Plotinus = Plotinus' (op. cit.: 177). The apriority of the vacuously true identity statement is thus argued to be a property of the informative identity statement, as well. A premise in the argument is that definite descriptions are non-referring, although – in free logic – still enable the sentences in which they figure to bear a positive, classical value. [See Evans (1979/1985: 167-169).] In free logic, closed formulas may receive a positive, classical semantic value when the terms therein have empty extensions (166). However, the informative identity statement is contingent. For example, it is metaphysically possible that the author of *The Enneads* is Plato, rather than Plotinus.

Evans distinguishes between a 'deep' type of contingency according to which a sentence is possibly true only if it is made true by a state of affairs (1979/1985: 185), and 'superficial' contingency which consists in that superficial contingency records the possible values of a formula when it embeds within the scope of a modal operator, e.g., 'possibly, x is red' and 'possibly, x is blue'. Two distinct propositions can express the same 'sense or content' (182), and are then epistemically equivalent and have the same truth conditions, although not the same truth conditions at a world, true$_w$, because they can have different values when embedded in distinct modal contexts (op. cit.: §III).[1] Evans' example for the two propositions is ϕ and Actually(ϕ) (210). In light of the approach to apriority which proceeds via targeting the content or epistemic equivalence of, for example, the propositions, ϕ and Actually(ϕ), or (i) and (ii) when both are true in the actual world, the content of formulas might thus be apriori and yet superficially contingent because 'there is no contingent feature of reality on which its truth depends' (211-212), and yet possibly the content is false.[2] Thus, distinct propositions embedding in distinct modal contexts can express the same content, and there is, too, an example of the contingent apriori.[3]

---

[1] Evans ties the notion of 'sense or content' (1979/1985: 208), i.e. epistemic equivalence, to belief. Evans writes: 'I shall not attempt to give an analysis of the notion of content here; I want to rely upon the intuitive sense according to which, if two sentences have the same content, then what is believed by one who understands and accepts the one sentence as true is the same as what is believed by one who understands and accepts the other sentence as true. On this, very strict, view of sameness of content, if two sentences have the same content, and a person understands both, then he cannot believe what one sentence says and disbelieve what the other sentence says. When two sentences meet this condition, I shall say that they are epistemically equivalent' (200).

[2] Kripke (1972/1980) anticipates Evans (1979/1985)'s example of the contingent apriori. Kripke (1980: 56, fn. 21) writes: 'What then, is the *epistemological* status of the statement "Stick S is one meter long at t_0", for someone who has fixed the metric system by reference to stick S? It would seem that he knows it *a priori*. For if he used stick S to fix the reference of the term "one meter", then as a result of this kind of "definition" (which is not an abbreviative or synonymous definition), he knows automatically, without further investigation, that S is one meter long.^{Footnote21} [Footnote 21: 'Since the truth he knows is contingent, I choose *not* to call it "analytic", stipulatively requiring analytic truths to be both necessary and *a priori*. See footnote 63.]'; 'I am presupposing that an analytic truth is one which depends on the *meanings* in the strict sense and therefore is necessary as well as *a priori*. If statements whose *a priori* truth is known via fixing of a reference are counted as analytic, then some analytic truths are contingent; this possibility is excluded in the notion of analyticity adopted here' (1980: 122, fn. 63).

See chapter **1**, fn. 4, for the relation between Kripke's definition of analyticity and super-rigidity.

[3] Evans' approach is defined within a single space of metaphysically possible worlds. See Davies (2004: 3.1). However, one could define the value of a formula relative to two spaces: A space of epistemic possibilities and a space of metaphysical possibilities. By contrast to securing apriority by (iii) eliding the values of informative and vacuous identity statements in a free logic within a single space of metaphysical possibilities, and then (iv) arguing that apriori identity statements are superficially contingent because possibly false, an alternative approach argues that an identity statement is contingent apriori if and only

Two-dimensional semantics provides a framework for regimenting the thought that the value of a formula relative to one parameter determines the value of the formula relative to another parameter. The semantics assigns truth-conditions to formulas, and semantic values to the formula's component terms. The conditions of the formulas and the values of their component terms are assigned relative to the array of intensional parameters. So, e.g., a term may be defined relative to a context; and the value of the term relative to the context will determine the value of the term relative to an index.

Primary, secondary, and 2D intensions can be defined as follows:

- Primary Intension:

  $\mathtt{pri}(x) = \lambda c. [\![x]\!]^{c,c}$.

  (The intension is a function mapping formulas, relative to two parameters ranging over possibilities from a first space, to truth-values.);

- Secondary Intension:

  $\mathtt{sec}_{v_@}(x) = \lambda w. [\![x]\!]^{v_@, w}$.

  (The intension is a function mapping formulas, relative to two parameters, where the first ranges over worlds, one of which is designated as actual, which determines the value of the formula relative to the second parameter ranging over worlds from a distinct space. The secondary intension picks out the semantic value of the formula relative to the second parameter.);

- 2D-Intension:

  $\mathtt{2D}(x) = \lambda c \lambda w [\![\mathrm{x}]\!]^{c,w} = 1$.

  (The intension determines a semantic value relative to two parameters, the first ranges over worlds from a first space and the second ranges over

---

if it is (v) apriori, because the statement is necessarily true in epistemic modal space, while the statement is (vi) contingent, because possibly the statement is false in metaphysical modal space. For further discussion, see Chapter **8**, fn. 2; Kripke (1980: 59, fn. 22); and Chalmers (2006b). Kripke writes: 'In the formal semantics of modal logic, the 'sense' of a term t is usually taken to be the (possibly partial) function which assigns to each possible world H the referent of t in H. For a rigid designator, such a function is constant. This notion of "sense" relates to that of "giving a meaning", not that of fixing a reference. In this use of "sense", "one meter" has a constant function as its sense, though its reference is fixed by "the length of S", which does not have a constant function as its sense' (1980: 59, fn. 22).

worlds from a distinct, second space. The value of the formula relative to the first parameter determines the value of the formula relative to the second.)

Interpretations of the intensions include the following. According to Kaplan (1979), an utterance's character is a mapping from the utterance's context of evaluation to the utterance's content. According to Stalnaker (op. cit.; 2004), having distinct functions associated with the value of an utterance provides one means of reconciling the necessity of a formula presupposed by speakers with the contingency of the values of assertions made about that formula.

According to Chalmers (op. cit.), there are cases in which the value of a formula relative to a first parameter, which ranges over epistemically possible worlds, determines the value of a formula relative to a second parameter, which ranges over metaphysically possible worlds. The dependence is recorded by 2D-intensions. Epistemic possibility entails metaphysical possibility in cases in which terms or formulas are, furthermore, 'super-rigid' (2012: 474), i.e. have a 'constant two-dimensional intension' (370), i.e. map to the same truth-value in all epistemically possible worlds and all metaphysically possible worlds (369), and with regard to which $\blacksquare(\square\phi)$.[4]

According to Lewis (op. cit.), the context may be treated as a concrete situation ranging over individuals, times, locations, and worlds; and the index may be treated as ranging over shiftable parameters of the context. According to MacFarlane (op. cit.), formulas may receive their value relative to a context of use, a context of evaluation, and the value of the formula may yet be defined relative to a third context ranging over the states of an independent, third assessor. Finally, in decision theory, the value of a formula relative to a context, which ranges over a time, location, and agent, constrains the value of the formula relative to a first index on which a space of the agent's possible acts is built, and the latter will subsequently constrain the value of the formula relative to a second index on which a space of possible outcomes may be built.

---

[4]See Chapter **1**, fn. 4, for further discussion. Thanks here to Dave Chalmers (private correspondence).

4.2.2 Truthmaker Semantics

Truthmaker semantics has been applied, in order to explain the conditions under which parts of worlds, rather than worlds in their entirety, verify propositions.

Truthmaker semantics is defined over a state space, $F = \langle S, \sqsubseteq \rangle$, where S is a set of states which are parts of a world, and $\sqsubseteq$ is a parthood relation on S which is a partial order, such that it is reflexive, anti-symmetric, and transitive (2017a: 19).

A proposition $P \subseteq S$ is verifiable if P is non-empty, and is otherwise unverifiable (20).

A model, M, over F is a tuple, $M = \langle F, D, V \rangle$, where D is a domain of closed formulas (i.e. propositions), and V is an assignment function mapping propositions $P \in D$ to pairs of subsets of S, $\{1, 0\}$, i.e. the verifier and falsifier of P, such that $[\![P]\!]^+ = 1$ and $[\![P]\!]^- = 0$ (35).

s exactly verifies p if and only if $s \vdash p$ if $s \in [\![p]\!]$;

s inexactly verifies p if and only if $s \vartriangleright p$ if $\exists s' \leqslant S$, $s' \vdash p$; and

s loosely verifies p if and only if, $\forall v$, s.t. $s \sqcup v \vdash p$, where $\sqcup$ is the relation of compatibility (35-36);

Differentiated contents may be defined as follows.[5] A state $s \sqsubseteq S$ is differentiated only if s is the fusion of distinct parts, s.t. $s = s_1 \sqcup s_2$. There is thus an initial state, s_1; an additional state, s_2; and a total state, s. The three states correspond accordingly to three contents: The initial content $s_1 \vdash P_1$; the additional content, $s_2 \vdash P_2$; and the total content, $s \vdash P_{1,2}$ (2017b: 15).

Finally, subject matters may be defined as follows.

A verifiable proposition, $[\![P]\!]^+$, is about a positive subject matter, $\mathbf{p}^+$ (20-21).

A falsifiable proposition, $[\![P]\!]^-$ is about a negative subject matter, $\mathbf{p}^-$ (21).

The intersection of the subject matters both verified and falsified by the fusion of a number of states comprise a comprehensive subject matter:

$\mathbf{p}_{1,+,-} = \mathbf{p}_{1,+} \sqcap \mathbf{p}_{1,-} = \langle s \vdash P \text{ and } s \dashv P \rangle$;
$\mathbf{p}_{2,+,-} = \mathbf{p}_{2,+} \sqcap \mathbf{p}_{2,-} = \langle s \vdash P_2 \text{ and } s \dashv P_2 \rangle$; s.t.,
$\mathbf{p}_{1,2,+,-} = \mathbf{p}_{1,2,+} \sqcap \mathbf{p}_{1,2,-} = \langle s \vdash P_{1,2} \text{ and } s \dashv P_{1,2} \rangle$ (op. cit.).

---

[5] Fine (op. cit.: 8, 12) avails of product spaces in his discussion of content and subject matter, though we continue here to work with a single space for ease of exposition.

The union of the subject matters that are either verified or falsified by the fusion of a number of states comprise a differentiated subject matter:
$\mathbf{p}_{1,+/-} = \mathbf{p}_{1,+} \sqcup \mathbf{p}_{1,-} = \langle s \vdash P \text{ or } s \dashv P \rangle$;
$\mathbf{p}_{2,+,-} = \mathbf{p}_{2,+} \sqcup \mathbf{p}_{2,-} = \langle s \vdash P_2 \text{ or } s \dashv P_2 \rangle$; s.t.,
$\mathbf{p}_{1,2,+/-} = \mathbf{p}_{1,2,+} \sqcup \mathbf{p}_{1,2,-} = \langle s \vdash P_{1,2} \text{ or } s \dashv P_{1,2} \rangle$ (op. cit.).
Informally, propositions P and Q are about
P is exactly about Q if $\mathbf{p} = \mathbf{q}$;
P is partly about Q if $\mathbf{p}$ and $\mathbf{q}$ overlap, such that $\exists u \sqsubset S(u \vdash R)$; $\forall s_1, s_2 \subseteq S$, $s_1 \vdash P$, $s_2 \vdash Q$; and $u = s_1 \sqcap s_2$, such that $R = P \cap Q$;
P is entirely about Q if $\mathbf{p} \subseteq \mathbf{q}$; and
P is about Q in its entirety if $\mathbf{p} \supseteq \mathbf{q}$ (5).

4.2.3 Two-dimensional Truthmaker Semantics

In order to account for two-dimensional indexing, we augment the model, M, with a second state space, S*, on which we define both a new parthood relation, $\sqsubset^*$, and partial function, V*, which serves to map propositions in D to pairs of subsets of S*, $\{1, 0\}$, i.e. the verifier and falsifier of P, such that $[\![P]\!]^+ = 1$ and $[\![P]\!]^- = 0$. Thus, $M = \langle S, S^*, D, \sqsubset, \sqsubset^*, V, V^* \rangle$. The two-dimensional hyperintensional profile of propositions may then be recorded by defining the value of P relative to two parameters, c,i: c ranges over subsets of S, and i ranges over subsets of S*.

(*) M, s∈S, s*∈S* $\vdash$ P iff:
(i) $\exists c_s [\![P]\!]^{c,c} = 1$ if s∈$[\![P]\!]^+$; and
(ii) $\exists i_{s*} [\![P]\!]^{c,i} = 1$ if s*∈$[\![P]\!]^+$

(Distinct states, s, s*, from distinct state spaces, S, S*, provide a two-dimensional verification for a proposition, P, if the value of P is provided a truthmaker by s. The value of P as verified by s determines the value of P as verified by s*).

We say that P is hyper-rigid iff:

(**) M, s∈S, s*∈S* $\vdash$ P iff:
(i) $\forall c'_s [\![P]\!]^{c,c'} = 1$ if s∈$[\![P]\!]^+$; and
(ii) $\forall i_{s*} [\![P]\!]^{c,i} = 1$ if s*∈$[\![P]\!]^+$
Hyper-rigidity is the analogue of super-rigidity in the hyperintensional setting.

The foregoing provides a two-dimensional hyperintensional semantic framework within which to interpret the values of a proposition. Two-dimensional truthmakers can further be exact, inexact, or loose:

s is a two-dimensional exact truthmaker of P if and only if (*);

s is a two-dimensional inexact truthmaker of P if and only if $\exists s' \sqsubset S$, s $\rightarrow$ s', s' $\vdash$ P and such that
$\exists c_{s'} \llbracket P \rrbracket^{c,c} = 1$ if s'$\in \llbracket P \rrbracket^+$, and
$\exists i_{s*} \llbracket P \rrbracket^{c,i} = 1$ if s*$\in \llbracket P \rrbracket^{+6}$;

s is a two-dimensional loose truthmaker of P if and only if, $\exists s'$, s.t. s $\sqcup$ s' $\vdash$ P:
$\exists c_{s \sqcup t} \llbracket P \rrbracket^{c,c} = 1$ if s'$\in \llbracket P \rrbracket^+$, and
$\exists i_{s*} \llbracket P \rrbracket^{c,i} = 1$ if s*$\in \llbracket P \rrbracket^+$.

- $\llbracket P \rrbracket^{c,i}$ is exactly about $\llbracket Q \rrbracket^{c,i}$ if $f_{1-1}[\mathbf{p}^{c,i} \iff \mathbf{q}^{c,i}]$

  (Suppose that the values of P and of Q are two-dimensionally determined, as above. Then P is exactly about Q if there is a bijection between the two-dimensionally individuated subject matters that they express)[7];

---

[6]'x $\rightarrow$ x" is read as claiming that the state, x, is extended by the state, x', while not forming a fusion of states, rather than as entailment or containment. Thanks here to Francesco Berto, for discussion.

[7]Fritz and Goodman (2017) write 'For a given sentence S involving a modalized use of a generalized quantifier, we will count a sentence P as a paraphrase of S just in case P expresses the proposition conveyed by S' (1067). Fritz and Goodman take the standard of content-preserving paraphrase to be what they term 'metaphysical equivalence' (op. cit.). Metaphysical equivalence assumes a unique, coarse-grained approach to delineating propositions, according to which the following sentences express the same proposition:
(α) 'Hesperus is a planet',
(β) 'Hesperus is a planet $\wedge$ Hesperus is a planet',
(γ) 'Phosphorus is a planet' (6).
A Russellian approach to propositions would distinguish, by contrast, the propositions expressed by α and β while identifying α and γ. A Fregean approach to propositions would distinguish each of α, β, and γ.

Metaphysical equivalence is taken, further, to be distinct from logical and interpretational equivalence. Two languages are logically equivalent iff they are expressively equivalent, and a language can be interpreted in another iff the languages prove the same theorems. Fritz and Goodman note, however, that it is sufficient for the paraphrase strategy that a language with modalized quantifiers can be translated into one without such quantifiers (1070).

Finally, Fritz and Goodman rule-out modal approaches to propositions – according to which two sentences express the same proposition in virtue of being necessarily equivalent,

- $\llbracket P \rrbracket^{c,i}$ is partly about $\llbracket Q \rrbracket^{c,i}$ if **p** and **q** overlap, s.t. $\exists u \sqsubseteq S$, s.t. $u \vdash R$, and $\forall s_1, s_2 \sqsubseteq S$, $s_1 \vdash P$, $s_2 \vdash Q$, and $u = s_1 \sqcap s_2$ such that $R^{c,c} = P \cap Q$. A function, A, maps u to a state s* in i where $s^* \vdash R^{c,i}$.

- $\llbracket P \rrbracket^{c,i}$ is entirely about $\llbracket Q \rrbracket^{c,i}$ if $\mathbf{p}^{c,i} \Leftarrow \mathbf{q}^{c,i}$

  (Suppose that the values of P and of Q are two-dimensionally determined. Then P is entirely about Q if there is a surjection from the subject matter of Q onto the subject matter of P);

- $\llbracket P \rrbracket^{c,i}$ is about $\llbracket Q \rrbracket^{c,i}$ in its entirety if $\mathbf{p}^{c,i} \Rightarrow \mathbf{q}^{c,i}$

  (Suppose that the values of P and of Q are two-dimensionally determined. Then P is about Q in its entirety if there is an injection from the subject matter of P onto the subject matter of Q).

4.3 Topic-sensitive Two-dimensional Truthmaker Semantics

Following the presentation of topic models in Berto (2018; 2019), Canavotto et al (2022), and Berto and Hawke (2021), the diamond, box, and least and greatest fixed point operators can be sensitive to topics, i.e. hyperintensional atomic subject matters. Topic fusion is a binary operation, such that for all x, y, z∈T, where T is a set of topics, the following properties are satisfied: idempotence ($x \oplus x = x$), commutativity ($x \oplus y = y \oplus x$), and associativity

---

such that all necessary sentences express the same proposition and would thus define the same classes of models (1.4). One issue for the modal approach is that Fritz and Goodman assume a variable domain semantics for their universe of models. So – given the variable domain semantics, s.t. the cardinality of worlds can vary across models – whether formulas are true and thus necessarily so will vary in different models, s.t. necessary formulas will define different classes of models. Another issue for the modal approach is Fritz and Goodman's ultimate appeal to intensions with impossible values.

A class of models is *probative*, only if: 'for any two sentences interpreted over those models, the two sentences convey the same proposition, only if *they define the same class of those models*' [my emphasis - D.E.] (op. cit.).

Fritz and Goodman note that the probativity condition places a necessary condition which can act as a heuristic negative constraint on successful paraphrases. They concede, however, that probativity is itself insufficient to account for successful, systematic paraphrase, given that probative classes of models have to be augmented by 'independent judgments of propositional granularity' (1072).

$[(x \oplus y) \oplus z = x \oplus (y \oplus z)]$ (Berto, 2018: 5). Topic parthood is a partial order, $\leqslant$, defined as $\forall x,y \in T(x \leqslant y \iff x \oplus y = y)$ (op. cit.: 5-6). Atomic topics are defined as follows: $\text{Atom}(x) \iff \neg \exists y < x$, with $<$ a strict order. Topic parthood is thus a partial ordering such that, for all x, y, z∈T, the following properties are satisfied: reflexivity ($x \leqslant x$), anti-symmetry ($x \leqslant y \wedge y \leqslant x \to x = y$), and transitivity ($x \leqslant y \wedge y \leqslant z \to x \leqslant z$) (6). For formulas, ϕ, atomic formulas, p, q, r ($p_1, p_2, \ldots$), and a set of atomic topics, $Ut\phi = \{p_1, \ldots p_n\}$, the topic of ϕ, $t(\phi) = \oplus Ut\phi = t(p_1) \oplus \ldots \oplus t(p_n)$ (op. cit.).

If a formula is two-dimensional and the two parameters for the formula range over distinct spaces, then there won't be only one subject matter for the formula, because total subject matters are construed as sets of verifiers and falsifiers and there will be distinct verifiers and falsifiers relative to each space over which each parameter ranges. This is especially clear if one space is interpreted epistemically and another is interpreted metaphysically. Availing of topics, however, and assigning the same topics to each of the states from the distinct spaces relative to which the formula gets its value is one way of ensuring that the two-dimensional formula has a single subject matter.

A topic frame can then be defined as $\{W, R, T, \oplus, t\}$, with t a valuation function assigning atomic topics to atomic formulas, and the diamond, box, and least and greatest fixed point operators can then be defined relative to topics:

$\langle M, w \rangle \Vdash \Diamond^t \phi$ iff $\langle R_{w,t} \rangle(\phi)$
$\langle M, w \rangle \Vdash \Box^t \phi$ iff $[R_{w,t}](\phi)$, with
$\langle R_{w,t} \rangle(\phi) := \{w' \in W t' \in T \mid R_{w,t}[w', t'] \cap \phi \neq \emptyset$ and $t'(\phi) \leqslant t(\phi)\}$
$[R_{w,t}](\phi) := \{w' \in W t' \in T \mid R_{w,t}[w', t'] \subseteq \phi$ and $t'(\phi) \leqslant t(\phi)\}$
$\langle M, w \rangle \Vdash \mu x.\phi^t$ iff $\bigcap \{U \subseteq W \mid \phi^t \subseteq U\}$
$\langle M, w \rangle \Vdash \nu x.\phi^t$ iff $\bigcup \{U \subseteq W \mid U \subseteq \phi^t\}$.

Hyperintensions can then be defined as functions from world, topic pairs to extensions.

- Epistemic Hyperintension:

  $\mathtt{pri}_t(x) = \lambda c \lambda t. \llbracket x \rrbracket^{c \cap t, c \cap t}$,

- Subjunctive Hyperintension:

  $\mathtt{sec}_{v @ \cap t}(x) = \lambda w \lambda t. \llbracket x \rrbracket^{v @ \cap t, w \cap t}$

- 2D-Hyperintension:

$$2D(x) = \lambda c \lambda w \lambda t [\![x]\!]^{c \cap t, w \cap t} = 1.$$

We can also combine topics with truthmakers rather than worlds, thus countenancing multi-hyperintensional semantics, i.e. topic-sensitive epistemic two-dimensional truthmaker semantics:

- Topic-Sensitive Epistemic Hyperintension:
$\text{pri}_t(x) = \lambda s \lambda t.[\![x]\!]^{s \cap t, s \cap t}$, with s a truthmaker from an epistemic state space.

- Topic-Sensitive Subjunctive Hyperintension:
$\text{sec}_{v_@ \cap t}(\text{x}) = \lambda w \lambda t.[\![x]\!]^{v_@ \cap t, w \cap t}$, with w a truthmaker from a metaphysical state space.

- Topic-Sensitive 2D-Hyperintension:
$2D(x) = \lambda s \lambda w \lambda t [\![x]\!]^{s \cap t, w \cap t} = 1.$

In Chapter **2**, my dynamic logic for conceptual engineering accounts for topic-preservation at the level of variables.[8]

4.4 New Interpretations

The two-dimensional account of truthmaker semantics provides a general framework in which a number of interpretations of the state spaces at issue can be defined. The framework may accommodate, e.g., the 'metasemantic' and 'epistemic' interpretations of the framework. The metasemantic interpretation accommodates the update effects of contingently true assertions on a context set with regard to necessary propositions (see Stalnaker, op. cit.). The framework may further be provided an epistemic interpretation, in order to countenance hyperintensional distinctions in the relations between conceivability, i.e. the space of an agent's epistemic states, and metaphysical possibility, i.e. the state space of facts (see Chalmers, op. cit.). Chapter **2** outlines an epistemic two-dimenisonal truthmaker semantics in detail, and epistemic two-dimensional semantics, both intensional and truthmaker, are

---

[8]The issue of topic-preservation in conceptual engineering is discussed by Haslanger (2000, 2021), Sawyer (2018; 2020), Cappelen (2018; 2020), Prinzing (2018), Pinder (2020), Nado (2021), and Koch (2023).

applied in Part III. In this section, I advance three novel interpretations of two-dimensional semantics, as witnessed by the new relations induced by the interaction between two-dimensional indexing and hyperintensional value assignments. The three interpretations concern (i) the distinction between fundamental and derivative truths; (ii) probabilistic grounding in the setting of decision theory; and (iii) the structural contents of the types of intentional action.

4.4.1 Fundamental and Derivative Truths

The first novel interpretation concerns the distinction between fundamental and derivative truths. In the foregoing model, the value of the subject matter expressed by a proposition may be verified by states in a first space, which determine, then, whether the proposition is verified by states in a second space. Allowing the first space to be interpreted so as to range over fundamental facts and the second space to be interpreted so as to range over derivative facts permits a precise characterization of the determination relations between the fundamental and derivative grounds for a truth.

Suppose, e.g., that the fundamental facts concern the computational characterization of a subject's mental states, and let the fundamental facts comprise the first state space. Let the derivative facts concern states which verify whether the subject is consciously aware of their mental representations, and let the derivative facts comprise the second state space. Finally, let ϕ be a psychological formula, e.g. a characterization of a mental state in an experimental task where there is a particular valence for the contrast-level of a stimulus. The formula's having a truthmaker in the first space – where the states of which range, as noted, over the subject's psychofunctional facts – will determine whether the formula has a truthmaker in the second space – where the states of which range over the mental representations of which the subject is consciously aware. If the deployment of some attentional functions provides a necessary condition on the instantiation of phenomenal awareness, then the role of the state of the attentional function in the first space in verifying ϕ will determine whether ϕ is subsequently verified relative to the second space. Intuitively: Attending to a stimulus with a particular value will constrain whether a truthmaker can be provided for being consciously aware of the stimulus. If the computational facts at issue are fundamental, and the phenomenal facts at issue are derivative, then a precise characterization may be provided of the multi-dimensional relations between the verifiers which target fundamental and derivative truths.

4.4.2 Decision Theory

A second novel interpretation of two-dimensional truthmaker semantics concerns the types of intentional action, and the interaction of the latter with decision theory. As noted in the foregoing, two-dimensional semantics may be availed of in order to explain how the value of a formula relative to a context ranging over an agent and time will determine the value of the formula relative to an index ranging over a space of admissible actions made on the basis of the formula, where the value of the formula relative to the context and first index will determine the value of the formula relative to a second index, ranging over a space of outcomes.

One notable feature of the decision-theoretic interpretation is that it provides a natural setting in which to provide a gradational account of truthmaking. A proposition and its component expressions are true, just if they are verified by states in a state space, such that the state and its parts fall within the proposition's extension. In decision theory, a subject's expectation that the proposition will occur is recorded by a partial belief function, mapping the proposition to real numbers in the $\{0, 1\}$ interval. The subject's desire that the proposition occurs is recorded by a utility function, the quantitative values of which – e.g., 1 or 0 – express the qualitative value of the proposition's occurrence. The evidential expected utility of a proposition's occurrence is calculated as the probability of its obtaining conditional on an agent's action, as multiplied by the utility to the agent of the proposition's occurrence. The causal expected utility of the proposition's occurrence is calculated as the probability of its obtaining, conditional on both the agent's acts and the causal efficacy of their actions, multiplied by the utility of the proposition's occurrence.

There are three points at which a probabilistic construal of the foregoing may be defined. One point concerns the objective probability that the proposition will be verified, i.e. the chance thereof. The second point concerns subjective probability with which a subject partially believes that the proposition will obtain. A third point concerns the probability that an outcome will occur, where the space of admissible outcomes will be constrained by a subject's acts. An agent's actions will, in the third case, constrain the admissible verifiers in the space of outcomes, and thus the probability that the verifier for the proposition will obtain as an outcome. A proponent of metaphysical indeterminacy might further suggest that the verifiers are themselves gradational; thus, rather than target the probability of a verifier's realization, the proponent of metaphysical indeterminacy will suggest that a proposition P is made true only to a certain degree, such that both of the proposition's extension and anti-extension will have non-negative, real values. One objection to the foregoing account of metaphysical indeterminacy for truthmakers is, however, that the metalogic for many-valued logic is classical (see Williamson, 2014a). A distinct approach to metaphysical indeterminacy is proffered by Barnes and Williams (2011), who argue that metaphysical indeterminacy consists in persistently unpointed models, i.e. a case in which it is unclear which among a set of worlds is actual, even upon precisifying the set with precisifications. A proponent of metaphysical indeterminacy for probabilistic truthmaker semantics might then argue both that the realization of a verifier has a gradational value and that it is indeterminate which of the states which can verify a given formula is actual.

In order formally to countenance the foregoing, we define a probability measure on a state space, such that the probability measure satisfies the Kolmogorov axioms: normality $[\Pr(T) = 1]$; non-negativity $[\Pr(\phi) \geq 0]$; additivity [For disjoint ϕ and $\psi [\Pr(\phi \cup \psi) = \Pr(\phi) + \Pr(\psi)]]$; and conditionalization $[\Pr(\phi \mid \psi) = \Pr(\phi \cap \psi) / \Pr(\psi)]$. In order to account for the interaction between objective probability and the verification-conditions in truthmaker semantics, we avail, then, of a regularity condition in our earlier model, M, in which the assignment function, V, maps propositions P∈D to pairs of subsets of S, $\{1, 0\}$, i.e. the verifier and falsifier of P, such that $[\![P]\!]^+ = \{0, 1\}$ and $[\![P]\!]^- = 1 - P$. In our gradational truthmaker semantics, a state, s, verifies a proposition, P, if the probability that s is in P's extension is greater than or equal to .5:

s ⊢ P if $\Pr(s \in [\![P]\!]^+) \geq .5$.

A state, s, falsifies a proposition P if the probability that s is in P's extension is less than .5 iff the probability that s is in P's anti-extension is greater than or equal to .5

s ⊣ P if $\Pr(s \in [\![P]\!]^-) \geq .5$
iff $\Pr(s \in [\![P]\!]^+) < .5$.

The subjective probability with regard to the proposition's occurrence is expressed by a probability measure satisfying the Kolmogorov axioms as defined on a second state space, i.e., a space whose points are interpreted as concerning the subject's states of information. The formal clauses for partial belief in truthmaker semantics are the same as in the foregoing, save that the probability measures express the mental states of an agent, by being defined on the space of their states of information.

Finally, the interaction between objective and subjective probability measures in hyperintensional semantics may be captured in two ways.

One way to countenance the foregoing is via the interaction between the chance of a proposition's occurrence, the subject's partial belief that the proposition will occur, and the spaces for the subjects actions and outcomes. The formal clause for the foregoing will then be as follows:

$M,s \vdash [\![P]\!]^{c(c',a,o)} > .5$,

where c ranges over the space of physical states, and a probability measure recording objective chance is defined thereon; c' ranges over the space of an agent's states of information, and the value of P relative to c' determines the value of P relative to the space of the agent's acts, a, where the latter determines the space of admissible outcomes concerning P's occurrence, o. Thus, the parameters, c', a, o possess a hyperintensional two-dimensional profile, and the space of physical states, c, determines the values of the subject's partial beliefs and their subsequently conceivable actions and outcomes.

Accounting for the relation between c and c' – i.e., specifying a norm on the relation between chances and credences – provides one means by which to account for how objective gradational truthmakers interact with a subject's partial beliefs about whether propositions are verified. Following Lewis (1980,b/1987), a candidate chance-credence norm may be what he refers to as the 'principal principle'.[9] The principal principle states that an agent's partial belief that a proposition will be verified, conditional on the objective chance of the proposition's occurrence and the admissible evidence, will be equal to the objective chance of the proposition's occurrence itself:

$Pr_s(P \mid ch(P) \wedge E) = ch(P)$.

---

[9]See Pettigrew (2012), for a justification of a generalized version of the principal principle based on Joyce (1998)'s argument for probabilism. Probabilism provides an accuracy-based account of partial beliefs, defining norms on the accuracy of partial beliefs with reference only to worlds, metric ordering relations, and probability measures thereon. The proposal contrasts to pragmatic approaches, according to which a subject's probability and utility measures are derivable from a representation theorem, only if the agent's preferences with regard to a proposition's occurrence are consistent (see Ramsey, 1926). Probabilism states, in particular, that, if there is an ideal subjective probability measure, the ideality of which consists e.g. in its matching objective chance, then one's probability measure ought to satisfy the Kolmogorov axioms, on pain of there always being a distinct probability measure which will be metrically closer to the ideal state than one's own.

4.4.3 Intentional Action

A third novel interpretation of two-dimensional hyperintensional semantics provides a natural setting in which to delineate the structural content of the types of intentional action. For example, the mental state of intending to pursue a course of action may be categorized as falling into three types, where intending-that is treated as a two-dimensional hyperintensional state. One type targets a unique structural content for the state of acting intentionally, such that an agent intends to bring it about that ϕ just if the intention satisfies a clause which mirrors that outlined in the last paragraph:

- $[\![\text{Intenton-in-Action}(\phi)]\!]_w = 1$ only if $\exists w' [\![\phi]\!]^{w',c(=t,l),a,o} = 1$.

A second type of intentional action may be recorded by a future-directed state, such that an agent intends to ϕ only if they intend to pursue a course of action in the future, only if there is a state and a future time relative to which the agent's intention is satisfied:

- $[\![\text{Intention-for-the-future}(\phi)]\!]_w = 1$ only if $\exists w' \forall t \exists t' [t < t' \land [\![\phi]\!]^{w',t'} = 1]$.

Finally, a third type of intentional action concerns reference to the intention as an explanation for one's course of action. Chapter **14** regiments the structural content of this type of intention as a state which receives its value only if a hyperintensional grounding operator which takes scope over a proposition and an action, receives a positive semantic value.

- $[\![\text{Intention-with-which}(\phi)]\!]_w = 1$ only if $\exists w' [[\![\psi]\!]^{w'} = 1 \land [\![G(\phi,\psi)]\!] = 1]$,

where G(x,y) is a grounding operator encoding the explanatory connection between ϕ and ψ.

The varieties of subject matter, as defined in two-dimensional truthmaker semantics, can be availed of in order to enrich the present approach. Having multiple state spaces from which to define the verifiers of a proposition enables a novel solution to issues concerning the interaction between action and explanation. The third type of intentional action may be regimented, as noted, by the agent's reference to an intention as an explanation for her course of action.

The foregoing may also be availed of, in order to provide a novel solution to an issue concerning the interaction between involuntary and intentional action. The issue is as follows. Wittgenstein (1953/2009; 621) raises the inquiry: 'When I raise my arm, my arm goes up. Now the problem arises: what is left over if I subtract the fact that my arm goes up from the fact that I raise my arm?' Because the arm's being raised has at least two component states, namely, the arm's going up and whatever the value of the variable state might be, the answer to Wittgenstein's inquiry is presumably that the agent's intentional action is the value of the variable state, such that a combination of one's intentional action and one's arm going up is sufficient for one's raising one's arm. The aforementioned issue with the foregoing concerns how precisely to capture the notion of partial content, which bears on the relevance of the semantics of the component states and the explanation of the unique state entrained by their combination.

Given our two-dimensional truthmaker semantics, a reply to Wittgenstein's inquiry which satisfies the above desiderata may be provided. Let W express a differentiated subject matter, whose total content is that an agent's arm is raised. W expresses the total content that an agent's arm is raised, because W is comprised of an initial content, U (that one's arm goes up), and an additional content, R (that one intends to raise one's arm).

The verifier for W may be interpreted as a two-dimensional loose truthmaker. Let c range over an agent's motor states, S. Let i range over an agent's states of information, S*. We define a state for intentional action in the space of the agent's motor actions. The value of the state is positive just if a selection function, f, is a mapping from the power set of motor actions in S to a unique state s' in S. This specifies the initial, partial content, U, that one's arm goes up. An intention may then be defined as a unique state, s*, in the agent's state of information, S*. The state, s*, specifies the additional, partial content R, that one intends to raise one's arm.

Formally:

$s \vdash U$ only if $\exists s' \sqsubset S$, such that f: $s \rightarrow s'$, s.t. $s' \vdash U$,

$\exists s^*$, $s^* \vdash R$, and

$W = U \sqcup R$.

The two-dimensional loose truthmaker for one's arm being raised may then be defined as follows:

$\exists c_{s \rightarrow s'} [\![W]\!]^{c,c} = 1$ if $s' \in [\![W]\!]^+$, and

$\exists i_{s*} [\![W]\!]^{c,i} = 1$ if $s^* \in [\![W]\!]^+$.

Intuitively, the value of the total content that one's arm is raised is defined relative to a set of motor states – where a first intentional action selects a series of motor states which partly verify that one's arm goes up. The value of one's arm being raised, relative to (the intentionally modulated) motor state of one's arm possibly going up, determines the value of one's arm being raised relative to the agent's distinct intention to raise their arm. The agent's first intention selects among the admissible motor states, and – all else being equal – the motor states will verify the fact that one's arm goes up.[10] The fusion of (i) the state corresponding to the initial partial content that one's arm goes up, and (ii) the state corresponding to the additional partial content that one intends to raise one's arm, is sufficient for the verification of (iii) the state corresponding to the total content that one's arm is raised.

---

[10]The role of the first intention in acting as a selection function on the space of motor actions corresponds to the comparator functions stipulated in the cognitive science of action theory. For further discussion of the comparator model, see Frith et al. (2000) and Pacherie (2012).

4.5 Concluding Remarks

In this essay, I have endeavored to establish foundations for the interaction between two-dimensional indexing and hyperintensional semantics. I examined, then, the philosophical significance of the framework by developing three, novel interpretations of two-dimensional truthmaker semantics, in light of the new relations induced by the model.

The first interpretation enables a rigorous characterization of the distinction between fundamental and derivative truths. The second interpretation evinces how the elements of decision theory are definable within the two-dimensional hyperintensional setting, and a novel account was then outlined concerning the interaction between probability measures and hyperintensional grounds. The third interpretation of two-dimensional hyperintensional semantics concerns the structural content of the types of intentional action. Finally, I demonstrated how the hyperintensional array of state spaces, relative to which propositions may be verified, may serve to resolve a previously intransigent issue concerning the role of intention in action.

Chapter 5

Fixed Points in the Epistemic Hyperintensional μ-Calculus and the KK Principle

This essay provides a novel account of self-knowledge, which avoids the epistemic indeterminacy witnessed by the invalidation of modal axiom 4 in epistemic logic; i.e. the KK principle: $\Box \phi \to \Box \Box \phi$. The essay argues, by contrast, that – despite the invalidation of modal axiom 4 on its epistemic interpretation – states of epistemic determinacy might yet be secured by countenancing self-knowledge on the model of fixed points in the modal μ-calculus.

Counterinstances to modal axiom 4 – which records the property of transitivity in labeled transition systems[1] – have been argued to occur within various interpretations of the sorites paradox. Suppose, e.g., that a subject is presented with a bounded continuum, the incipient point of which bears a red color hue and the terminal point of which bears an orange color hue. Suppose, then, that the cut-off points between the points ranging from red to orange are indiscriminable, such that the initial point, a, is determinately red, and matches the next apparent point, b; b matches the next apparent point, c; and thus – by transitivity – a matches c. Similarly, if b matches c, and c matches d, then b matches d. The sorites paradox consists in that iterations of transitivity would entail that the initial and terminal points in the bounded continuum are phenomenally indistinguishable. However, if one takes transitivity to be the culprit in the sorites, then eschewing the principle

---

[1] See Kripke (1963).

would entail a rejection of the corresponding modal axiom (4), which records the iterative nature of the relation.[2] Given the epistemic interpretation of the axiom – namely, that knowledge that a point has a color hue entails knowing that one knows that the point has that color hue – a resolution of the paradox which proceeds by invalidating axiom 4 subsequently entrains the result that one can know that one of the points has a color hue, and yet not know that they know that the point has that color hue (Williamson, 1990: 107-108; 1994: 223-244; 2001: chs. 4-5). The non-transitivity of phenomenal indistinguishability corresponds to the non-transitivity of epistemic accessibility. As Williamson (1994: 242) writes: 'The example began with the non-transitive indiscriminability of days in the height of the tree, and moved on to a similar phenomenon for worlds. It seems that this can always be done. Whatever x, y and z are, if x is indiscriminable from y, and y from z, but x is discriminable from z, then one can construct miniature worlds w_x, w_y and w_z in which the subject is presented with x, y and z respectively, everything else being relevantly similar. The indiscriminability of the objects is equivalent to the indiscriminability of the corresponding worlds, and therefore to their accessibility. The latter is therefore a non-transitive relation too.' The foregoing result holds, furthermore, in the probabilistic setting, such that the evidential probability that a proposition has a particular value may be certain – i.e., be equal to 1 – while the iteration of the evidential probability operator – recording the evidence with regard to that evidence – is yet equal to 0. Thus, one may be certain on the basis of one's evidence that a proposition has a particular value, while the higher-order evidence with regard to one's evidence adduces entirely against that valuation (Williamson, 2014).

In the foregoing argument, 'safety' figures as a necessary condition on knowledge, and is codified by margin-for-error principles of the form: '$\forall x \forall \phi [K^{m+1}\phi(x) \to K^m\phi(x+1)]$', with m ranging over the natural numbers (Williamson, 2001: 128; Gómez-Torrente, 2002: 114). Intuitively, the safety condition ensures that if one knows that a predicate is satisfied, then one knows that the predicate is satisfied in relevantly similar worlds. Williamson targets the inconsistency of margin-for-error principles, the luminosity principle ['$\forall x \forall \phi [\phi(x) \to K\phi(x)]$'], and the characterization of the sorites as occurring when an initial state satisfies a condition, e.g. being red, and a terminal state satisfies a distinct condition, e.g. being orange. As Srinivisan (2013: 4) writes: 'By

---

[2] For more on non-transitivist approaches to the sorites, see Zardini (2019).

[the luminosity principle], if C obtains in α_0, then S knows that C obtains in α_0. By [margin-for-error principles], if S knows that C obtains in α_0, then C obtains in α_1. By [the characterization of the sorites], C does obtain in α_0; therefore, C obtains in α_1. Similarly, we can establish that C also obtains in $\alpha_2, \alpha_3, \alpha_4, \ldots, \alpha_n$. But according to [the characterization of the sorites] C doesn't obtain in α_n. Thus we arrive at a contradiction'. The triad evinces that the luminosity principle is false, given the plausibility of margin-for-error principles and the characterization of the sorites. In cases, further, in which conditions on knowledge are satisfied, epistemic indeterminacy is supposed to issue from the non-transitivity of the accessibility relation on worlds (1994: 242).

The anti-luminosity argument can be availed of to argue against the KK principle. If states are not luminous, then knowing that ϕ will not entail that one knows that one knows that ϕ. A different argument is presented, as well, in Williamson (2001: ch. 5, pp. 115-116). Suppose the following:

(1_i) If K that x is i+1 inches tall, then ¬K¬x is i inch tall

(If an agent knows that some object is i+1 inches tall, then for all the agent knows the object is i inch tall); and

(C) 'If p and all members of the set X are pertinent propositions, p is a logical consequence of X, and [an agent] knows each member of X, then he knows p' (op. cit.: 116).

Suppose that:

(2_i) An agent knows that the object is not i inch tall.

By the KK principle, (3_i) follows from (2_i).

(3_i) An agent knows that she knows that the object is not i inch tall.

Suppose a proposition (q) which states that the object is i+1 inches tall. By (1), then the agent knows that ¬(2_i). However, if (3_i), then the agent knows (2_i). Thus, (q) → (2_i) ∧ ¬(2_i). Thus – by (C) – (1_i) and (3_i) imply that the agent knows ¬(q):

(2_{i+1}) the agent knows that the object is not i+1 inches tall.

Thus, from (KK), (C), and (2_i), we can infer (2_{i+1}).

Repeating the argument for values of i ranging from 0 to 664, we have

(2_0) An agent knows that the object is not 0 inches tall.

(2_{664}) An agent knows that the object is not 664 inches tall.

However, suppose that the object is in fact 664 inches tall and grant the factivity of knowledge (modal axiom T: $\Box\phi \to \phi$). Then (2_{664}) is false. So, from (1), (2_0), (C), and (KK), we can derive a false conclusion, (2_{664}).

(C) is a principle of deductive closure, and thus arguably ought to be preserved. Williamson takes (2_i) to be a truism, and (1) to be defensible. He thus argues that we ought to reject the KK principle.

In this essay, I endeavor to provide a novel account which permits the retention of both classical logic as well as a modal approach to the phenomenon of vagueness, while salvaging the ability of subjects to satisfy necessary conditions on there being iterated epistemic states. I will argue that – despite the invalidity of modal axiom 4 – a distinct means of securing an iterated state of knowledge concerning one's first-order knowledge that a particular state obtains is by availing of fixed point, non-deterministic automata in the setting of coalgebraic modal logic.

The modal μ-calculus is equivalent to the bisimulation-invariant fragment of monadic second-order logic.[3] $\mu(x)$. is an operator recording a least fixed point. Despite the non-transitivity of sorites phenomena – such that, on its epistemic interpretation, the subsequent invalidation of modal axiom 4 entails structural, higher-order epistemic indeterminacy – the modal μ-calculus provides a natural setting in which a least fixed point can be defined with regard to the states instantiated by non-deterministic modal automata. In virtue of recording iterations of particular states, the least fixed points witnessed by non-deterministic modal automata provide, then, an escape route from the conclusion that the invalidation of the KK principle provides an exhaustive and insuperable obstruction to self-knowledge. Rather, the least fixed points countenanced in the modal μ-calculus provide another conduit into subjects' knowledge to the effect that they know that a state has a determinate value. Thus, because of the fixed points definable in the modal μ-calculus, the non-transitivity of the similarity relation is yet consistent with necessary conditions on epistemic determinacy and self-knowledge, and the states at issue can be luminous to the subjects who instantiate them. Gödel (1972a: p. 272, fn. c) writes, too, of the iterative nature of rational intuition that 'In order to make the concept of accessibility ['concrete intuition of certain infinite procedures' (op. cit.) - D.E.] *fruitful, abstract conceptions* are always necessary, e.g., *insights about infinitely many possible insights* [my emphasis - D.E.] in Gentzen's original definition, which is somewhat different from that given above (see his 1936, p. 555, line 7)'.[4]

In the remainder of the essay, we introduce labeled transition systems,

---

[3]See Janin and Walukiewicz (1996).
[4]See chapter **11**, for further discussion of Gödelian intuition.

the modal μ-calculus, and non-deterministic Kripke (i.e., μ-) automata. We recount then the sorites paradox in the setting of the modal μ-calculus, and demonstrate how the existence of fixed points enables there to be iterative phenomena which ensure that – despite the invalidation of modal axiom 4 – iterations of mental states can be secured, and can thereby be luminous.

A labeled transition system is a tuple comprised of a set of worlds, S; a valuation, V, from S to its power set, $\wp(S)$; and a family of accessibility relations, R. So LTS = $\langle S, V, R \rangle$ (see Venema, 2012: 7). A Kripke coalgebra combines V and R into a Kripke functor, σ; i.e. the set of binary morphisms from S to $\wp(S)$ (op. cit.: 7-8). Thus for an s∈S, $\sigma(s) := [\sigma_V(s), \sigma_R(s)]$ (op. cit.). Satisfaction for the system is defined inductively as follows: For a formula ϕ defined at a state, s, in S,

$[\![\phi]\!]^S = V(s)$
$[\![\neg\phi]\!]^S = S - V(s)$
$[\![\bot]\!]^S = \varnothing$
$[\![\top]\!]^S = S$
$[\![\phi \vee \psi]\!]^S = [\![\phi]\!]^S \cup [\![\psi]\!]^S$
$[\![\phi \wedge \psi]\!]^S = [\![\phi]\!]^S \cap [\![\psi]\!]^S$
$[\![\Diamond_s\phi]\!]^S = \langle R_s \rangle [\![\phi]\!]^S$
$[\![\Box_s\phi]\!]^S = [R_s][\![\phi]\!]^S$, with
$\langle R \rangle(\phi) := \{s \in S \mid R[s] \cap \phi \neq \varnothing\}$ and
$[R](\phi) := \{s \in S \mid R[s] \subseteq \phi\}$ (9)
$[\![\mu x.\phi]\!] = \bigcap\{U \subseteq S \mid [\![\phi]\!] \subseteq U\}$ (Fontaine, 2010: 18)
$[\![\upsilon x.\phi]\!] = \bigcup\{U \subseteq S \mid U \subseteq [\![\phi]\!]\}$ (op. cit.; Fontaine and Place, 2010),

A Kripke coalgebra can be represented as the pair (S, σ: S $\to$ **K**A) (Venema, 2020: 8.1)

In our Kripke coalgebra, we have M,s ⊩ $\langle \pi^* \rangle \phi \iff (\phi \vee \Diamond_s \langle \pi^* \rangle \phi)$ (Venema, 2012: 25). $\langle \pi^* \rangle \phi$ is thus said to be the *fixed point* for the equation, $x \iff \phi \vee \Diamond x$, where the value of the formula is a function of the value of x conditional on the constancy in value of ϕ (38). The smallest solution of the formula, $x \iff \phi \vee \Diamond x$, is written $\mu x.\phi \vee \Diamond x$ (25). The value of the least fixed point is, finally, defined more specifically thus:

$[\![\mu x.\phi \vee \Diamond x]\!] = V(\phi) \cup \langle R \rangle([\![\mu x.\phi \vee \Diamond x]\!])$ (38).

A non-deterministic automaton is a tuple $\mathbb{A} = \langle A, \Xi, Acc, a_I \rangle$, with A a finite set of states, a_I being the initial state of A; Ξ is a transition function s.t. $\Xi: A \to \wp(A)$; and $Acc \subseteq A$ is an acceptance condition which specifies admissible conditions on Ξ (60, 66).

Let two Kripke models $\mathbb{A} = \langle A, a \rangle$ and $\mathbb{S} = \langle S, s \rangle$, be bisimilar if and only if there is is a non-empty binary relation, $Z \subseteq \mathbb{A} \times \mathbb{S}$, which is satisfied, if:

(i) For all $a \in \mathbb{A}$ and $s \in \mathbb{S}$, if aZs, then a and s satisfy the same proposition letters;
(ii) *The forth condition.* If aZs and $R_\triangle a, v_1 \ldots v_n$, then there are $v'_1 \ldots v'_n$ in $\mathbb{S}$, s.t.
- for all i ($1 \leqslant i \leqslant n$) $v_i Z v'_i$, and
- $R'_\triangle s, v'_1 \ldots v'_n$;

(iii) *The back condition.* If aZs and $R'_\triangle s, v'_1 \ldots v'_n$, then there are $v_1 \ldots v_n$ in $\mathbb{A}$, s.t.
- for all i ($1 \leqslant i \leqslant n$) $v_i Z v'_i$ and
- $R_\triangle a, v_1 \ldots v_n$ (see Blackburn et al, 2001: 64-65).

Bisimulations may be redefined as *relation liftings*. We let, e.g., a Kripke functor, **K**, be such that there is a relation $\overline{\mathbf{K}} \subseteq \mathbf{K}(A) \times \mathbf{K}(A')$ (Venema, 2020: 81). Let Z be a binary relation s.t. $Z \subseteq A \times A'$ and $\wp \overline{Z} \subseteq \wp(A) \times \wp(A')$, with
$\wp \overline{Z} := \{(X, X') \mid \forall x \in X \exists x' \in X'$ with $(x, x') \in Z \wedge \forall x' \in X' \exists x \in X$ with $(x, x') \in Z\}$ (op. cit.). Then, we can define the relation lifting, $\overline{\mathbf{K}}$, as follows:
$\overline{\mathbf{K}} := \{[(\pi, X), (\pi', X')] \mid \pi = \pi'$ and $(X, X') \in \wp \overline{Z}\}$ (op. cit.).

Finally, given the Kripke functor, **K**, **K** can be defined as the μ-automaton, i.e., the tuple $\mathbb{A} = \langle A, \Xi, a_I \rangle$, with $a_I \in A$ defined again as the initial state in the set of states A; and Ξ defined as a mapping such that $\Xi : A \to \wp_\exists(\mathbf{K}A)$, where the $\exists$ subscript indicates that $(a, s) \in A \times S \to \{(a', s) \in \mathbf{K}(A) \times S \mid a' \in \Xi(a)\}$ (93). The duality between the categories of coalgebras, A, and algebras, S, and the definition of the functor, **K**, as an expression relation expressed by μ-automata, provide an account of expressivism for self-knowledge.[5]

The philosophical significance of the foregoing can now be witnessed by defining the μ-automata on an alphabet; in particular, a non-transitive set

---

[5] See chapter **2**, for further discussion of modal and hyperintensional expressivism. For further discussion of the convergence between expressivism and self-knowledge, see Wittgenstein (1953/2009: 310-326); Bar-On (2004); Bar-On (2012); Wright (2012c); and Bar-On and Wright (2023). For discussion of the convergence between self-knowledge and Wright (2001)'s intention-based view of rule-following, see Coliva (2012), §**4.4.3** and chapter **14**.

comprising a bounded real-valued, ordered sequence of terms for chromatic properties. Although the non-transitivity of the ordered sequence of terms for color hues belies modal axiom 4, such that one can know that a particular point in the sequence has a particular value although not know that one knows that the point satisfies that value, terms for chromatic values, ϕ, in the non-transitive set of color terms nevertheless permits every sequential input state in the μ-automaton to define a fixed point. In order for there to be least and greatest fixed points, there must be monotone operators defined on complete lattices. As Venema (2020: A-2) writes: '*A partial order* is a structure $\mathbb{P} = \langle P, \leqslant \rangle$ such that $\leqslant$ is a reflexive, transitive and anti-symmetric relation on P. Given a partial order $\mathbb{P}$, an element p∈P is an *upper bound* (*lower bound, respectively*) of a set $X \subseteq P$ if $p \geqslant x$ for all x∈X ($p \leqslant x$ for all x∈X). If the set of upper bounds of X has a minimum, this element is called the *least upper bound, supremum,* or *join* of X, notation: $\bigvee X$. Dually, the *greatest lower bound, infimum,* or *meet* of X, if existing, is denoted as $\bigwedge X$ ... A partial order $\mathbb{P}$ is called a *lattice* if every two-element subset of P has both an infimum and a supremum; in this case, the notation is as follows: $p \wedge q := \bigwedge \{p, q\}$, $p \vee q := \bigvee \{p, q\}$ ... A partial order $\mathbb{P}$ is called a *complete lattice* if every subset of P has both an infimum and a supremum ... A complete lattice will usually be denoted as a structure $\mathbb{C} = \langle C, \bigvee, \bigwedge \rangle$.' 'Let $\mathbb{P}$ and $\mathbb{P}'$ be two partial orders and let $f: P \to P'$ be some map. Then f is called *monotone* or *order preserving* if $f(x) \leqslant' f(y)$ whenever $x \leqslant y$ ...' (3.1). 'Let $\mathbb{P} = \langle P, \leqslant \rangle$ be a partial order, and let $f: P \to P$ be some map. Then an element p∈P is called a *prefixpoint* of f if $f(p) \leqslant p$, a *postfixpoint* of f if $p \leqslant f(p)$, and a *fixpoint* if $f(p) = p$. The sets of prefixpoints, postfixpoints, and fixpoints of f are denoted respectively as PRE(f), POS(f) and FIX(f). / In case the set of fixpoints of f has a least (respectively greatest) member, this element is denoted LFP.f (GFP.f, respectively)' (3-2). The Knaster-Tarski Theorem says, then, that, for a complete lattice, $\mathbb{C} = \langle C, \bigvee, \bigwedge \rangle$, with $f: C \to C$ being monotone, f has both a least and greatest fixpoint, LFP.$f = \bigwedge$PRE(f), and GFP.$f = \bigvee$POS(f) (op. cit.).[6]

The epistemicist approach to vagueness relies, as noted, on the epistemic interpretation of the modal operator, such that the invalidation of transitivity and modal axiom 4 ($\Box\phi \to \Box\Box\phi$) can be interpreted as providing a barrier to a necessary condition on self-knowledge.[7] Crucially, μ-automata

---

[6]Knaster (1928); Tarski (1955).
[7]Williamson (private correspondence) writes: 'My argument against KK is compatible

can receive a similar epistemic interpretation.[8] An epistemic interpretation of a μ-automaton is just such that the automaton operates over epistemically possible worlds. The automaton can thus be considered a model for an epistemic agent. The transition function accounts for the transition from one epistemic state to another, e.g. as one proceeds along the stages of a continuum. A fixed point operator on a given epistemic state, e.g. $\Box(\phi)$ where $\Box$ is interpreted so as to mean knowledge-that, amounts to one way to iterate the state. If one knows a proposition, ϕ, the least fixed point operation, $\mu x.(\Box(\phi))$, records an iteration of the epistemic state, knowledge of knowledge, and similarly for belief. Thus, interpreting the μ-automaton epistemically permits the fixed points relative to the arbitrary points in the ordered continuum to provide a principled means – distinct from the satisfaction of the KK principle – by which to account for the pertinent iterations of epistemic states unique to an agent's self-knowledge.

Fixed points for the iteration of epistemic states can, further, be obtained, as with the Knaster-Tarski Theorem, by Lambek (1968)'s Theorem according to which 'C denotes a category and F an endofunctor on it [...] a fixed point of an endofunctor F consists of an object X and an isomorphism between FX and X. Hence, it can be viewed as an algebra or a coalgebra for F. Fundamental examples are

$F(\mu F) \to_\iota \mu F$

the initial algebra for F, whose algebra structure ι is an isomorphism by Lambek's Lemma [...] and dually

$F \to_\tau F(F)$

the terminal coalgebra for F. Of course, both μF and F are unique up to isomorphism, if they exist' (Adámek et al., 2018).[9] Fixed points for the iteration of epistemic states can, too, be obtained from

(i) Brouwer (1911)'s fixed point theorem, which states: '[E]very continuous function from a compact convex set to itself has at least one fixed

---

with there being lots of cases of knowledge that one knows. Indeed, the picture of knowledge it suggests (with margins for error) suggests that there is plenty of knowledge, as I think there is. Right now, I know that I know that I'm typing on my laptop. The failure of KK is a limitation on self-knowledge but not a very severe one'. See Gertler (2021), for further discussion.

[8]For more on the epistemic μ-calculus, see Bulling and Jamroga (2011); Bozianu et al (2013); and Dima et al (2014). For an examination of the modal μ-calculus and common knowledge, see Alberucci (2002).

[9]Lambek (1968: §2). See Adámek et al. (forthcoming).

point',[10] defining fixed points for real values, for example the real numbered magnitudes of a continuum.

'Let (X, τ) be a topological space. Then an open cover is a set $\{U_i \subset X\}_{i \in I}$ of open subsets (i.e. $(U_i \subset X) \in \tau \subset \wp(X)$) such that their union is all of X

'$\bigcup_{i \in I} U_i = X$. This is called a finite open cover if I is a (Kuratowski-)finite set.

'A *subcover* of an open cover as above is a subset $J \subset I$ of the given open subsets, such that their union still exhausts X i.e. $\bigcup_{i \in J \subset I} U_i = X$.
[...]
'A topological space is called *compact* if every open cover has a finite subcover'.[11]

'A subset S of a real affine space X is **convex** if for any two points x, $y \in S$, the straight line segment connecting x with y in X is also contained in S. In other words, for any x, $y \in S$, and any $t \in [0, 1]$, we have also $tx + (1 - t)y \in S$'.[12]

(ii) the Kleene fixed point theorem (Tarski, 1955), which states: 'Let $f: P \to P$ be a monotone function on a poset [i.e. partially ordered set: reflexive; transitive; anti-symmetric - D.E.] P. If P has a least element $\perp$ and joins of increasing sequences, and if f preserves joins of increasing sequences, then a least fixed point of f can be constructed as the join of the increasing sequence:
$\perp \leq f(\perp) \leq f^2(\perp) \leq ...$' (https://ncatlab.org/nlab/show/Kleene[poundsymbol]27s+fixed+point+theorem),[13] and

(iii) the Pohlová-Adámek fixed point theorem, which states: 'Let C be a category with an initial object 0 and transfinite composition of length ω, hence colimits [i.e. sums] of sequences $\omega \to C$ (where ω is the first infinite ordinal), and suppose $F: C \to C$ preserves colimits of $C\omega$-chains. Then the colimit γ of the chain
$0 \to_i F(0) \to_{F(i)} ... F^{(n)}(0) \to_{F^{(n)}(0)} F^{(n+1)}(0) \to ...$
carries a structure of the initial F-algebra' (https://ncatlab.org/nlab/show/Adámek[poundsymbol]27s+fixed+point+theorem).[14]

The fixed point operators in the modal μ-calculus can be rendered hyperintensional, by defining the elements in the sets in the semantics for the

---

[10]https://ncatlab.org/nlab/show/Brouwer[poundsymbol]27s+fixed+point+theorem .
[11]https://ncatlab.org/nlab/show/compact+space .
[12]https://ncatlab.org/nlab/show/convex+set .
[13]Kleene (1952/1971); Cousot and Cousot (1979).
[14]Pohlová (1973); Adámek (1974).

operators above, such that they are hyperintensional parts of epistemically possible worlds, rather than whole epistemically possible worlds. The semantics for each operator can then remain as presented in the foregoing, while changing the sets and their subsets to hyperintensional epistemic states or verifiers instead of worlds.

The fixed point approach to iterated epistemic states will provide a compelling alternative to the KK principle, if Williamson's argument against the KK principle does not hold for all ancestral relations of knowledge but rather only for specific applications of luminosity and modal axiom 4.[15] If Williamson's argument does not generalize to all ancestral relations of knowledge, then one can avoid the objection that the fact that $\mu x.(\Box(\phi))$ entails that one knows that one knows that ϕ is such that the state collapses just to KK such that the state would rarely be satisfied in light of the argument against the KK principle. An iteration procedure via a fixed point operation on a knowledge state is distinct from an application of modal axiom 4 i.e. the KK principle, and provides a novel formal method for accounting for the iteration of epistemic states.

---

[15] Thanks here to Jon Litland for the objection.

Part II: The Relation between Hyperintensional Conceivability and Metaphysical States

Chapter 6

Conceivability, Haecceities, and Essence

6.1 Introduction

In this essay, I endeavor to provide an account of how the epistemic interpretation of two-dimensional semantics can be sensitive to haecceities and essential properties more generally. Let a model, M, be comprised of a set of epistemically possible worlds C; a set of metaphysically possible worlds W; a domain, D, of terms and formulas; binary relations defined on each of C and W; and a valuation function mapping terms and formulas to subsets of C and W, respectively. So, M = $\langle C, W, D, R_C, R_W, V \rangle$. A term or formula is epistemically necessary or apriori iff it is inconceivable for it to be false ($\Box \iff \neg \Diamond \neg$). A term or formula is negatively conceivable iff nothing rules it out apriori ($\Diamond \iff \neg \Box \neg$). A term or formula is positively conceivable only if the term or formula can be perceptually imagined. According to the epistemic interpretation of two-dimensional semantics, the semantic value of a term or formula can then be defined relative to two parameters.[1] The first parameter ranges over the set of epistemically possible worlds, and the second parameter ranges over the set of metaphysically possible worlds. The value of the term or formula relative to the first parameter determines the value of the term or formula relative to the second parameter. Thus, the epistemically possible value of the term or formula constrains the metaphysically possible value of the term or formula; and so conceivability might, given

---
[1]Chalmers and Rabern (2014: 211-212).

the foregoing, serve as a guide to metaphysical possibility.

Roca-Royes (2011) and Chalmers (2010a; 2011; 2014) note that, on the above semantics, epistemic possibility cannot track the difference between the metaphysical modal profile of a non-essential proposition – e.g., that there is a shooting star – and the metaphysical modal profile of an essential definition, such as a theoretical identity statement – e.g., that water = H2O. Another principle of modal metaphysics to which epistemic possibilities are purported to be insensitive is haecceity comprehension; namely, that $\Box \forall x,y \Box \exists \Phi (\Phi x \iff x = y)$.

The aim of this note is to redress the contention that epistemic possibility cannot be a guide to the principles of modal metaphysics. I will argue that the interaction between the two-dimensional framework and the mereological parthood relation, which is super-rigid, enables the epistemic possibility of parthood to be a guide to its metaphysical profile. Further, if essential properties are haecceitistic properties, then the super-rigidity of haecceitistic properties entrains that the epistemic possibility of their obtaining entails the metaphysical possibility of their obtaining.

In Section **2**, I examine a necessary condition on admissible cases of conceivability entailing metaphysical possibility in the two-dimensional framework, focusing on the property of super-rigidity. I argue that – despite the scarcity of properties which satisfy the super-rigidity condition – metaphysical properties such as the parthood relation do so. In Section **3**, I address objections to two dogmas of the semantic rationalism underpinning the epistemic interpretation of two-dimensional semantics. The first dogma of semantic rationalism mirrors Quine's (1951) contention that one dogma of the empiricist approach is the distinction that it records between analytic and synthetic claims. The analogous dogma in the semantic rationalist setting is that a distinction can be drawn between contextual intensions by contrast to epistemic intensions. The second dogma states that there are criteria on the basis of which formal from informal domains, unique to the extensions of various concepts, can be distinguished, such that the modal profiles of those concepts would thus be determinate. I examine the Julius Caesar problem as a test case. I specify, then, a two-dimensional formula encoding the relation between the epistemic possibility of essential properties obtaining and its metaphysical possibility, and I generalize the approach to haecceitistic properties. In Section **4**, I address objections from the indeterminacy of ontological principles relative to the space of epistemic possibilities, and from the consistency of epistemic modal space. Section **5** provides concluding remarks.

6.2 Super-rigidity and Hyper-rigidity

Mereological parthood satisfies a crucial condition in the epistemic interpretation of two-dimensional semantics. The condition is called super-rigidity, and its significance is that, unless the semantic value for a term is super-rigid, i.e. maps to the same extension throughout the classes of epistemic and metaphysical possibilities, the extension of the term in epistemic modal space risks diverging from the extension of the term in metaphysical modal space.[2] Chalmers provides two other conditions for the convergence between the epistemic and metaphysical profiles of expressions. In his (2002), epistemically possible worlds are analyzed as being centered metaphysically possible worlds, such that conceivability entails metaphysical (1-)possibility. In his (2010), the epistemic and metaphysical intensions of terms for physics and consciousness are argued to coincide, such that the conceivability of physics without consciousness (i.e. zombies) entails the metaphysical possibility of physics without consciousness. Thus, the 1- and 2-intensions of an expression can converge without super-rigidity. In this chapter, however, I focus just on the role of the super-rigidity condition in securing epistemic possibility as a guide to modal metaphysics. Super-rigidity ought to be replaced by the hyper-rigidity condition specified below, in hyperintensional contexts.

There appear to be only a few expressions which satisfy the super-rigidity condition. Such terms include those referring to the properties of phenomenal consciousness, to the parthood relation, and perhaps to the property of friendship (Chalmers, 2012: 367, 374). Other candidates for super-rigidity are taken to include metaphysical terms such as 'cause' and 'fundamental', and logical constants such as '∧' (Chalmers, op. cit.). However, there are counterexamples to each of the foregoing proposed candidates.

Against the super-rigidity of 'fundamental', Fine (2001: 3) argues that a proposition is fundamental if and only if it is real, while Sider (2011: 112, 118) argues that a proposition is fundamental iff it possesses a truth-condition (in a 'metaphysical semantics', stated in perfectly joint-carving terms) for the sub-propositional entities – expressed by quantifiers, functions, predicates – comprising the target proposition. The absolute joint-carving terms are taken

---

[2]See chapter **7**.

to include logical vocabulary (including quantifiers), metaphysical predicates such as mereological parthood, and physical predicates.

Against the super-rigidity of 'cause', Sider (op. cit.: 8.3.5) notes that a causal deflationist might argue that causation is non-fundamental. By contrast, a causal nihilist might argue that causation is non-fundamental as well, though for the distinct reason that there is no causation. So, while both the deflationist and nihilist believe that 'cause' does not carve at the joints – the nihilist can still state that there is a related predicate, 'cause*', such that they can make the joint-carving claim that 'Nothing causes* anything', whereas the deflationist will remain silent, and maintain that no broadly causal locutions carve at the joints.

Against the super-rigidity of the logical connective, $\wedge$, the proponent of model-theoretic validity will prefer a definition of the constant according to which, for propositions ϕ and ψ and a model, M, M validates $\phi \wedge \psi$ iff M validates ϕ and M validates ψ. By contrast, the proponent of proof-theoretic validity will prefer a distinct definition which makes no reference to truth, according to which $\wedge$ is defined by its introduction and elimination rules: $\phi, \psi \vdash \phi \wedge \psi$; $\phi \wedge \psi \vdash \phi$; $\phi \wedge \psi \vdash \psi$.

Finally, referring expressions for physical entities, such as 'metric tensor field', might have a rigid intension mapping the referring expression to the same extension in metaphysical modal space, and a non-rigid intension mapping the referring expression to distinct extensions in epistemically possible space, such that what is known about the referring expression is contingent and might diverge from its necessary metaphysical profile.[3] That physical terms are not super-rigid, i.e. epistemically and metaphysically rigid, might be one way to challenge the soundness of the conceivability argument to the effect that, if it is epistemically possible that truths about consciousness cannot be derived from truths about physics, then the dissociation between phenomenal and physical truths is metaphysically possible (see Chalmers, 2010: 151).

Crucially for the purposes of this chapter, there appear to be no clear counterexamples to the claim that mereological parthood is super-rigid. If this is correct, then mereological parthood in the space of epistemic modality can serve as a guide to the status of mereological parthood in metaphysical modal space. The philosophical significance of the foregoing is that it belies the contention proffered by Roca-Royes (op. cit.) and Chalmers (op.

---

[3] See Arntzenius (2012: 72-73), for the definition, and discussion, of metric tensor fields.

cit.) concerning the limits of conceivability-based modal epistemology. The super-rigidity of the parthood relation permits the conceivability of mereological parthood to bear on its metaphysical profile. I argue further that – supposing essential properties are haecceitistic properties, and essential and haecceitistic properties are super-rigid – the conceivability of haecceitistic properties obtaining can be a guide to the metaphysical possibility of haecceitistic properties obtaining.

In the hyperintensional setting, the super-rigidity property is replaced by a hyper-rigidity property, which is defined as follows:

(*) $M, s \in S, s^* \in S^* \vdash p$ iff:
(i) $\forall c'_s \llbracket p \rrbracket^{c,c'} = 1$ if $s \in \llbracket p \rrbracket^+$; and
(ii) $\forall i_{s*} \llbracket p \rrbracket^{c,i} = 1$ if $s^* \in \llbracket p \rrbracket^+$

6.3 Two Dogmas of Semantic Rationalism

The tenability of the foregoing depends upon whether objections to what might be understood as the two dogmas of semantic rationalism can be circumvented.[4]

6.3.1 The First Dogma

The first dogma of semantic rationalism mirrors Quine (1951)'s contention that one dogma of the empiricist approach is the distinction that it records between analytic and synthetic claims. The analogous dogma in the semantic rationalist setting is that a distinction can be drawn between contextual intensions – witnessed by differences in the cognitive significance of two sentences or terms which have the same extension, e.g., with x = 2, 'x^2' and '2x' – by contrast to epistemic intensions. Chalmers (2006) delineates orthographic, linguistic, semantic, and cognitive (including conceptual) contextual intensions, and argues that they are all distinct from epistemic intensions in light of apriority figuring in the definition of the epistemic possibilities which are input to the latter functions. The distinction coincides with two interpretations of two-dimensional semantics. As noted, the epistemic interpretation of two-dimensional semantics takes the value of a formula relative to a first

---
[4]Thanks to Josh Dever for the objections.

parameter ranging over epistemically possible worlds to determine the extension of the formula relative to a second parameter ranging over metaphysically possible worlds (see Chalmers, op. cit.). According to the metasemantic interpretation, a sentence, such as that 'water = H_2O', is metaphysically necessary, whereas assertions made about metaphysically necessary sentences record the non-ideal epistemic states of agents and are thus contingent (see Stalnaker, 1978, 2004). The first dogma is thus to the effect that there are distinct sets of worlds – sets of either epistemic possibilities or of contextual presuppositions, respectively – over which the context ranges in the epistemic and metasemantic interpretations.

If no conditions on the distinctness between contextual and epistemic intensions can be provided, then variance in linguistic intension might adduce against the uniqueness of the epistemic intension. Because of the possible proliferation of epistemic intensions, conditions on the super-rigidity of the formulas and terms at issue might thereby not be satisfiable. The significance of the first dogma of semantic rationalism is that it guards against the collapse of epistemic and linguistic intensions, and thus the collapse of language and thought.

A defense of the first dogma of semantic rationalism might, in response, be proffered, in light of the status of higher-order distributive plural quantification in natural language semantics. Plural quantifiers are distributive, if the individuals comprising the plurality over which the quantifier ranges are conceived of singly, rather than interpreting the quantifier such that it ranges over irreducible collections. Natural language semantics permits plural quantification into both first and second-level predicate position. However, there are no examples of plural quantification into third-level predicate position in empirical linguistics, despite that examples thereof can be readily countenanced in intended models of formal languages. As follows, higher-order plural quantification might adduce in favor of the first dogma of semantic rationalism, to the effect that contextual linguistic and epistemic intensions can be sufficiently distinguished.

6.3.2 The Second Dogma

The second dogma of semantic rationalism mirrors Quine (op. cit.)'s contention that another dogma of empiricism is the reduction of the meaning of a sentence to the empirical data which verifies its component expressions. The analogous dogma in the semantic rationalist setting states that individuation-conditions on concepts can be provided in order to distinguish between concepts unique to formal and informal domains. The significance of the second dogma of semantic rationalism is that whether the objects falling under a concept belong to a formal domain of inquiry will subsequently constrain its modal profile.

The analogous dogma in the semantic rationalist setting states that individuation-conditions on concepts can be provided in order to distinguish between concepts unique to formal and informal domains. The significance of this dogma of semantic rationalism is that whether the objects falling under a concept belong to a formal domain of inquiry will subsequently constrain its modal profile.

In the space of epistemic possibility, it is unclear, e.g., what reasons there might be to preclude implicit definitions such as that the real number of the x's is identical to Julius Caesar (see Frege, 1884/1980: 56; Clark, 2007) by contrast to being identical to a unique set of rational numbers as induced via Dedekind cuts. It is similarly unclear how to distinguish, in the space of epistemic possibility, between formal and informal concepts, in order to provide a principled account of when a concept, such as the concept of 'set', can be defined via the axioms of the language in which it figures, by contrast to concepts such as 'water', where definitions for the latter might target the observational, i.e. descriptive and functional, properties thereof.

The notion of scrutability concerns 'suppositional' inferences from a base class of truths, $PQTI$ – i.e. physical, phenomenal, and indexical truths and a 'that's-all' truth – which determine canonical specifications, A_{1-n}, of epistemically possible worlds, to other truths (Chalmers, 2010b: 3). Scrutability from a canonical description of an epistemically possible world i.e. scenario, characterized by the set of truths, $PQTI$, to an arbitrary sentence, fixes an epistemic intension. Chalmers (2012: 245) is explicit about this: 'The intension of a sentence S (in a context) is true at a scenario w iff S is a priori scrutable from [A] (in that context), where [A] is a canonical specification of w (that is, one of the epistemically complete sentences in the equivalence class of w) ... A Priori Scrutability entails that this sentence S is a priori scrutable (for me) from a canonical specification [A] of my actual scenario, where [A] is something along the lines of $PQTI$'. However, physical, phenomenal, and indexical truths are orthogonal to truths about necessarily non-concrete objects such as abstracta.[5] How then are the epistemic intensions for abstracta

---

[5]For challenges to the indexing account of mathematical explanation, see Baker and

fixed? The most obvious maneuver would be to add mathematical truths to the scrutability base from which sentences about mathematical objects can be inferred.[6] It is not obvious, however, which mathematical, or perhaps logical, truths would be necessary to add in order to capture all truths about formal domains. In this section, I thus provide an explanation of how formal and informal domains can be distinguished which departs from this suggestion, and where the distinction can thereby serve to determine the modal profiles of the relevant domain classes.

The concept of mereological parthood provides a borderline case. While the parthood relation can be axiomatized so as to reflect whether it is non-reflexive, non-symmetric, and transitive, its status as a formal property is more elusive. The fact, e.g., that an ordinal is part of the sequence of ordinal numbers impresses as being necessary, while yet the fact that a number of musicians comprise the parts of a chamber ensemble might impress as being contingent.

The Julius Caesar Problem

The Julius Caesar problem, and the subsequent issue of whether there might be criteria for delineating formal from informal concepts in the space of epistemic modality, may receive a unified response. The ambiguity with regard to whether the parthood relation is formal – given that its relata can include both formal and informal objects – is similar to the ambiguity pertaining to the nature of real numbers. As Frege (1893/2013: §162) notes: 'Instead of asking which properties an object must have in order to be a magnitude, one needs to ask: how must a concept be constituted in order for its extension to be a domain of magnitudes? [...] Let us, for short, say "*Relation*" instead of "extension of a relation", then we can say: the magnitudes considered by us are Relations. Accordingly, the ratios of magnitudes, or real numbers, will be regarded as Relations on Relations. Our domains of magnitudes are classes of Relations, namely extensions of concepts that are subordinate to the concept *Relation*' (op. cit.). The interest of Frege's definition of the concept of real number is that explicit mention must be made therein to an

---

Colyvan (2011). For more on mathematical explanation and its relation to scientific truths, see Mancosu (2008); Pincock (2012); Lange (2017); and Baron et al (2020).

[6]Chalmers (2012: 388) suggests this maneuver with regard to the problem of the scrutability of mathematical truths in general. See Prelević (2021), for further discussion.

131

infinite domain of entities to which the number is supposed, as a type of measurement, to be applied (§163).

In response: The following implicit definitions – i.e., abstraction principles – can be provided for the concept of real number, where the real numbers are defined as sets, or Dedekind cuts, of rational numbers. Following Shapiro (2000), let F, G, and R denote rational numbers, such that concepts of the reals can be specified as follows: $\forall F,G[\mathbf{C}(F) = \mathbf{C}(G) \iff \forall R(F \leq R \iff G \leq R)]$. Concepts of rational numbers can themselves be obtained via an abstraction principle in which they are identified with quotients of integers – $[\mathbf{Q}\langle m, n\rangle = \mathbf{Q}\langle p, q\rangle \iff n = 0 \land q = 0 \lor n \neq 0 \land q \neq 0 \land m \times q = n \times p]$; concepts of the integers are obtained via an abstraction principle in which they are identified with differences of natural numbers – $[\mathbf{Diff}(\langle x, y\rangle) = \mathbf{Diff}(\langle z, w\rangle) \iff x + w = y + z]$; concepts of the naturals are obtained via an abstraction principle in which they are identified with pairs of finite cardinals – $\forall x,y,z,w[\langle x, y\rangle(=\mathbf{P}) = \langle z, w\rangle(=\mathbf{P}) \iff x = z \land y = w]$; and concepts of the cardinals are obtained via Hume's Principle, to the effect that cardinals are identical if and only if they are equinumerous – $\forall \mathbf{A} \forall \mathbf{B}[[\text{Nx: } \mathbf{A} = \text{Nx: } \mathbf{B} \equiv \exists R[\forall x[\mathbf{A}x \rightarrow \exists y(\mathbf{B}y \land Rxy \land \forall z(\mathbf{B}z \land Rxz \rightarrow y = z))] \land \forall y[\mathbf{B}y \rightarrow \exists x(\mathbf{A}x \land Rxy \land \forall z(\mathbf{A}z \land Rzy \rightarrow x = z))]]]$.

Frege notes that 'we can never [...] decide by means of [implicit] definitions whether any concept has the number Julius Caesar belonging to it, or whether that same familiar conqueror of Gaul is a number or not' (1884/1980: 56). A programmatic line of response endeavors to redress the Julius Caesar problem by appealing to sortal concepts, where it is an essential property of objects that they fall in the extension of the concept (see Hale and Wright, 2001: 389, 395). In order further to develop the account, I propose to avail of recent work in which identity conditions are interpreted so as to reflect relations of essence and explanatory ground. The role of the essentiality operator will be to record a formal constraint on when an object falls under a concept 'in virtue of the nature of the object' (Fine, 1995: 241-242). The role of the grounding operator will be to record a condition on when two objects are the same, entraining a hyperintensional type of implicit definition for concepts which is thus finer-grained and less susceptible to error through misidentification.

In his (2015a), Fine treats identity criteria as generic statements of ground. By contrast to **material** identity conditions which specify when two objects are identical, **criterial** identity conditions explain in virtue of what the two objects are the same. Arbitrary, or generic, objects are then argued to be constitutive of criterial identity conditions. Let a model, M, for a first-order language, L, be a tuple, where $M = \langle I, A, R, V \rangle$, with I a domain of concrete and abstract individuals, A a domain of arbitrary objects, R a dependence relation on arbitrary objects, and V a non-empty set of partial functions from A to I (see Fine, 1985). The arbitrary objects in A are reified variables. The dependence relation between any a and b in A can be interpreted as a relation of ontological dependence (op. cit.: 59-60). Informally, from $a \in A$ s.t. $F(a)$, one can infer $\forall x. F(x)$ and $\exists x. F(x)$, respectively (57). Then, given two arbitrary objects, x and y, with an individual i in their range, '$[(x = i \land y = i) \rightarrow x = y]$', such that x and y mapping to a common individual explains in virtue of what they are the same (Fine, 2015b).

Abstraction principles for, e.g., the notion of set, as augmented so as to record distinctions pertaining to essence and ground, can then be specified as follows:

- Given x,y, with $\text{Set}(x) \land \text{Set}(y)$: $[\forall z(z \in x \equiv z \in y) \leftarrow_{x,y} (x = y)]$

(Intuitively, where the 'given' expression is a quantifier ranging over the domain of variables-as-arbitrary objects: Given x, y, whose values are sets, it is essential to x and y being the same that they share the same members); and

- Given x,y, with $\text{Set}(x) \land \text{Set}(y)$: $[\forall z(z \in x \equiv z \in y) \rightarrow_{x,y} (x = y)]$

(Intuitively: Given arbitrary objects, x, y, whose values are sets, the fact that x and y share the same members grounds the fact that they are the same).

Combining both of the above directions yields the following hyperintensional, possibly asymmetric, biconditional:

- Given x,y, with $\text{Set}(x) \land \text{Set}(y)$: $[\forall z(z \in x \equiv z \in y) \leftrightarrow_{x,y} (x = y)]$.

A reply to the Julius Caesar problem for real numbers might then avail of the foregoing metaphysical implicit definitions, such that the definition would record the essentiality to the reals of the property of being necessarily non-concrete as well as provide a grounding-condition:

- Given F,G$[\mathbf{C}(F) = \mathbf{C}(G) \leftrightarrow_{F,G} \forall R(F \leqslant R \iff G \leqslant R)]$, and
- $\Box \forall X_{X/F} \Box \exists Y [\neg C(Y) \land \Box(X = Y)]$

(Intuitively: Given arbitrary objects, F, G, whose values are the real numbers: It is essential to the F's and the G's that the concept of the Fs is identical to the concept of the G's iff (i) F and G are identical subsets of a limit rational number, R, and (ii) with C(x) a concreteness predicate, necessarily for all real numbers, X, necessarily there is a non-concrete object Y, to which necessarily X is identical; i.e., the reals are necessarily non-concrete. The foregoing is conversely the ground of the identification.)

Heck (2011: 129) notes that the Caesar problem incorporates an epistemological objection: 'Thus, one might think, there must be more to our apprehension of numbers than a mere recognition that they are the references of expressions governed by HP [Hume's Principle - D.E.]. Any complete account of our apprehension of numbers as objects must include an account of what distinguishes people from numbers. But HP alone yields no such explanation. That is why Frege writes: "Naturally, no one is going to confuse [Caesar] with the [number, zero]; but that is no thanks to our definition of [number]"' (Gl, §62).

The condition of being necessarily non-concrete in the metaphysical definition for real numbers, as well as the conditions of essence and ground therein, provide a reply to the foregoing epistemological objection, i.e. the required explanation, by contrast to the insufficient definition, of what distinguishes people from numbers.

6.3.3 Mereological Parthood

The above proposal can then be generalized, in order to countenance the abstract profile of the mereological parthood relation. By augmenting the axioms for parthood in, e.g., classical mereological parthood with a clause to the effect that it is essential to the parthood relation that it is necessarily non-concrete, parthood can thus be understood to be abstract; and truths in which the relation figures would thereby be necessary.

- Given x: $\Phi(x) \land \Box\forall x \Box\exists y\, [\neg C(y) \land \Box(x = y)] \leftrightarrow_x \Gamma(x)$ where

- $\Gamma(x) :=$ x is the parthood relation, $<$, which is non-reflexive, asymmetric, and transitive, and where the relation satisfies the axioms of classical extensional mereology codified by the predicate, $\Phi(x)$ (see Cotnoir, 2014):

Weak Supplementation: x < y → ∃z[¬∃w[(w < z ∨ w = z) ∧ (w < x ∨ w = x)] ∧ (z < y ∨ z = y)], and

Unrestricted Fusion: ∀xx∃y[F(y,xx)],

with the axiom of Fusion defined as follows:

Fusion: F(t,xx) := (xx < t ∨ xx = t) ∧ ∀y[(y < t ∨ y = t) → (y < xx ∨ y = xx)]

As with sets, members of which can be concrete yet for which membership is necessary, fusions are themselves abstracta, formed by a fusion-abstraction principle. The abstraction principle states that two singular terms – in which an abstraction operator, σ, from pluralities to fusions figures as a term – are identical, if and only if the fusions overlap the same locations (see Cotnoir, ms). Let a topological model be a tuple, comprised of a set of points in topological space, μ; a domain of individuals, D; an accessibility relation, R; and a valuation function, V, assigning distributive pluralities of individuals in D to subsets of μ:

M = ⟨μ, D, R, V⟩;

R = R(xx, yy)$_{xx,yy \in \mu}$ iff R$_{xx}$ ⊆ μ$_{xx}$ × μ$_{xx}$, s.t. if R(xx, yy), then ∃o⊆μ, with xx∈o s.t. ∀yy∈oR(xx, yy), where the set of points accessible from a privileged node in the space is said to be open; and V = $f(ii \in D, m \in \mu)$.[7] Necessity is interpreted as an interiority operator on the space:

M,xx ⊩ □φ iff ∃o ⊆ μ, with xx∈o, such that ∀yy∈o M,yy ⊩ φ.

The following fusion abstraction principle can then be specified:

Given xx,yy,F[σ(xx, F) = σ(yy, F) ↔$_{xx,yy}$ [f(xx, m_1) ∩ f(yy, m_1) (≠ ∅)]].

(Intuitively, given arbitrary objects whose values are the pluralities, xx,yy: It is essential to xx and yy that fusion-abstracts – formed by mapping the pluralities to the abstracta – are identical, because the fusions overlap the same nonstationary – i.e., ≠ ∅ – locations. The converse is the determinative ground of the identification.)

The foregoing constraints on the formality of the parthood relation – both being necessarily non-concrete and figuring in pluralities which serve to individuate fusions as abstract objects – are sufficient then for redressing the objections to the second dogma of semantic rationalism; i.e., that individuation-conditions are wanting for concepts unique to formal and informal domains, which would subsequently render the modal profile of such

---

[7] μ is further Alexandrov; i.e., closed under arbitrary unions and intersections. Thanks again to Peter Milne.

concepts indeterminate. That relations of mereological parthood are abstract adduces in favor of the claim that the values taken by the relation are necessary. The significance of both the necessity of the parthood relation, as well as its being abstract rather than concrete, and thus being in some sense apriori, is that there are thus compelling grounds for taking the relation to be super-rigid, i.e., to be both epistemically and metaphysically necessary.

Finally, a third issue related to the dogma is that, following Dummett (1963/1978: 195-196), the concept of mereological parthood might be taken to exhibit a type of 'inherent vagueness', in virtue of being indefinitely extensible. Dummett (1996: 441) defines an indefinitely extensible concept as being such that: 'if we can form a definite conception of a totality all of whose members fall under the concept, we can, by reference to that totality, characterize a larger totality all of whose members fall under it'.[8] It will thus be always possible to increase the size of the domain of elements over which one quantifies, in virtue of the nature of the concept at issue; e.g., the concept of ordinal number is such that ordinals can continue to be generated, despite the endeavor to quantify over a complete domain, in virtue of iterated applications of the successor relation, and the concept of cardinal number is such that the cardinals can continue to be generated via elementary embeddings. Bernays (1942)'s theorem states that class-valued functions from classes to sub-classes are not onto, where classes are non-sets (see Uzquiano, 2015a: 186-187). A generalization of Bernays' theorem can be recorded in plural set theory,[9] where the cardinality of the sub-pluralities of an incipient plurality

---

[8]The concept of indefinite extensibility is introduced by Cantor's proof of the uncountability of the real numbers (1874); Cantor (1882; 1883: §11)'s two principles of generation, successor and limits, and third principle of cumulative structure; and Cantor's Theorem (1891). See Zermelo (1904). Russell (1905: 36) writes: 'The above considerations point to the conclusion that the contradictions result from the fact that, according to current logical assumptions, there are what we may call *self-reproductive processes* and classes. That is, there are some properties such that, given any class of terms all having such a property, we can always define a new term also having the property in question. Hence we can never collect *all* the terms having the said property into a whole; because, whenever we hope we have them all, the collection which we have immediately proceeds to generate a new term also having the said property'. The concept is reintroduced by Dummett (1963/1978), in the setting of a discussion of the philosophical significance of Gödel (1931/1986)'s first incompleteness theorem. See Shapiro and Wright (2006: 255-260) and the essays in Rayo and Uzquiano (2006); Section **6.3.3**; Chapter **12**; and Studd (op. cit.), for further discussion.

[9]See Burgess (2004/2008), for an axiomatization of 'Boolos-Bernays' plural set theory, so named after the contributions of Bernays (op. cit.) and Boolos (1984, 1985). See

will always be greater than the size of that incipient plurality. If one takes the cardinal height of the cumulative hierarchy to be fixed, then one way of tracking the variance in the cardinal size falling in the extension of the concept of mereological parthood might be by redefining the intension thereof (Uzquiano, 2015b). Because it would always be possible to reinterpret the concept's intension in order to track the increase in the size of the plural universe, the intension of the concept would subsequently be non-rigid; and the concept would thus no longer be super-rigid.

One way in which the objection might be countered is by construing the variance in the intension of the concept of parthood as tracking interpretational modalities, rather than alethic modal properties. Fine (2005c: 547) has a similar approach to the concept of set, and writes: 'On the usual conception of the cumulative hierarchy of ZF, we think of the membership predicate as given and of the ontology of sets or classes as something to be made out. Thus given an understanding of membership, we successively carve out the ontology of sets by using the membership-predicate to specify which further sets should be added to those that are already taken to exist. Under the present approach, by contrast, we think of the ontology of classes as given and of the membership predicate as something to be made out. Thus given an understanding of the ontology of classes, we successively carve out extensions of the membership-predicate by using conditions on the domain of classes to specify which further membership-relationships should obtain'. The parthood relation can then be necessary while the modality satisfies full S5 – i.e., modal axioms K $[(\Box\phi \to \psi) \to (\Box\phi \to \Box\psi)]$, T $(\Box\phi \to \phi)$, and E $(\neg\Box\phi \to \Box\neg\Box\phi)$ – and there can be variations in the size of the quantifier domains over which the relation and its concept are defined. Let $\uparrow$ be an intensional parameter which indexes and stores the relevant formulas at issue to a particular world (see Vlach, 1973). The $\downarrow$-symbol is an operator which serves to retrieve, as it were, that indexed information. These arrow-operators are referred to as Vlach-operators. Adding Vlach-operators is then akin to multiple-indexing: The value of a formula, as indexed to a particular world, will then constrain the value of that formula, as indexed – via the addition of the new arrows – to different worlds. Interpreting the operators interpretationally permits there to be multiple-indexing in the array of intensional parameters relative to which a formula gets its value, while the underlying logic for metaphysical modal operators can be S5, partitioning

---

Linnebo (2007), for critical discussion.

the space of worlds into equivalence classes. Formally:
$\uparrow_1 \forall x \exists \phi \uparrow_2 \exists y [\phi(x) \downarrow_1 \land \phi(y) \downarrow_2]$.

The clause states that, relative to a first interpretational parameter in which all of the x's satisfying the parthood relation are quantified over, there is – relative to a distinct interpretational parameter – another element which satisfies that predicate. Crucially, differences in the interpretational indices, as availed of in order to record variance in the size of the cumulative hierarchy of elements falling in the range of the parthood relation, is yet consistent with the cardinality of the elements in the domain being fixed, such that the valuation of the relation can yet be metaphysically necessary.

6.3.4 Summary

In this section, I addressed objections to a dogma of the semantic rationalism underpinning the epistemic interpretation of two-dimensional semantics. In response to the objections to the dogma – according to which criteria on distinguishing formal from informal domains unique to the extensions of various concepts are lacking, which subsequently engenders indeterminacy with regard to the modal profiles of those concepts – I availed of generic criterial identity conditions, in which it is essential to identical arbitrary representatives of objects that they satisfy equivalence relations which are conversely ground-theoretically determinative of the identification, and further essential thereto that they satisfy the predicate of being necessarily non-concrete. The extensions of indefinitely extensible concepts can further be redefined relative to distinct hyperintensional parameters. Thus, parthood can be deemed a necessary, because abstract, relation, despite (i) hyperintensional variance in the particular objects on which the parthood relation is defined; and (ii) variance in the cardinality of the domain in which those objects figure, relative to which the concept's intensions are defined.

My strategy in what follows will be to provide two-dimensional formulas for essential properties. The first dimension is interpreted epistemically and the second dimension is interpreted metaphysically. Then, supposing essential properties are haecceitistic properties (see Korbmacher, 2016), I will generalize the formula to account for the interaction between epistemic and metaphysical profiles of haecceities.

Suppose that essential properties either are super-rigid or ground super-rigidity.[10] Following Fine (2000), suppose there is an operator, $\Box_F$, where $\Box_F A$ is read 'it is true in virtue of the nature of (some or all) of the F's that A' where 'each of the objects mentioned in A is involved in the nature of one of the F's' (op. cit.: 543). $\Box_F$ satisfies the axioms KTE and necessitation:

$\Box_F A \to A$,
$\Box_F(A \to B) \to (\Box_F A \to \Box_F B)$,
$\neg \Box_F A \to \Box_{F,|A|} \neg \Box_F A$, F rigid, where

F is rigid 'if it is a rigid predicate symbol or is of the form $\lambda x \bigvee_{1 \leq i \leq n} A_i$, $n \geq 0$, where each formula A_i, $i = 1, \ldots, n$, is either of the form Px or of the form $x = y$ for some variable y distinct from x' (545), and

$|E|$ stands for $\lambda x (x \eta E)$ x the first variable not free in E, where $x \eta E$ stands for $\bigvee_{1 \leq i \leq m} x = x_i \vee \bigvee_{1 \leq i \leq m} P_i x$,

$A \vdash \Box_{|A|} A$, and
$F \subset G \to (\Box_F A \to \Box_G A)$ (546).

A model M is a quadruple $\langle W, I, \leq, \phi \rangle$, where

W is a non-empty set of worlds, I is a function taking each w∈W into a non-empty set of individuals$_w$, $\leq$ is a reflexive transitive dependence relation on $\bigcup_{w \in W}$ with respect to which each world is closed (a∈I_w and a $\leq$ b implies b∈I_w), and ϕ is a valuation function taking each constant a into an individual $\phi(a)$ of some I_w (w∈W), each rigid predicate symbol H into a subset $\phi(H)$ of some I_w, and each world w and pure n-place predicate symbol F into a set $\phi(F, w)$ of n-tuples of I_w, where a pure predicate involves no reference to any object (544, 547-548).

For a subset J of $\bigcup I_w$, the closure c(J) of J in M is {b: a $\leq$ b for some a∈J} (548).

M is a model with E a sentence or closed predicate whose constants are $a_1, \ldots, a_m$ and whose rigid predicate symbols are $P_1, \ldots, P_n$ (op. cit.). The objectual content $[E]^M$ of E in M is then $\{\phi(a_1), \ldots, \phi(a_m)\} \cup \{\phi(P_1), \ldots, \phi(P_n)\}$ and E is defined in M at w∈W if $[E]^M \subseteq I_w$ (op. cit.).

Then the semantics for $\Box_F$ can be defined as follows:

$w \Vdash \Box_F A$ iff (i) $[A]^M \subseteq c(F_w)$, and (ii) $v \Vdash A$ whenever $I_v \supseteq F_w$, where F_w is $\phi(w, F)$ (op. cit.).

$\Box_F$ can the be defined relative two parameters, the first ranging over epistemically possible worlds or truthmakers considered as actual, and the second ranging over metaphysically possible worlds or truthmakers, such that

---

[10] See Fine (1994), for the locus classicus of accounts according to which essence grounds metaphysical necessity.

the conceivability of it being true in virtue of the nature of the nature of (some or all) of the F's that A entails the metaphysical possibility or verification of it being true in virtue of the nature of the nature of (some or all) of the F's that A:

$\forall c \in C, w \in W [\![\Box_F A]\!]^{c,w} = 1$ iff $\exists c' \in C, w' \in W [\![\Box_F A]\!]^{c',w'} = 1$.

Korbmacher (2016) argues that essential properties are haecceitistic properties. If so, then the following two-dimensional formula can be specified. If it is epistemically possible that Φx, then it is metaphysically possible that Φx. Formally:

$\forall c \in C, w \in W [\![\Phi x]\!]^{c,w} = 1$ iff $\exists c' \in C, w' \in W [\![\Phi x]\!]^{c',w'} = 1$.

Thus, the epistemic possibility of haecceity comprehension constrains the value of the metaphysical possibility of haecceity comprehension, and – in response to Roca-Royes and Chalmers – there is a case according to which conceivability is a guide to a principle of modal metaphysics.

Conceivability is not a fail-safe method of alighting upon haecceities or essential properties. However, evidence about the haecceitistic or essential properties of objects can play a role in ascertaining which of a number of epistemic possibilities or truthmakers is actual.[11] The epistemic two-dimensional method countenanced in the foregoing is such that – because haecceities and essential properties either are super-rigid or entail super-rigidity – epistemic truthmakers or possibilities about essential properties considered as actual will determine the values of their metaphysical truthmakers or possibilities. An accidental property might mistakenly be thought to be essential, in which case conceivability would not be an adequate guide to metaphysical verification or possibility. However, once essential properties are discovered in the actual world, the actuality of the epistemic verification or possibility thereof can serve as a guide to their metaphysical verification or possibility. Another way that evidence might bear on the actuality of epistemic truthmakers is via the role of apriori scrutability in defining primary intensions. Chalmers writes that '[t]he primary intension of [a sentence] S is true at a scenario [i.e. epistemically possible world] w iff [A] epistemically necessitates S, where [A] is a canonical specification of w', where '[A] epistemically necessitates S iff a [material] conditional of the form "[A] → S" is apriori' and the apriori entailment is the relation of scrutability (Chalmers, 2006; see also 2012: 245, quoted above). Because physical, phenomenal, and indexical truths are built into the scrutablity base, and scrutability plays a central role in the defi-

---
[11] See Lowe (2008; 2012).

nition of primary intensions, there is thus at least one viable route to the epistemology of essence via conceivability as constrained by actual evidence.

In the remainder of the chapter, I will examine issues pertaining to the determinacy of epistemic possibilities.

6.4 Determinacy and Consistency

In his (2014), Chalmers argues for the law of excluded middle, such that it is either apriori derivable – modeled via a strict conditional with an epistemic necessity i.e. apriority opterator, i.e. 'scrutable' – that p or scrutable that ¬p, depending on the determinacy of p. Chalmers refers to the case in which p must be determinate, entailing determinate scrutability, as the Hawthorne model, and the case in which it can be indeterminate, entailing indeterminate scrutability, as the Dorr model (259).[12] Chalmers argues that, for any p, one can derive 'p iff it is scrutable that p' from 'p iff it is true that p' (262). However, 'p iff it is scrutable that p' is unrestrictedly valid only on Dorr's, and not Hawthorne's, model (op. cit.).[13]

Turner (2014) argues that Chalmers needs 'p ∧ it is indeterminate that p' to be inconsistent, otherwise both p and ¬p would be epistemically possible such that one could then scry whether p or ¬p. One compelling maneuver might be to restrict the valid apriori material entailments to determinately true propositions; and to argue, against Chalmers (2009)'s preferred ontological anti-realist methodology, that the necessity of parthood is both epistemically and metaphysically determinately true, if true at all. Restricting scrutability to determinate truths is Chalmers' maneuver in his (2012: 31).

More generally, however, there are barriers to establishing the consistency of the space of epistemic modality. Because, on this approach, conceivability is a guide to metaphysical possibility, inconsistency in epistemic modal space might then entrain inconsistency or indeterminacy in metaphysical modal space, despite the foregoing cases of conceivability being a guide to principles of modal metaphysics. One might reject explosion (p ∧ ¬p → q, for any q), such that inconsistencies do not entail everything, in order to mitigate the

---

[12] See Dorr (2003: 103-4) and Hawthorne (2005: sec. 2).

[13] Chalmers rejects the epistemicist approach to indeterminacy, which reconciles the determinacy in the value of a proposition with the epistemic indeterminacy concerning whether the proposition is known (op. cit.: 288). For further discussion, see Williamson (1994).

issue, though p ∧ ¬p in epistemic space might yet entrain indeterminacy whether p in metaphysical space. This might be an insuperable issue for conceivability-based accounts of modal epistemology.

One route to securing the epistemic interpretation of consistency is via Chalmers' conception of idealized epistemic possibility. Conceivability is ideal if and only if nothing rules it out apriori upon unbounded rational reflection (2012: 143). The rational reflection pertinent to idealized conceivability can be countenanced modally, normatively, and so as to concern the notion of epistemic entitlement. An idealization is (i) modal iff it concerns what it is metaphysically possible for an agent to know or believe; (ii) normative iff it concerns what agents ought to believe; and (iii) warrant-involving iff it concerns the propositions which agents are implicitly entitled to believe (2012: 63).

More general issues for the consistency of epistemically possible worlds, even assuming that the idealization conditions specified in (i)-(iii) are satisfied, include Yablo (1993)'s paradox, and Gödel (1931/1986)'s incompleteness theorems. Yablo's paradox is as follows:

(S1) For all k > 1, S_k is false;
(S2) For all k > 2, S_k is false;
...
(Sn) For all k > n, S_k is false;
(Sn+1) For all k > n+1, S_k is false.

(Sn) says that (Sn+1) is false. Yet (Sn+1) is true. Contradiction.[14]

Gödel's incompleteness theorems can be thus outlined.[15] 'A numeral canonically denoting a natural number **n** is abbreviated as $\overline{n}$. A formalized theory F is ω-consistent if it is not the case that for some formula A(x), both $F \vdash \neg A(\overline{n})$, and for all **n**, $F \vdash \exists x A(x)$. A set S of natural numbers is strongly representable in F if there is a formula A(x) of the language of F with one free variable x such that for every natural number **n**:

'n∈S ⇒ $F \vdash A(\overline{n})$;
'n∉S ⇒ $F \vdash \neg A(\overline{n})$.

'A set S of natural numbers is weakly representable in F if there is a formula A(x) of the language of F such that for every natural number **n**:

'n∈S ⟺ $F \vdash A(\overline{n})$.

---

[14]For further discussion, see Cook (2014a).

[15]The presentation follows that of Raatikainen (2022). I will quote the entire text, because the definitions and characterizations are mostly owing to Raatikainen.

'The representability theorem says then that in any consistent formal system which contains Robinson Arithmetic i.e. **Q**:[16]:

'1. A set (or relation) is strongly representable if and only if it is recursive;

'2. A set (or relation) is weakly representable if and only if it is recursively enumerable. [See §**2.3**.]

'Suppose that there is a coding of symbols and formulas by the natural numbers. The Gödel number of a formula A is denoted as ⌜A⌝.

'Suppose that the diagonalization lemma holds, such that $F \vdash Q \iff A(⌜Q⌝)$.

'For the first incompleteness theorem, the diagonalization lemma is applied to the negation of the provability predicate, $\neg Prov_F(x)$, which yields the following sentence:

'(Z) $F \vdash M_F \iff \neg Prov_F(⌜M_F⌝)$.

'Assume that M_F is provable. By the weak representability of provability-in-F by $Prov_F(x)$, F would also prove $Prov_F(M_F)$. Because F proves Z – i.e. $F \vdash M_F \iff \neg Prov_F(⌜M_F⌝)$ – F would then prove $\neg M_F$. So F would be inconsistent. Thus, if F is consistent, then M_F is not provable in F.

'Assume that F is ω-consistent. Assume, then, that $F \vdash \neg M_F$. Then F cannot prove M_F, because it would then be ω-inconsistent. Thus no natural number **n** is the Gödel number of a proof of M_F. Because the proof relation is strongly representable, for all **n**, $F \vdash \neg Prf_F(\overline{n}, ⌜M_F⌝)$. If $F \vdash \exists x Prf_F(x, ⌜M_F⌝)$, F is not ω-consistent. Thus F does not prove $\exists x Prf_F(x, ⌜M_F⌝)$, i.e. F does not prove $Prov_F(⌜M_F⌝)$. By the equivalence recorded in (Z), F does not prove $\neg M_F$.

'For the second incompleteness theorem: Suppose that consistency, Con(F), is defined as $\neg Prov_F(⌜\bot⌝)$, where $\bot$ expresses an inconsistent formula such as $\overline{0} = \overline{1}$. Formalizing the proof of the first incompleteness theorem in F yields $F \vdash Cons(F) \to M_F$. If Cons(F) were provable in F, so would be M_F.

---

[16]The signature of **Q** is first-order Peano Arithmetic without the induction schema, with 0 a constant for zero, a unary function symbol s for successor, and binary function symbols + and • for addition and multiplication. The axioms of **Q** are:
1. $\forall x \neg s(x) = 0$
2. $\forall_{x,y} s(x) = s(y) \to x = y$
3. $\forall_x x = 0 \lor \exists_y x = s(y)$
4. $\forall_x x + 0 = x$
5. $\forall_{x,y} x + s(y) = s(x + y)$
6. $\forall_x X • 0 = 0$
7. $\forall_{x,y} x • s(y) = x • y + x$

(https://ncatlab.org/nlab/show/Robinson+arithmetic).

Cons(F) is thus unprovable, given the first incompleteness theorem.'

6.5 Concluding Remarks

One of the primary objections to accounting for the relationship between conceivability and metaphysical possibility via the epistemic interpretation of two-dimensional semantics is that epistemic possibilities are purportedly insensitive to modal metaphysical propositions, concerning, e.g., parthood and the haecceitistic properties of individuals. In this chapter, I have endeavored to redress the foregoing objection. Further objections, from both the potential indeterminacy in, and inconsistency of, the space of epistemic possibilities, were then shown to be readily answered. In virtue of the super-rigidity of the parthood relation and essential properties, conceivability can thus serve as a guide to to principles of modal metaphysics.

Chapter 7

Grounding, Conceivability, and the Mind-Body Problem

This essay argues that Chalmers (1996; 2010)'s two-dimensional conceivability argument against the derivation of phenomenal truths from physical truths risks being obviated by a hyperintensional regimentation of the ontology of consciousness.

Chalmers (2010a) provides the following argument against the identification of phenomenal truths with physical and functional truths. Let M be a model comprised of a domain D of formulas; C a set of epistemic possibilities; W a set of metaphysical possibilities; R_c and R_w, accessibility relations on C and W, respectively; and V a valuation function assigning formulas to subsets of C and W. So, M = $\langle$D, C, W, R_c, R_w, V$\rangle$. Let P denote the subset of formulas in the domain concerning fundamental physics, as well as both neurofunctional properties such as oscillations of neural populations, and psychofunctional properties such as the retrieval of information from memory stores. Let Q denote the subset of formulas in the domain concerning phenomenal consciousness. A formula is epistemically necessary or apriori ($\Box$), if and only if it has the same value at all points in C, if and only if it is impossible, i.e. inconceivable, for the formula to a variant value ($\neg \Diamond \neg$). A formula is negatively conceivable ($\Diamond$) if and only if nothing rules it out apriori ($\neg \Box \neg$) (144). A formula is metaphysically necessary if and only if it has the same value at all points in W. A formula is said to be 'super-rigid', if and only if it is both epistemically and metaphysically necessary, and thus has the same value at all points in epistemic and metaphysical modal space (2012: 474).

The two-dimensional conceivability argument against physicalism proceeds as follows.

The physicalist thesis states that:
P → Q.
Suppose, however, that the physicalist thesis is false. Thus,
1. ¬(P → Q).
By the definition of the material conditional,
2. ¬(¬P ∨ Q).
By the De Morgan rules for negation,
3. ¬¬P ∧ ¬Q.
By double negation elimination,
4. P ∧ ¬Q.[1]

'P ∧ ¬Q' can receive a truth value relative to two parameters, C and W. In two-dimensional semantics, the value of the formula relative to C determines the value of the formula relative to W. Let C range over a space of epistemic possibilities and let W range over a space of metaphysical possibilities. Then,

⟦P ∧ ¬Q⟧^{c,w} = 1 iff ∃c'∈C∃w'∈W⟦P ∧ ¬Q⟧$^{c',w'}$ = 1.

The foregoing clause codifies the thought that, if it is epistemically possible that the truths about physics and functional organization obtain while the truths about consciousness do not, then the dissociation between P and Q is metaphysically possible as well.

Chalmers' informal characterization of the argument proceeds as follows:
1. P ∧ ¬Q is conceivable.
2. If P ∧ ¬Q is conceivable, P ∧ ¬Q is [epistemically, i.e.] 1-possible.
3. If P ∧ ¬Q is 1-possible, P ∧ ¬Q is [metaphysically, i.e.] 2-possible.
4. If P ∧ ¬Q is 2-possible, then materialism is false.
Thus,
5. Materialism is false (2010: 149).

The thesis of 'weak modal rationalism' states that conceivability can be a guide to 1-possibility, i.e. conceivability entails 1-possibility or truth at a centered metaphysically possible world (2002). Thus conceivability can be a

---

[1] For the formal equivalence, given the definition of the material conditional, see Chalmers (2010a: 169).

guide to metaphysical possibility on the metaphysical construction of epistemically possible worlds. In the hyperintensional setting, epistemic states might be analyzed as centered metaphysical states.

However, in his (2002) and (2010), Chalmers argues for line (3) of the argument to the effect that 1-, i.e. epistemic, possibility entails 2-, i.e. metaphysical, possibility, in the case when the primary and secondary intensions for physics and consciousness coincide. Thus, there is no gap between the epistemic and metaphysical profiles for expressions involving physics or consciousness, and the conceivability about scenarios concerning them will entail the 1-possibility and the 2-possibility of those scenarios. In the hyperintensional setting, one works with hyperintensions, i.e. topic-sensitive truthmakers, rather than intensions.

Finally, in his (2012), Chalmers defines a notion which he refers to as super-rigidity: 'When an expression is epistemically rigid and also metaphysically rigid (metaphysically rigid *de jure* rather than *de facto*, in the terminology of Kripke 1980), it is *super-rigid*' (Chalmers, 2012: 239). He writes: 'I accept Apriority/Necessity and Super-Rigid Scrutability. (Relatives of these theses play crucial roles in "The Two-Dimensional Argument against Materialism")' (241). The Apriority/Necessity Thesis is defined as the 'thesis that if a sentence S contains only super-rigid expressions, s is a priori iff S is necessary' (468), and Super-Rigid Scrutability is defined as the 'thesis that all truths are scrutable from super-rigid truths and indexical truths' (474). This is thus a third way for conceivability to be a guide to metaphysical possibility. The epistemic necessity i.e. apriority of a sentence involving only super-rigid expressions is such that it converges with the metaphysical necessity of that sentence. In the hyperintensional setting, super-rigidity is replaced by a hyper-rigidity condition defined in chapter **4**.

If the conceivability argument is sound, then the physicalist thesis – that all phenomenal truths are derivable from physical and functional truths – is possibly false. The foregoing argument entrains, thereby, the metaphysical possibility of a property-based version of dualism between phenomenal consciousness and fundamental physics.

One of the standard responses to Chalmers' conceivability argument is to endeavor to argue that there are 'strong' necessities, i.e. cases according to which the necessity of the physical and phenomenal formulas throughout epistemic and metaphysical modal space is yet consistent with the epistemic possibility that the formulas have a different value.[2] Note, however, that strong necessities are ruled-out, just if one accepts the normal duality axioms for the modal operators: i.e., it is necessary that ϕ if and only if it is impossible for ϕ to be false: $\Box\phi$ iff $\neg\Diamond\neg\phi$. Thus, the epistemic necessity of ϕ rules out the epistemic possibility of not-ϕ by fiat. So, proponents of the strong necessity strategy are committed to a revision of the classical duality axioms.

Another line of counter-argument proceeds by suggesting that the formulas and terms at issue are not super-rigid. Against the super-rigidity of physical truths, one might argue, for example, that our knowledge of fundamental physics is incomplete, such that there might be newly discovered phenomenal or proto-phenomenal truths in physical theories from which the truths about consciousness might be derived.[3] More contentiously, the epistemic profile of consciousness – as recorded by the concepts comprising our thoughts thereof, or by the appearance of its instantiation – might be dissociable from its actual instantiation. A variation on this reply takes our concepts of phenomenal consciousness still to refer to physical properties (see Block, 2006). A related line of counter-argument relies on the assumption that phenomenal concepts are entities which are themselves physically reducible (see Balog, 1999).

Finally, a counter-argument to the conceivability argument that has yet to be advanced in the literature is that its underlying logic might be non-classical. Thus, for example – by relying on double negation elimination in the inference from line 3 to 4 above – the two-dimensional conceivability ar-

---

[2] As Chalmers (2010a: 166-167) writes, 'Before proceeding, it is useful to clarify [the general conceivability-possibility thesis] CP by making clear what a counterexample to it would involve ... Let us say that a *negative strong necessity* is a statement S such that S is [epistemically]-necessary and [metaphysically]-necessary but ¬S is negatively conceivable'. For a case-by-case examination of purported examples of strong necessities, see Chalmers (op. cit.: 170-184; 2014a). Because it is epistemically possible for there to be scenarios in which there is no consciousness, the target neighborhood of epistemically possible worlds is that in which the conditions on there being phenomenal consciousness are assumed to obtain. [Thanks here to Dave Chalmers (private correspondence).] Thus, the notion of epistemic necessity will satisfy conditions on real world validity, rather than general validity. In the latter case, a formula is necessary if and only if it has the same value in all worlds in a model. In the former case, the necessity at issue will hold throughout the neighborhood, where a neighborhood function assigns the subset of worlds in which consciousness obtains to a privileged world in the model.

[3] See Seager (1995) and Strawson (2006) for the panpsychist proposal. Proponents of the pan-protopsychist approach include Stoljar (2001, 2014) and Montero (2010).

gument is intuitionistically invalid. A novel approach might further consist in arguing that epistemic modality might be governed by the Routley-Meyer semantics for relevant logic.[4] Relevant validity can be defined via a ternary relation, such that $[\![\phi \to \psi]\!]^\alpha = 1$ iff $[\![\phi]\!]^\beta \leq [\![\psi]\!]^\gamma$ and $R(\alpha,\beta,\gamma)$, where the parameters, α, β, and γ, range over epistemic possibilities. The philosophical interest of relevant logic is that it eschews the principle of disjunctive syllogism; i.e., $\forall \phi,\psi [[(\phi \vee \psi) \wedge \neg \phi] \to \psi]$ and $\forall \phi,\psi[[\phi \wedge (\neg \phi \vee \psi)] \to \psi]$. Without disjunctive syllogism, logical entailment can no longer be identified with the material conditional, and this would block the derivation of line 2 from line 1 in the two-dimensional conceivability argument.

In this essay, I will pursue a line of argument which is novel and distinct from the foregoing. I argue, in turn, that the conceivability argument can be circumvented, when the relationship between the truths about fundamental physics and the truths about phenomenal consciousness is analyzed in a classical, hyperintensional setting. Suppose, for example, that the physicalist thesis is defined using hyperintensional, grounding operators rather than metaphysical necessitation.[5] Then, the epistemic and metaphysical possibility that $\neg(P \to Q)$ is classically valid, although targets a less fine-grained metaphysical connection between physical and phenomenal truths. Even if P's grounding Q still entails the metaphysical necessitation of Q by P, the epistemic-intensional value of '$\neg(P \to Q)$' – will be an insufficient guide to the metaphysical-hyperintensional value of the proposition. So, even if the intension for 'consciousness' is rigid in both epistemic and metaphysical modal space, the epistemic intension recording the value of the proposition will be blind to its actual metaphysical value, because the latter will be hyperintensional.

In the remainder of this essay, I will outline the regimentation of the proposals in the ontology of consciousness using hyperintensional grounding operators, rather than the resources of modality and identity.[6] By contrast to the modal approach underlying the conceivability argument, the hyperintensional regimentation targets the properties of reflexivity and bijective mappings, in order to countenance novel, ontological dependence relations between the properties of consciousness and physics, which are finer-grained

---

[4]See Routley and Meyer (1972a,b; 1973).

[5]For the logic and operator-based semantics for the notion of explanatory ground, see Fine (2012c,d).

[6]See Elohim (2018), for the regimentation and for further discussion.

than necessitation.[7]

Following Fine (2012c,d), let a polyadic operator have a *ground-theoretic* interpretation, only if the profile induced by the interpretation concerns the hyperintensional *truth-making* connection between an antecedent set of truths or properties and the relevant consequent. Let a grounding operator be *weak* if and only if it induces reflexive grounding; i.e., if and only if it is sufficient for the provision of its own ground. A grounding operator is *strict* if and only if it is not weak. A grounding operator is *full* if and only if it uniquely provides the explanatory ground for a fact. A grounding operator is *part* if and only if it - along with other facts - provide the explanatory ground for a fusion of facts.

Combinations of the foregoing explanatory operators may also obtain: $x < y$ iff ϕ is a *strict full* ground for ψ; $x \leqslant y$ iff ϕ is a *weak full* ground for ψ; $x \prec y$ iff ϕ is a *strict part* ground for ψ; $x \preceq y$ iff ϕ is a *weak part* ground for ψ; $x \leq y \land \neg(y \leq x)$ iff ϕ is a *strict partial* ground for ψ; $x \prec^* y$ iff $x_1, ..., x_n \leqslant y$, iff ϕ is a *partial strict* ground for ψ; $x \prec' z$ iff $[\phi \prec^* \psi \land \psi \leq \mu]$ iff

---

[7]The claim that necessitation must be present in cases in which there is grounding is open to counterexample. Because, e.g., hyperintensional dependencies can obtain in only parts of, rather than entirely within, a world, the hyperintensional dependencies need not reflect necessitation. For further discussion of the grounding-necessitation thesis, see Rosen (2010) and Skiles (2015).

ϕ is a *part strict* ground for some further fact, μ.[8]

The proposals in the metaphysics of consciousness can then be regimented in the hyperintensional framework as follows.

- Functionalism (modally: truths about consciousness are identical to truths about neuro- or psychofunctional role):

  Functional truths (F) ground truths about consciousness (Q) if and only if the grounding operator is:

  -strict full, s.t. $F < Q$

  -distributive (i.e. bijective between each truth-ground and grounded truth), s.t. $\exists f_{1-1} \langle F, Q \rangle$

- Phenomenal Realist Type Identity (modally: truths about consciousness are identical to truths about biological properties, yet phenomenal

---

[8]The derivation is induced by the following proof-rules:

- Subsumption

  $(<, \leqslant)$:
  $$[(x_1, \ldots, x_n < y)] \rightarrow (x \leqslant y)$$

  $(<, <)$:
  $$[(x_1, \ldots x_n) < y] \rightarrow (x < y)$$

  $(\prec, \leq)$:
  $$(x \prec y) \rightarrow (x \leq y)$$

  $(\leqslant, \leq)$:
  $$(x \leqslant y) \rightarrow (x \leq y)$$

- Distributivity/Bijection:

  $$\forall x \in X, y \in Y$$
  $$[G[(\ldots x \ldots)(\ldots y \ldots)], \text{ s.t.}$$
  $$f_{1-1}: [x_1 \rightarrow y_1], \ldots, f_{1-1}: [x_n \rightarrow y_n]].$$

properties are – in some sense – non-reductively real).[9]

Biological truths (B) ground truths about consciousness (Q) if and only if the grounding operator is:

-strict partial, s.t. $B \leq Q \wedge \neg Q \leq B$;

-distributive, s.t. $\exists f_{1-1}\langle B, Q\rangle$; and

-truths about consciousness are weak part (i.e. the set partly reflexively grounds itself), s.t. $Q \leq Q$

- Property Dualism (modally: truths about consciousness are identical neither to functional nor biological truths, yet are necessitated by physical truths):

Physical truths (P) ground truths about consciousness (Q) if and only if the grounding operator is:

-$P \leq Q$;

-non-distributive, s.t. $\neg \exists f_{1-1}\langle P, Q\rangle$; and

-truths about consciousness are weak part, s.t. $Q \leq Q$

- Panpsychism (in Non-constitutive guise: Phenomenal properties are the intrinsic realizers of extrinsic functional properties and their roles; in Constitutive guise: (i) fundamental microphysical entities are functionally specified and they instantiate microphenomenal properties, where microphenomenal properties are the realizers of the fundamental microphysical entity's role/functional specification; and (ii) microphenomenal properties constitute the macrophenomenal properties of macrophysical entities):

Truths about consciousness (Q) ground truths about functional role (F) if and only if the grounding operator is:

-strict full, s.t. $Q < F$; and

-non-distributive, s.t. $\neg \exists f_{1-1}\langle Q, F\rangle$

The philosophical significance of the hyperintensional regimentation of the ontology of consciousness is at least three-fold. First, the regimentation

---

[9]See, e.g., Smart (1959: 148-149), for an attempt to account for how phenomenal properties and biological properties can be identical, while phenomenal properties might yet have distinct higher-order properties.

permits one coherently to formulate Phenomenal Realist Type Identity. Leibniz's law states that for all propositional variables x,y and for all properties R, $x = y$ iff $(Rx \iff Ry)$. According to the Phenomenal Realist Type Identity proposal, phenomenal properties are identical to biological properties, while phenomenal properties are in some sense non-reductively real. Thus, in the modal setting, Phenomenal Realist Type Identity belies Leibniz's law, on the assumption that the latter can be applied to intensional entities. One virtue of the hyperintensional regimentation is thus that it avoids this result, by providing a framework with the expressive resources sufficient to formulate the non-reductive Type Identity proposal.

Second, the hyperintensional grounding regimentation evinces how functionalist approaches to the ontology of consciousness can be explanatory, because the identification of phenomenal properties with functional organization can be defined via the foregoing ground-theoretic explanatory properties. Block (2015) suggests that – by contrast to Phenomenal Realist Type Identity – identifying phenomenal properties with functional roles cannot sufficiently account for the ground-theoretic explanation of the identity. Block distinguishes between metaphysical and ontological versions of physicalism. Block's 'ontological physicalism' is a reductive, functionalist theory, and eschews of explanation by restricting the remit of its theory to 'what there is'; i.e. to specifying identity statements between entities in the domain of quantification (114). By contrast, Block's 'metaphysical physicalism' – namely, Phenomenal Realist Type Identity – purports to account for the nature of the entities figuring in theoretical identity statements via availing of relations of explanatory, ontological dependence (op. cit.).

Block poses the following consideration against the functionalist (117). Suppose that there is a counterpart of a human organism with isomorphic functional properties, but comprised of distinct biological properties. Suppose that the functional isomorph instantiates phenomenal properties. Block argues that the functional isomorph 'is like us superficially, but not in any deep property that can plausibly be one that scientists will one day tell us is the physical ground of consciousness [...] So there is a key question that that kind of reductive physicalism – ontological physicalism – does not ask nor answer: what is it that creatures with the same phenomenology share that grounds that phenomenology' (op. cit.)? The foregoing does not provide an argument that the neuro- and psycho-functionalist must provide an account of in virtue of what phenomenal properties are instantiated. Rather, Block suggests only that functionalist proposals do not sufficiently inquire into the realizers of the functional roles that they specify. He suggests that this theoretical approach would be insufficient, if one were to seek an explanation of the psychofunctional correlations between phenomenal property types and the relevant functional roles.

The second theoretical virtue of the hyperintensional regimentation is thus that it demonstrates how Block's analysis might be circumvented. Functionalism can be regimented within the logic of hyperintensional ground; and can therefore satisfy the formal requirements on explaining in virtue of what phenomenal truths ontologically depend upon functional truths.[10]

Third, and most crucially: The regimentation demonstrates how metaphysically possible relations between consciousness and physics cannot be witnessed by epistemic constraints, when the latter are recorded by the conceivability – i.e., the epistemic possibility – thereof. Propositional epistemic modality is blind to the hyperintensional, metaphysical dependencies holding between phenomenal and physical truths. Thus, the two-dimensional conceivability argument against the derivation of phenomenal truths from physical truths risks being obviated by a hyperintensional regimentation of the ontology of consciousness.

One way to resolve the third issue is to provide a hyperintensional semantics for epistemic space, such that epistemic space can track metaphysical space when the latter is itself hyperintensionally defined via e.g. grounding operators. Following chapter **2**, hyperintensional epistemic two-dimensional truthmaker semantics permits conceivability to be a guide to metaphysical possibility in the hyperintensional setting.

---

[10] Of pertinence to the foregoing is another distinction drawn by Fine (2015b), between material and criterial identity conditions. A crucial point of departure between the foregoing and the approach proffered in this essay is Fine's ontology of arbitrary objects, to which the present proposal need make no appeal.

Part III: Epistemic Modality and Hyperintensionality in the Philosophy of Mathematics

Chapter 8

Epistemic Hyperintensionality and Absolute Decidability

This chapter aims to contribute to the analysis of the nature of mathematical modality and hyperintensionality, and to the applications of the latter to absolute decidability. I argue that mathematical modality falls under at least four types; the interpretational, the metaphysical, the non-maximally objective, and the logical. The interpretational type of mathematical modality has traditionally been taken to concern possible reinterpretations of quantifier domains (see Fine, 2006, 2007; Linnebo, 2009, 2010, 2013; Studd, 2013), and the possible reinterpretations of the intensions of the concept of set (Uzquiano, 2015a). The metaphysical type of modality concerns the ontological profile of abstracta and mathematical truth. Abstracta are thus argued to have metaphysically necessary being, and mathematical truths hold of metaphysical necessity, if at all (see Fine, 1981). Metaphysical modality is the maximal objective modality.[1] However, the phenomenon of indefinite extensibility of the natural numbers and the real numbers is, I argue, possessed of two modalities whose interaction is captured by a two-dimensional semantics, and which consist of an epistemic modality characterizing reinterpretations of quantifier domains, and a non-maximal, hence non-metaphysical, yet still objective modality characterizing ontological expansion. Another candidate for the non-maximal objective mathematical modality is the modal profile of forcing (see Kripke 1965; Hamkins and Löwe, 2008). Instances, finally,

---

[1]For endorsements of this contention, see Kripke (1980: 99), Lewis (1986), Stalnaker (2003: 203), and Williamson (2016b: 459-460). For an argument in opposition, see Clarke-Doane (2021).

of the logical type of mathematical modality might concern the property of consistency (see Field, 1989: 249-250, 257-260; Rayo, 2013: 50; Leng, 2007; 2010: 258; and Berry, 2022), and can perhaps be further witnessed by the logic of provability (see Boolos, 1993).

The significance of the present contribution is as follows. (i) Rather than countenancing the interpretational type of mathematical modality as a primitive, I argue that the interpretational type of mathematical modality is a species of epistemic modality. (ii) I argue, then, that the framework of two-dimensional hyperintensional semantics ought to be applied to the mathematical setting. The framework permits of a formally precise account of the priority and relation between epistemic mathematical modality and metaphysical mathematical modality.[2] I target, in particular, the modal axioms that the respective interpretations of the modal operator ought to satisfy. The discrepancy between the modal systems governing the parameters in the two-dimensional setting provides an explanation of the difference between the metaphysical possibility of absolute decidability and our knowledge thereof. (iii) I examine the application of the mathematical modalities beyond the issue of indefinite extensibility. As a test case for the two-dimensional approach, I investigate the interaction between epistemic and metaphysical mathematical modalities and both large cardinal axioms and Orey sentences which are undecidable relative to the axioms of ZFC, such as the Generalized Continuum Hypothesis. The two-dimensional framework permits of a formally precise means of demonstrating how the metaphysical possibility of

---

[2]Kripke (1980: 140-141) treats mathematical modality as a for all one knows operator, by contrast to epistemic possibilities concerning theoretical identity statements for rigid designators for objects. Epistemic possibilities for theoretical identities are primary conceivable - Kripke's phrase for primary conceivability is 'logically possible' (141); see Chalmers (2010: ch. 6, §10) for a discussion of modal monism and primary, i.e. epistemic, and secondary, i.e. subjunctive, intensions defined on 'logically possible worlds' - not secondary conceivable. See, further, **§1.1.2** and **§1.1.3**, fn. 12. Kripke writes in the guise of an 'objector' that: 'Nor can you evade the difficulty by declaring the "might have" of "might have turned out otherwise" to be merely epistemic, in the way that "Fermat's Last Theorem might turn out to be true and might turn out to be false" merely expresses our present ignorance, and "Arithmetic might have turned out to be complete" signals our former ignorance. In these mathematical cases, we may have been ignorant, but it was in fact mathematically impossible for the answer to turn out otherwise than it did' (141), and responds: 'The reason the example of Fermat's Last Theorem gives a different impression is that here no analogue suggests itself, except for the extremely general statement that, in the absence of proof or disproof, it is possible for a *mathematical conjecture* to be either true or false' (143).

absolute decidability and the Continuum Hypothesis can be accessed by their epistemic-modal-mathematical profile. I argue that, in the absence of disproof, large cardinal axioms are epistemically possible, and thereby provide a sufficient guide to the objective mathematical possibility of determinacy claims and the Continuum Hypothesis. (iv) Finally, I apply the hyperintensional, topic-sensitive epistemic two-dimensional truthmaker semantics from chapters **2** and **4**. I examine the relation between epistemic truthmakers and the axioms of epistemic set theory, large cardinal axioms, the Epistemic Church-Turing Thesis, as well as the verification-profile of Ω-logical consequence.

In Section **1-2**, I discuss how the properties of the epistemic mathematical modality and objective mathematical modality converge and depart from previous attempts to delineate the contours of similar notions. I define the formal clauses and modal axioms governing the epistemic and metaphysical types of mathematical modality. I also advance a topic-sensitive epistemic two-dimensional truthmaker semantics, if hyperintensional approaches are to be preferred to possible worlds semantics. I examine the relation between epistemic truthmakers and the axioms of epistemic set theory, large cardinal axioms, the Epistemic Church-Turing Thesis, as well as the verification-profile of Ω-logical consequence. Section **3** extends the two-dimensional framework to the issue of mathematical knowledge; in particular, to the hyperintensional profile of large cardinal axioms and to the absolute decidability of the Continuum Hypothesis. Section **4** provides concluding remarks.

8.1 Mathematical Modality

8.1.1 Metaphysical Mathematical Modality

A formula is a logical truth if and only if the formula is true in an intended model structure, M = <W, D, R, V>, where W designates a space of metaphysically possible worlds; D designates a domain of entities, constant across worlds; R designates an accessibility relation on worlds; and V is an assignment function mapping elements in D to subsets of W.

Metaphysical Mathematical Possibility
$$[\![\Diamond\phi]\!]^{v,w} = 1 \iff \exists w' [\![\phi]\!]^{v,w'} = 1$$

Metaphysical Mathematical Necessity
$$[\![\Box\phi]\!]^{v,w} = 1 \iff \forall w' [\![\phi]\!]^{v,w'} = 1,$$

with $\Diamond := \neg\Box\neg$

8.1.2 Epistemic Mathematical Modality

In order to accommodate the notion of epistemic possibility, we enrich M with the following conditions: M = <C, W, D, R, V>, where C is a set of epistemically possibilities.

The interpretation of epistemic possibility which will here be at issue defines the notion as conceivability, the dual of epistemic necessity i.e. apriority (see Chalmers, 2006, 2011), instead of consistent logical reasoning (Jago, 2009; Bjerring, 2012) and a for all one knows operator (see MacFarlane, 2011). In the hyperintensional setting outlined below, the box and diamond operators are replaced by necessary and possible truthmakers which serve as verifiers for propositions. On the consistent logical reasoning interpretation of epistemic possibility, necessary truthmakers receive the same interpretation as $\Box_n$, i.e. that a proposition A is provable in n steps of logical reasoning using the rules in R. On their metaphysical interpretation, truthmakers verify the truth values of propositions and are orthogonal to the logical reasoning which figures in the interpretation of the epistemic truthmakers. The consistent logical reasoning interpretation ties truthmaking to provability and is of relevance to the discussion of epistemic possibility and hyperintensionality and their bearing on absolute decidability, but will not be here examined.[3]

---

[3]Gödel (1931/1986: 195) writes that 'I wish to note expressly that Theorem XI ["*Let K be any recursive consistent class of* FORMULAS; *then the* SENTENTIAL FORMULA *stating that K is consistent is not K-PROVABLE;* in particular, the consistency of P is not provable in P, provided P is consistent (in the opposite case, of course, every proposition is provable [in P])" (op. cit.: 193)] (and the corresponding results for ["set theory" (op. cit.: 195),] M and ["classical mathematics" (op. cit.),] A) do not contradict Hilbert's formalistic viewpoint. For this viewpoint presupposes only the existence of a consistency proof in which nothing but finitary means of proof is used, and it is *conceivable* [my emphasis - D.E.] that there exist finitary proofs that *cannot* be expressed in the formalism of P (or of M or A)' (op. cit.; Kleene in Gödel, 1986: pp. 138-139); and Herbrand's and Gödel's correspondence in Gödel (2003). See chapter **8.3**, for further discussion. Thanks here to John Goodridge and to Øystein Linnebo. See Linnebo and Shapiro (2020), for an account of intensional realizability semantics as a kind of truthmaking.

8.1.3 Modal Axioms

- Metaphysical mathematical modality is governed by the modal system KTE, as augmented by the Barcan formula and its Converse (see Fine, 1981).

K: $\Box[\phi \to \psi] \to [\Box\phi \to \Box\psi]$
T: $\Box\phi \to \phi$
E: $\neg\Box\phi \to \Box\neg\Box\phi$
Barcan: $\Diamond \exists x Fx \to \exists x \Diamond Fx$
Converse Barcan: $\exists x \Diamond Fx \to \Diamond \exists x Fx$

- Epistemic mathematical modality is governed by the modal system, K4+GL.[4]

K: $\blacksquare[\phi \to \psi] \to [\blacksquare\phi \to \blacksquare\psi]$
4: $\blacksquare\phi \to \blacksquare\blacksquare\phi$
GL: $\blacksquare[\blacksquare\phi \to \phi] \to \blacksquare\phi$

Note that, if one prefers a hyperintensional semantics to an intensional semantics, one can avail of the definitions of hyperintensions as functions from states in a state space to extensions instead of from whole epistemically and metaphysically possible worlds. See chapters **2** and **4**, for the relevant clauses.

8.1.4 Hyperintensional Epistemic Set Theory

Following the presentation in Scedrov (1986: 104), an epistemic truthmaker set theory can be defined as follows. A(x) is defined as in chapter **2**, and is interpreted as an epistemically necessary truthmaker, though not as apriority.

---

[4] For further discussion of the properties of GL, see Löb (1955); Smiley (1963); Kripke (1965); and Boolos (1993). Löb's provability formula was formulated in response to Henkin (1952)'s problem concerning whether a sentence which ascribes the property of being provable to itself is provable. (See Halbach and Visser, 2014, for further discussion.) For an anticipation of the provability formula, see Wittgenstein (1933-1937/2005: 378). Wittgenstein writes: 'If we prove that a problem can be solved, the concept 'solution' must somehow occur in the proof. (There must be something in the mechanism of the proof that corresponds to this concept.) But the concept mustn't be represented by an external description; it must really be demonstrated. / The proof of the provability of a proposition is the proof of the proposition itself' (op. cit.). Wittgenstein contrasts the foregoing type of proof with 'proofs of relevance' which are akin to the mathematical, rather than empirical, propositions, discussed in Wittgenstein (2001: IV, 4-13, 30-31).

Logic

- Equality axioms, $x = y \wedge \phi(x) \to \phi(y)$

  All classical propositional tautologies

- From ϕ and $\phi \to \psi$ infer ψ

- $A(\phi) \to \phi$

- $A(\phi) \to AA(\phi)$

- $A(\phi) \wedge A(\phi \to \psi) \to A(\psi)$

- From ϕ infer $A(\phi)$

- $\forall \phi(x) \to \phi(y)$, where y is free for x in $\phi(x)$

- From $\phi \to \psi(x)$ infer $\phi \to \forall \psi(x)$, if x is not free in ϕ

- $\phi(y) \to \exists \phi(x)$, where y is free for x in $\phi(x)$

- From $\psi(x) \to \phi$ infer $\exists \psi(x) \to \phi$, if x is not free in ϕ

Non-logical Axioms

- Epistemic Extensionality: $A[\forall z(z \in x \to z \in y) \to x = y]$

- Foundation: $\forall x[\forall y \in x \phi(y) \to \phi(x)] \to \forall x \phi(x)$

- Epistemic Foundation: $A[[\forall x[A[\forall y \in x \phi(y) \to \phi(x)]] \to A[\forall x \phi(x)]]]$

- Pairing: $\exists z A(x \in z \wedge y \in z)$

- Union: $\exists z A[\forall w(\exists y \in x w \in y \to w \in z)]$

- Separation: $\exists z A[\forall y[y \in z \iff y \in x \wedge \phi(y)]]$, where z is not free in $\phi(y)$

- Epistemic Power Set: $\exists z A[\forall w[A(\forall y \in w y \in x \to w \in z)]]$

- Infinity: $\exists z A[\exists y A(y \in z) \wedge \forall u \in z \exists v A(v \in z \wedge u \in v)]$

- Collection: $\forall x \in u \exists y \phi(x, y) \to \exists z \forall x \in u \exists y \in z \phi(x, y)$ where z is not free in $\phi(x, y)$

161

- Epistemic Collection: A[∀x∈u∃yϕ(x, y) → ∃zA∀x∈u∃y[A(y∈z) ∧ ϕ(x, y)]], where z is not free in ϕ(x, y).

Two-dimensional hyperintensions can then be defined for each of the foregoing axioms, such that each axiom would be defined relative to two parameters, the first ranging over topic-sensitive epistemic truthmakers, which determines the value of the axiom relative to a second parameter ranging over either non-maximally objective or maximally objective i.e. metaphysical truthmakers.

8.1.5 Two-dimensional Hyperintensional Large Cardinals

A provisional definition of large large cardinal axioms is as follows.

∃xΦ is a large large cardinal axiom, because:

(i) Φx is a Σ_2-formula, where 'a sentence ϕ is a Σ_2-sentence if it is of the form: There exists an ordinal α such that $V_\alpha \Vdash \psi$, for some sentence ψ' (Woodin, 2019);

(ii) if κ is a cardinal, such that $V \models \Phi(\kappa)$, then κ is strongly inaccessible, where a cardinal κ is regular if the cofinality of κ – comprised of the unions of sets with cardinality less than κ – is identical to κ, and a strongly inaccessible cardinal is regular and has a strong limit, such that if $\lambda < \kappa$, then $2^\lambda < \kappa$ (see Kanamori, 2012: 360); and

(iii) for all generic partial orders $\mathbb{P} \in V_\kappa$, and all V-generics $G \subseteq \mathbb{P}$, $V[G] \models \Phi x$ (Koellner, 2006: 180).

The truthmaker 2D-hyperintension for large cardinal axioms is then $\forall s \in S, i \in I \llbracket \Phi x \rrbracket^{s,i} = 1$ iff $\exists s' \in S, i' \in I \llbracket \Phi x \rrbracket^{s',i'} = 1$.

The hyperintension states that the value of a large cardinal axiom relative to a topic-sensitive epistemic truthmaker determines the value of the axiom relative to a metaphysical truthmaker.

8.2 Departures from Precedent

Cantor (1897a/1996; 1897b/1996; 1899/1996) implicates conceivability in his definition of 'consistent' multiplicities i.e. sets or finite or transfinite numbers and his definition of 'inconsistent' i.e. 'absolute' infinities. Cantor (1899/1996: 931-935) writes: 'If we start from the notion of a definite multiplicity [Vielheit] (a system, a totality) of things, it is necessary, as I discovered, to distinguish two kinds of multiplicities (by this I always mean *definite* multiplicities).

'For a multiplicity can be such that the assumption that *all* of its elements "are together" leads to a contradiction, so that it is impossible to *conceive* [my emphasis - D.E.] of the multiplicity as a unity, as "one finished thing". Such multiplicities I call *absolutely infinite* or *inconsistent multiplicities*.

'As we can readily see, the "totality of everything thinkable", for example, is such a multiplicity; later still other examples will turn up.

'If on the other hand the totality of the elements of a multiplicity can be thought without contradiction as "being together", so that they can be gathered together into "*one* thing", I call it a *consistent multiplicity* or a "set".[5] Note that the two-dimensional intensions and hyperintensions of epistemic two-dimensional semantics account as well for the linking between what Cantor refers to as intrasubjective i.e. immanent reality and transsubjective i.e. transient reality (Cantor, 1883/1996: §8). Immanent reality concerns the reality of mathematical objects relative to the 'understanding', whereas transient reality concerns the reality of mathematical objects relative to the 'external world' (op. cit.). Cantor attributes the relation between the two realities as owing to the 'unity of the all to which we ourselves belong' (op. cit.). However, the existence of functions, i.e. hyperintensions, from topic-

---

[5]Cantor (1897a/1996: 926-927) writes: 'For the totality of all alephs is one that cannot be *conceived* [my emphasis - D.E.] as a determinate, well-defined, *finished* set. If this were the case, then this totality would be *followed* in size by a *determinate aleph*, which would therefore both *belong* to this totality (as an element) and *not belong*, which would be a contradiction.

[...]

'From this it is easy to show that, given the assumption (of a *determinate set* whose cardinal number is not an aleph), the totality of all alephs could then be grasped as a determinate, well-defined, finished set. But I have just proved that this is not so. Therefore, every a is always a determinate aleph.

'*Totalities* that cannot be grasped by us as "sets" (of which an example is the totality of alephs, as I just showed) I already called "absolutely infinite" many years ago, and distinguished them sharply from the *transfinite sets*'.

Cantor (1897b/1996: 927) writes: 'I say of a set that it can be thought of as *finished* (and call such a set, if it contains infinitely many elements, "transfinite" or "super-finite") if it is possible without contradiction (as can be done with finite sets) to think of *all its elements as existing together*, and so to think of the set itself as *a compounded thing for itself*; or (in other words) if it is *possible* to *imagine* [my emphasis - D.E.] the set as *actually existing* with the totality of its elements'.

sensitive epistemic state spaces to topic-sensitive objective or metaphysical state spaces to extensions provides a more illuminating explanation of the relation between concepts and metaphysics than does the contention that all entities can figure as members of sets or classes in set theory.

Gödel (1929/1986: 64-65) writes: 'On the other hand, the principle of the excluded middle seems to express nothing other than the decidability of every problem. To this, however, the following objections can be made:

'(1) The principle of the excluded middle is interpreted this way only by those of the intuitionistic persuasion;

'(2) Even if we accept this interpretation, what is affirmed is the solvability not at all through specified means but only through all means that are *in any way imaginable*,[4] [Footnote 4: 'It seems questionable, however, whether a notion of solvability that is so sweeping – and, consequently, the interpretation of the principle of the excluded middle that is at issue here – makes any sense at all'.] while what is shown below is precisely that every valid expression can be derived through completely *specified, concretely enumerated* inference rules'.[6]

Gödel (1932b/1986: 237) writes: 'If we imagine that the system Z is successively enlarged by the introduction of variables for classes of numbers, classes of classes of numbers, and so forth, together with the corresponding comprehension axioms, we obtain a sequence (continuable into the transfinite) of formal systems that satisfy the assumptions mentioned above, and it turns out that the consistency (ω-consistency) of any of those systems is provable in all subsequent systems'.

Gödel (1946/1990: p. 152) writes: 'But now, coming back to the definition of definability I suggested, it might be objected that the introduction of all ordinals as primitive terms is too cheap a way out of the difficulty, and that the concept thus obtained completely fails to agree with the intuitive concept we set out to make precise, because there exist undenumerably many sets definable in this sense. There is certainly some justification in this objection. For it has some plausibility that all things *conceivable* [my emphasis - D.E.] by us are denumerable, even if you disregard the question of expressibility in some language. But, on the other hand, there is much to be said in favor of the concept under consideration; namely, above all it is clear that, if the concept of mathematical definability is to be itself mathematically definable, it must necessarily be so that all ordinal numbers are

---

[6]See §**8.1.2**, fn. 3.

definable, because otherwise you could define the first ordinal number not definable, and | would thus obtain a contradiction. I think this does not mean that a concept of definability satisfying the postulate of denumerability is impossible, but only that it would involve some extramathematical element concerning the psychology of the being who deals with mathematics'.[7]

Gödel (1951/1995: 310) writes: 'For if the human mind were equivalent to a finite machine, then objective mathematics not only would be incompletable in the sense of not being contained in a well-defined axiomatic system, but moreover there would exist | *absolutely* unsolvable diophantine problems of the type described above, where the epithet "absolutely" means

---

[7]See Parsons in Gödel (1990: 147-148). Parsons writes: 'Gödel's philosophical remarks are prompted by the obvious objection that admitting all ordinals as primitive terms makes his notion no longer a notion of *definability*. He finds it plausible that "all things conceivable by us are denumerable". Does this mean that for any x, if x is conceivable by us, then x is denumerable, or that there are only denumerably many things conceivable by us? The latter reading seems more likely. Gödel remarks that, because of the paradox of the least indefinable ordinal, a notion of mathematical definability that makes the notion itself mathematically definable will have to have all ordinals definable. But what follows is that a notion satisfying the "postulate of denumerability" must involve some "extramathematical element concerning the psychology of the being which deals with mathematics" (page 4). It seems to me that the point is not just that to characterize such a notion will require some extramathematical vocabulary, for what will rule out a least ordinal not definable with the help of the extra vocabulary? Apparently the extra vocabulary must have the property that "definable with the help of the extra vocabulary" is no longer definable with the help of the extra vocabulary. Gödel finally argues that ordinal definability at least captures the notion of "being formed according to law" as opposed to "being formed by a random choice of the elements" (page 4). In particular, there is not a random element in the ordinals themselves.

'I would note further that, since the notion of definability contains a modal element, the question whether ordinal definability is a genuine notion of definability depends on the underlying modal notion, the "can" in "can be defined". Admitting all ordinals as definable might be viewed as an extreme extension of the notion of abstract mathematical possibility that arises in other contexts in the foundations of mathematics, such that of computable function, where complete abstraction is made both from the limitations of "hardware" and from feasibility in terms of the time required for a computation. It should be kept in mind that an ordinal definition of a set requires only finitely many ordinals (which can be reduced to one). To deny that an ordinal-definable set is "really" definable implies the existence of ordinals that are not really definable, no matter how our means of such definition might be extended. It is hard to see how a case could be made out for this so long as one stays on the abstract mathematical plane and does not introduce notions concerning "the psychology of the being which deals with mathematics", at least in a broad sense'.

that they would be undecidable, not just within some particular axiomatic system, but by *any* mathematical proof the human mind can conceive'.

Shoenfield (1967) writes: 'If we introduced symbols for new operations which cannot be defined in ZFC, we would increase our ability to describe sets and hence increase the power of the subset (and replacement) axioms. This appears to be a natural approach; but so far no one has been able to propose any suitable operations' (304). 'An approach which is more promising at the moment is to make fuller use of our principle of existence of stages: if we can imagine a situation in which all of the stages in a collection are completed, then there must be a stage after all the stages in the collection' (306).

Reinhardt (1974b) develops Shoenfield's comments, and countenances the following principle:

S 'If P is a property of stages, and if we can *imagine* a situation in which all the stages having P have been built up, then there *exists* a stage s beyond all the stages which have P' (5).

Imagination is defined thus: '[W]e choose here to pursue the sense of "imagine" according to which Existing = Real ⊆ Imaginary rather than that according to which Imaginable = Visualizable ⊆ (mathematically) Existing. To choose the latter would turn **S** into a tautology (or an exhortation to visualize!) and require for its usefulness a criterion of visualizability. One can read the axiom of **S** as such criteria, taking V to be the visualizable sets. Then [principle **S**] gives a condition under which ϕ will provide a "visualization" of a set. The axioms so read may be plausible for some kind of "visualizability" (such a concept of "visualizable" seems close to definable). However, I have trouble seeing that each $x \subseteq w$ is visualizable' (6).

Reinhardt writes of the relation between the imagination that a settheoretic rank at which a property is satisfied and the existence of the rank at which the property is satisfied that: 'We remark at the outset that one can read **S** in either (i) a more or (ii) a less constructive way, namely (i) that the stage s exists mathematically because of ... the act of imagination, which is thus a sort of construction of s, or (ii) that what can be imagined is but an indication of what has mathematical existence, so that the latter can retain changeless Platonic impregnability or Cantorian absoluteness' (op. cit.).[8]

---

[8]Reinhardt develops the following theory for principle **S**. 'The principle **S** distinguishes between "imagine" and "exist". We shall do this formally for sets by treating "imagine" as the quantifier and "exist" as quantification relativized to a certain predicate. (Consequently we should not call ∃ the existential quantifier, but a generalized existential quantifier. We keep the usual logic for ∃, however. We do not consider the possibility of a non-classical

Reinhardt (1974a: §6) proposes the use of imaginary sets and classes as 'imaginary experiments' (204), in order to define imaginary projections corresponding to the universe of sets which define Reinhardt cardinals.[9] An objection to the foregoing is advanced by Maddy (1988) who objects to the 'use of counterfactual situations to distinguish these new entities from sets' (754). Maddy writes: 'I think even those with strong modal intuitions will have trouble imagining how there might be more pure sets and ordinals than there are. After all, V is supposed to contain all the sets and ordinals there could possibly be' (op. cit.).

The approach to mathematical modality, according to which it yields a representation of the cumulative universe of sets, has been examined by Fine (2006) and Uzquiano (op. cit.). Fine argues that the mathematical modality ought to be postulational and interpretational; and thus taken to concern the reinterpretation of the domain over which the quantifiers range, in order to avoid inconsistency. Uzquiano argues similarly for an interpretational construal of mathematical modality, where the cumulative hierarchy of sets is fixed, yet what is possibly reinterpreted is the non-logical vocabulary of

---

logic for "imaginary" objects.) We suppose that we can imagine the set of all existing sets; V denotes this imaginary set. We have yet to explain the term "property". Generally, formulas correspond to properties of sets. Notice that the principle **S** only refers to existing properties, not imaginary ones (taking "is" not in the scope of "imagine" to have existential force). Formulas in which all parameters are existing sets will correspond to existing properties (we do not assume the correspondence is onto). Formulas involving merely imagined objects x, such as "$t \in x$", will not in general correspond to existing properties. Note in particular that we must not assume that V exists, and consequently the property P such that $t \in V \iff P(t)$ must not be assumed to exist either. / Our formalization of "there exists a set x such that ..." is "$\exists x(x \in V [\wedge] ...)$" (7).

'Imagine' and 'exist' can, too, be defined relative to ranks, with 'exists' to mean "exists at a higher level", and "imagine" to mean "exists, but possibly only at higher levels" (9).

The theory for principle **S** has the following axioms:

With $\exists x$ interpreted as "x can be imagined", "$x \in V$" as "x exists", and "$\phi(a_1, \ldots, a_n, t)$" as "$t$ has P", principle **S** is formalized as follows:

(i) $\langle a_1, \ldots, a_n \rangle \in V \wedge \exists x \forall t (\phi(a_1, \ldots, a_n, t) \to t \in x) \to \exists s [s \in V \wedge \forall t (\phi(a_1, \ldots, a_n, t) \to t \in s)]$, with ϕ a $\in$-formula '(i.e. any formula of L) with free variables $a_1, \ldots, a_n, t$, and x is distinct from all these' (8).

(ii) $\forall t (t \in x \iff t \in y) \to x = y$;

(iii) $t \in x \wedge x \in V \to t \in V$,
$t \subseteq x \wedge x \in V \to t \in V$,
$x \in V \wedge t \subseteq V \wedge t \equiv x \to t \in V$;

(iv) $\exists x \forall t (t \in x \iff \theta \wedge t \in x)$, where θ is a formula not involving x (7-8).

[9] $j: V \to V$.

167

the language, in particular the membership relation.[10]

On Fine (2005b)'s approach, there are simultaneously postulational, dynamic, and 'prescriptive' or imperatival modalities.[11] The prescriptive element consists in the rule:

'Introduction: !x.C(x)',

such that one is enjoined to postulate, i.e. to 'introduce an object x conforming to the condition C(x)' (2005b: 91; 2006: 38).

Fine clarifies that the postulational approach is consistent with a 'realist ontology' of the set of reals. He refers to the imperative to postulate new objects, and thereby reinterpret the domain for the quantifier, as the 'mechanism' by which epistemically to track the cumulative hierarchy of sets (2007: 124-125).

The present approach avoids a potential issue with Fine's account, with regard to the introduction of deontic modal properties of the prescriptive and imperatival rules that he mentions. It is sufficient that the interpretational modalities are a species of epistemic modality, i.e. possibilities that are relative to agents' spaces of states of information.

Developing Parsons (1983)'s program, Linnebo (2013) outlines a modalized version of ZF.[12] Similarly to the modal axioms for the epistemic math-

---

[10]Compare Gödel (1947/1990); Williamson (1998); and Fine (2005c).

[11]For the distinction between imperative rules: 'If C, do A!'; norms of requirement: 'At the beginning of the game, White must make the first move'; and norms of permission: 'If the configuration is C, you may castle', with regard to rule-following, see Boghossian (2008; 2012) and Wright (2012b).

[12]Linnebo (2018b) discusses the differences between Putnam's and Parsons' accounts of the role of modality in mathematics. Berry (2022) also discusses the differences between the foregoing. Linnebo (op. cit.: 265-266) avails of two-dimensional indexing for the relation between interpretational and circumstantial modalities. In Linnebo (2018a), he characterizes the relation between interpretational and circumstantial modalities via a bimodal product logic, rather than a two-dimensional semantics. He countenances two commutativity principles – '$\square\blacksquare\phi \iff \blacksquare\square$' and '$\lozenge\blacklozenge\phi \iff \blacklozenge\lozenge$', with $\square$ a circumstantial modality and $\blacksquare$ an interpretational modality – although details a counterexample to which they are susceptible. The present approach occurs in the setting of epistemic two-dimensional semantics, such that there is a one-way dependence of metaphysical profiles on epistemic profiles. Roberts (2019) countenances four interaction principles in a bimodal logic for interpretational and circumstantial modalities and applied to the indefinite extensibility of possibilia. Vlach-operators in two of the principles simulate two-dimensional indexing. The principles are a bimodal version of the converse Barcan formula: $\blacksquare\square\forall v\phi \to \forall v\blacksquare\square\phi$ (1159); $\blacksquare(\lozenge A \to \square\lozenge A)$ (1161); $\uparrow^{*1}\uparrow^1\blacklozenge\uparrow^{*2}\lozenge\uparrow^2\downarrow^{*1}\downarrow^1\square\forall x([E(x) \to \downarrow^{*2}\downarrow^2 E(x)])$ (1162); $\uparrow^{*1}\square\uparrow^1\blacklozenge\uparrow^{*2}\lozenge\uparrow^2[\downarrow^{*1}\downarrow^1\forall x[E(x) \to \downarrow^{*2}\downarrow^2 E(x)] \land \exists x[E(x) \land \downarrow^{*1}\downarrow^1 \neg E(x)]]$ (op. cit.). $\uparrow^*$A is a Vlach-operator on A selecting an interpretational modality from an ω-sequence

ematical modality specified in the previous section, Linnebo argues that his modal set theory ought to be governed by the system S4.2, the Converse Barcan formula, and (at least a restricted version of) the Barcan formula. However – rather than being either interpretational or epistemic – Linnebo deploys the mathematical modality in order to account for the notion of 'potential infinity', as anticipated by Aristotle.[13] The mathematical modality is thereby intended to provide a formally precise answer to the inquiry into the extent of the cumulative set-theoretic hierarchy; i.e., in order to precisify the answer that the hierarchy extends 'as far as possible' (2013: 205).[14]

Thus, Linnebo takes the modality to be constitutive of the actual ontology of sets; and the quantifiers ranging over the actual ontology of sets are claimed to have an 'implicitly modal' profile (2010: 146; 2013: 225). He suggests, e.g., that: 'As science progresses, we formulate set theories that characterize larger and larger initial segments of the universe of sets. At any one time, precisely those sets are actual whose existence follows from our strongest, well-established set theory' (2010: 159n21). However – despite his claim that the modality is constitutive of the actual ontology of sets – Linnebo concedes that the mathematical modality at issue cannot be interpreted metaphysically, because sets exist of metaphysical necessity if at all (2010: 158; 2013: 207). In order partly to allay the tension, Linnebo remarks, then, that set theorists 'do not regard themselves as located at some particular stage of the process of forming sets' (2010: 159); and this might

---

comprising a set thereof which validates A, and ↑A is a Vlach-operator on A selecting a metaphysical possibility from an ω-sequence comprising a set thereof which validates A. The appeal to epistemic two-dimensional semantics in order to account for interpretational as epistemic and circumstantial as objective modalities and their interaction in this essay was written in 2015 and pursued prior to knowledge of Linnebo's and Roberts' accounts. My approach differs, as well, by countenancing a hyperintensional, topic-sensitive epistemic two-dimensional truthmaker semantics and applying it to various phenomena in the philosophy of mathematics, as in chapters **8-12**.

[13] See Aristotle, *Physics*, Book III, Ch. 6.

[14] Precursors to the view that modal operators can be availed of in order to countenance the potential hierarchy of sets include Hodes (1984b). Intensional constructions of set theory are further developed by Reinhardt (1974a,b); Parsons (op. cit.); Myhill (1985); Scedrov (1985); Flagg (1985); Goodman (1985); Hellman (1990); Nolan (2002); and Studd (2013). (See Shapiro (1985) for an intensional construction of arithmetic.) Chihara (2004: 171-198) argues that 'broadly logical' conceptual possibilities can be used to represent imaginary situations relevant to the construction of open-sentence tokens. The open-sentences can then be used to define the properties of natural and cardinal numbers and the axioms of Peano arithmetic.

provide evidence that the inquiry – concerning at which stage in the process of set-individuation we happen to be, at present – can be avoided.

Another distinction to note is that both Linnebo (op. cit.) and Uzquiano (op. cit.) avail of second-order plural quantification, in developing their primitivist and interpretational accounts of mathematical modality. By contrast to their approaches, the epistemic and metaphysical modalities defined in the previous section are defined with second-order singular quantification over sets.

Linnebo and Uzquiano both suggest that their mathematical modalities ought to be governed by the G axiom; i.e. $\Diamond \Box \phi \to \Box \Diamond \phi$. The present approach eschews, however, the G axiom, in virtue of the following. Williamson (2009) demonstrates that an epistemic operator which validates the conjunction of the 4 axiom of positive introspection and the E axiom of negative introspection will be inconsistent with the condition of recursively enumerable 'conservativeness', although positive and negative introspection are individually consistent with r.e. (quasi-)conservativeness (453), where the conservativeness constraint defines recursively enumerable theories 'without reference to models' (442). 'A *theory* in [a language of propositional modal logic,] $L_\Box$ is a subset of $L_\Box$ containing all truth-functional tautologies and closed under modus ponens for $\supset$ (MP). A *model* M of $L_\Box$ induces a function M() : $L_\Box \to \{0, 1\}$ where $M(\bot) = 0$ and $M(\alpha \supset \beta) = 1$ if and only if $M(\alpha) \leq M(\beta)$. Intuitively, $M(\alpha) = 1$ if and only if α is true in M; $M(\alpha) = 0$ if and only if α is false in M. An *application* of epistemic logic determines a class of its intended models. The logic of the application is the set of formulas α such that $M(\alpha) = 1$ for every intended model M; thus the logic is a theory in $L_\Box$' (439). $\Box^{-1}X = \{\alpha \in L_\Box : \Box \alpha \in X\}$ (442). 'A model M is r.e. if and only if $\Box^{-1}M$ (which expresses what the agent cognizes in M) is an r.e. (recursively enumerable) theory in $L_\Box$. In that sense, the agent's cognition in an r.e. model does not exceed the computational capacity of a sufficiently powerful Turing machine / Consider the restriction of $\Box^{-1}M$ to the $\Box$-free sublanguage L, $L \cap \Box^{-1}M$. Let $\Box^{-1}M$ be an r.e. theory in $L_\Box$. Thus $L \cap \Box^{-1}M$ is an r.e. theory in L. It is the part of the agent's overall theory in M which is not specifically epistemic' (441-442). 'Σ is the logic of an application on which $\Box^{-1}M$ is a theory in $L_\Box$ for every intended model M' (440). 'Σ is r.e. conservative if and only if for every r.e. theory R in L, there is a maximal Σ-consistent set X such that $\Box^{-1}X$ is r.e. and $L \cap \Box^{-1}X = R$. Σ is r.e. quasi-conservative if and only if for every consistent r.e. theory R in L, there is a maximal Σ-consistent set X such that $\Box^{-1}X$ is r.e. and $L \cap \Box^{-1}X = R$' (442). S5 is not r.e. (quasi-)conservative (444). The G axiom is not r.e. quasi-conservative. K4 is r.e. conservative and GL is r.e. (quasi)-conservative (Williamson, 2009: 446, 449, 451-452). It is, however, an open question whether K4+GL is r.e. (quasi-)conservative.

Finally, my application of epistemic two-dimensional semantics to the epistemology of mathematics departs from full-blooded platonism, as well. According to full-blooded platonism, whatever mathematical objects can exist, do exist, and every consistent mathematical theory describes either a different part of the mathematical universe or distinct mathematical universes altogether (Balaguer, 1998). Thus, ZFC+CH and ZFC+¬CH both 'truly describe collections of mathematical objects', holding in distinct albeit equally real mathematical universes (Balaguer, 2001: 97: see also Hamkins, 2012).

Epistemic two-dimensionalism and full-blooded platonism differ on both the nature of their target possibilities and on the status of the actuality of the possibilities. Epistemic two-dimensionalism avails of epistemic possibilities, whereas full-blooded platonism avails of logical possibilities. Further, not all epistemic possibilities are actual according to epistemic two-dimensionalism, whereas the objects of any logically consistent theory actually exist according to full-blooded platonism. One reason to prefer epistemic two-dimensionalism to full-blooded platonism is that the former can be formalized, whereas Restall (2003) has shown that there are significant challenges to formalizing the latter. Another reason to prefer epistemic two-dimensionalism is that – unlike full-blooded platonism – it avoids commitment to the existence of inconsistent universes of sets where e.g. both ZFC+V=L and ZFC+V≠L would obtain.

8.3 Knowledge of Absolute Decidability

Williamson (2016a) examines the extension of the metaphysically modal profile of mathematical truths to the question of absolute decidability. A statement is decidable if and only if there is a mechanical procedure for deciding it or its negation. Statements are absolutely undecidable if and only if they are 'undecidable relative to any set of axioms that are justified' rather than just relative to a system (Koellner, 2006: 153), and they are absolutely decidable if and only if they are not absolutely undecidable. In this section, I aim to extend Williamson's analysis to the notion of epistemic mathematical modality that has been developed in the foregoing sections. The extension provides a crucial means of witnessing the significance of the two-dimensional approach for the epistemology of mathematics.

Williamson proceeds by suggesting the following line of thought. Suppose that A is a true interpreted mathematical formula which eludes present human techniques of provability; e.g. the Continuum Hypothesis (op. cit.). Williamson argues that mathematical truths are metaphysically necessary (op. cit.). Williamson then enjoins one to consider the following scenario: It is metaphysically possible that there is a species which finds A primitively compelling in virtue of their brain states and the evolutionary history thereof. Further, the species 'could not easily have come to believe ¬A or any other falsehood in a relevantly similar way'. He writes: 'In current epistemological terms, their knowledge of A meets the condition of safety: they could not easily have been wrong in a relevantly similar case. Here the relevantly similar cases include cases in which the creatures are presented with sentences that are similar to, but still discriminably different from, A, and express different and false propositions; by hypothesis, the creatures refuse to accept such other sentences, although they may also refuse to accept their negations' (11). Williamson writes that: 'The claim is not just that A *would* be absolutely provable *if* there were such creatures. The point is the stronger one that A is absolutely *provable* because there *could* in principle be such creatures.'

Williamson's scenario evinces one issue for the 'back-tracking' approach to modal epistemology, at least as it might be applied to the issue of possible mathematical knowledge. On the back-tracking approach, the method of modal epistemology is taken to proceed by first discerning the metaphysical modal truths – normally by natural-scientific means – and then working backward to the exigent incompleteness of an individual's epistemic states concerning such truths (see Stalnaker, 2003; Vetter, 2013).

The issue for the back-tracking method that Williamson's scenario illuminates is that the metaphysical mathematical possibility that CH is absolutely decidable must in some way converge with the epistemic possibility thereof. As with large cardinals, e.g., lack of inconsistency is a guide to metaphysical possibility. Woodin (2010, 2011b) and Bagaria et al. (2019) provide and discuss results with regard to the maximality of a core inner model for one supercompact cardinal, and avail of such results as evidence for the claim that the set-theoretic universe, V, is Ultimate-L.[15] The axiom for V = Ultimate-L implies the truth of CH, and states that '(i) There is a proper class of Woodin cardinals, and (ii) For each Σ_2-sentence ϕ, if ϕ holds in V then there is a universally Baire set $A \subseteq \mathbb{R}$ such that $\text{HOD}^{L(A,\mathbb{R})} \Vdash \phi$, where a set is universally Baire if for all topological spaces Ω and for all continuous functions $\pi : \Omega \to \mathbb{R}^n$, the preimage of A by π has the property of Baire in the space Ω' (Woodin, 2019). Such evidence might comprise an epistemic possibility with regard to the truth of CH, which can thus be a guide to its metaphysical mathematical possibility.[16]

The significance of the two-dimensional hyperintensional framework outlined in the foregoing is that it provides an explanation of the discrepancy between metaphysical mathematical modality and epistemic mathematical modality. Metaphysical mathematical modality is governed by the system S5, the Barcan formula, and its Converse, whereas epistemic mathematical modality is governed by K4+GL. Thus, epistemic mathematical modality figures as the mechanism, such that it can provide a guide to the metaphysical possibility of mathematical truth.

The relation between the Epistemic Church-Turing Thesis and absolute undecidability is complicated, however, by there being results pointing to two opposing conclusions.

The first result is by Leitgeb (2009). Leitgeb endeavors similarly to argue for the convergence between the notion of informal provability – countenanced as an epistemic modal operator, K – and mathematical truth. Availing of Hilbert (1923/1996: ¶18-42)'s epsilon terms for propositions, such that, for an arbitrary predicate, **C**(x), with x a propositional variable, the term 'ϵp.**C**(p)' is intuitively interpreted as stating that 'there is a proposition, x(/p), s.t. the formula, that p satisfies **C**, obtains' (op. cit.: 290). Leitgeb purports to demonstrate that $\forall p(p \to Kp)$, i.e. that informal provability is absolute; i.e. truth and provability are co-extensive. He argues as follows. Let Q(p) abbreviate the formula 'p $\land \neg K(p)$', i.e., that the proposition, p, is true while yet being unprovable. Let K be the informal provability operator reflecting knowability or epistemic necessity, with $\langle K \rangle$ its dual. Then:

1. $\exists p(p \land \neg Kp) \iff \epsilon p.Q(p) \land \neg K\epsilon p.Q(p)$.

---

[15]'For λ an ordinal' and κ a cardinal, 'κ being λ-supercompact is equivalent to there being an elementary embedding j: V $\to$ M such that j(a) = a for all a $< \kappa$ and j(κ) $> \lambda$, where M is an inner model such that $\{f \mid \lambda \to M\} \subset M$, i.e. every λ-sequence of elements of M is an element of M'. https://ncatlab.org/nlab/show/supercompact+cardinal .

[16]See §10.2.2, for further discussion.

By necessitation,
2. $K[\exists p(p \wedge \neg Kp)] \iff K[\epsilon p.Q(p) \wedge \neg K\epsilon p.Q(p)]$.
Applying modal axioms, KT, to (1), however,
3. $\neg K[\epsilon p.Q(p) \wedge \neg K\epsilon p.Q(p)]$.
Thus,
4. $\neg K\exists p(p \wedge \neg Kp)$.
Leitgeb suggests that (4) be rewritten
5. $\langle K\rangle \forall p(p \to Kp)$.
Abbreviate $\forall p(p \to Kp)$ by B. By existential introduction and modal axiom K, both
6. $B \to \exists p[K(p \to B) \vee K(p \to \neg B) \wedge p]$, and
7. $\neg B \to \exists p[K(p \to B) \vee K(p \to \neg B) \wedge p]$.
Thus,
8. $\exists p[K(p \to B) \vee K(p \to \neg B) \wedge p]$.
Abbreviate (8) by $C(p)$. Introducing epsilon notation,
9. $[K(\epsilon p.C(p) \to B) \vee K(\epsilon p.C(p) \to \neg B)] \wedge \epsilon p.C(p)$.
By K,
10. $[K(\epsilon p.C(p) \to KB) \vee K(\epsilon p.C(p) \to K\neg B)]$.
From (9) and necessitation, one can further derive
11. $K\epsilon p.C(p)$.
By (10) and (11),
12. $KB \vee K\neg B$.
From (5), (12), and K, Leitgeb derives
13. KB.
By, then, the T axiom,
14. $\forall p(p \to Kp)$ (291-292).

Rather than accounting for the coextensiveness of epistemic provability and truth, Leitgeb interprets the foregoing result as cause for pessimism with regard to whether the formulas countenanced in epistemic logic and via epsilon terms are genuinely logical truths if true at all (292). By contrast to this response, Leitgeb's proof might be thought to provide independent justification in favor of the epistemic two-dimensional approach pursued in this chapter, according to which the epistemic possibility or verification of abstraction principles, large cardinal axioms, and Orey sentences such as the Continuum Hypothesis is a guide to the metaphysical possibility or verification thereof. The notion of epistemic possibility at issue converges with Leitgeb's notion of informal provability according to which it has semantic and intuitive aspects (274) and is not exhaustively determined by syntax and logic (268). Epistemic states and possibilities comprise what is conceivable, where what is conceivable might best be countenanced by what Azzouni (2013: 73) refers to as 'inference packages'. Azzouni (op. cit.) defines inference packages as follows: 'Inference packages are topic-specific, bundled, sets of principles naturally applied to certain areas: various visualization capabilities, language-manipulation capacities, kinesthetic abilities, and so on'. If epistemic states and possibilities are countenanced via inference packages, then the relevant notion of conceivability would be prima facie, non-ideal conceivability. Ideal conceivability targets the limit of apriori reflection unconstrained by finite limitations, whereas non-ideal conceivability is hostage to the feasibility of computability and the psychological limitations of finite knowers.[17]

The second result is by Marfori and Horsten (2016: 260-261), who prove that if the Epistemic Church-Turing Thesis is true, then there are absolutely undecidable propositions in the language of Epistemic Arithmetic. They prove the following theorem:

'If ECT restricted to Π_1 arithmetical relations $\phi(x, y)$ holds, then there are absolutely undecidable Π_3 sentences of L_{EA}'.

They proceed by proving the contrapositive: If there are no Π_3 absolute undecidable sentences of L_{EA}, then ECT restricted to Π_1 arithmetical relations is false. They write: 'Suppose that there are no absolutely undecidable Π_3 sentences in L_{EA}:

'$\Box\Psi \iff \Psi$ for all Π_3 $\Psi \in L_{EA}$.

'Choose a Turing-uncomputable total functional Π_1 arithmetical relation $\phi(x, y)$; from elementary recursion theory we know that such $\phi(x, y)$ exist.

'Then, $\forall x \exists y \phi(x, y)$. But then we also have that $\forall x \exists y \Box \phi(x, y)$. The reason is that $\Pi_1 \subset \Pi_3$, so for every m and n, $\phi(m, n)$, being a Π_1 statement, entails $\Box\phi(m, n)$. However, $\forall x \exists y \Box \phi(x, y)$ is now a Π_3 statement of L_{EA}, so again from our assumption it follows that $\Box \forall x \exists y \Box \phi(x, y)$.

'Therefore, for the chosen $\phi(x, y)$ the antecedent of ECT is true whereas its consequent is false. Therefore, for the chosen $\phi(x, y)$, ECT is false.'

Leitgeb's result demonstrates that informal provability converges with truth, and thus corroborates that mathematical truths are absolutely decidable, whereas Marfori and Horsten's result demonstrates the inconsistency of the Epistemic Church-Turing Thesis and absolute decidability. The consis-

---

[17] For the distinction between ideal and prima facie (i.e. non-ideal) conceivability, see Chalmers (2002).

tency of these results is innocuous, and vindicates Gödel (1951/1995)'s Disjunction: '*Either mathematics is incompletable in this sense, that its evident axioms can never be comprised in a finite rule, that is to say, the human mind (even within the realm of pure mathematics) infinitely surpasses the powers of any finite machine, or else there exist absolutely unsolvable diophantine problems of the type specified* (where the case that both terms of the disjunction are true is not excluded, so that there are, strictly speaking, three alternatives)' (Gödel, 1951/1995: 310; see §2.3). When epistemic possibility is interpreted as informal provability rather than as a type of mechanism, mathematical truths are absolutely decidable.[18]

A final question is whether Orey sentences have a determinate epistemic intension given that there are currently models in which CH is true and models in which CH is false, such that it is not determinate which epistemic possibility is actual. The epistemic intensions of Orey sentences are arguably indeterminate for non-ideal reasoners, yet determinate for ideal ones. This optimism about the determinate truth of CH is corroborated by Woodin (2019), who demonstrates that the Ultimate-L conjecture is an existential number theoretic statement, such that it 'must be either true or false; it cannot be meaningless'.[19]

8.4 Concluding Remarks

In this essay, I have endeavored to delineate the types of mathematical modality, and to argue that the epistemic interpretation of two-dimensional semantics can be applied in order to explain, in part, the epistemic status of large cardinal axioms and the decidability of Orey sentences. The formal constraints on mathematical conceivability adumbrated in the foregoing can therefore be considered a guide to our possible knowledge of unknown mathematical truth.

---

[18] See Section **8.1.2**, fn. 3, Chapter **11**.
[19] See chapter **10**, for the definitions relevant to the Ultimate-L Conjecture.

Chapter 9

Hyperintensional Foundations of Mathematical Platonism

9.1 Introduction

Modal notions have been availed of, in order to argue in favor of nominalist approaches to mathematical ontology. Field (1989) argues, for example, that mathematical modality can be treated as a logical consistency operator on a set of formulas comprising an empirical theory, such as Newtonian mechanics, in which the mathematical vocabulary has been translated into the vocabulary of physical geometry.[1] Hellman (1993) argues that intensional models both of first- and second-order arithmetic and of set theory motivate an eliminativist approach to mathematical ontology. On this approach, reference to mathematical objects can be eschewed, and possibly the mathematical structures at issue are nothing.[2]

This essay aims to provide modal foundations for mathematical platonism, i.e., the proposal that mathematical terms for sets; functions; and the natural, rational, real, and complex numbers refer to abstract – necessarily non-concrete – objects. Intensional constructions of arithmetic and set theory have been proposed by, inter alia, Putnam (1967a), Fine (1981); Parsons (1983); Shapiro (1985); Myhill (1985); Reinhardt (1988); Chihara (1990);

---

[1]For a generalization of Field's nominalist translation scheme to the differential equations in the theory of General Relativity, see Arntzenius and Dorr (2012).

[2]For further discussion of modal approaches to nominalism, see Burgess and Rosen (1997: II, B-C) and Leng (2007; 2010: 258).

Nolan (2002); Linnebo (2013; 2018a); and Studd (2013; 2019). Williamson (2013a) emphasizes that mathematical languages are extensional, although in Williamson (2016a) he argues that Orey sentences, such as the Generalized Continuum Hypothesis – $2^{\aleph_\alpha} = \aleph_{\alpha+1}$ – which are currently undecidable relative to the axioms of Zermelo-Fraenkel Set Theory with choice as augmented by large cardinal axioms, are yet possibly decidable.[3] This chapter argues that the epistemic interpretation of two-dimensional semantics provides a novel approach to the epistemology of mathematics, such that if the decidability of mathematical axioms is epistemically possible, then their decidability is metaphysically possible.[4] Epistemic mathematical modality, suitably constrained, can thus serve as a guide to metaphysical mathematical modality.[5] Hamkins and Löwe (2007; 2013) argue that the modal logic of set-forcing extensions of ground models satisfies at least S4.2, i.e., axioms K $[\Box(\phi \to \psi) \to (\Box\phi \to \Box\psi)]$; T $(\Box\phi \to \phi)$; 4 $(\Box\phi \to \Box\Box\phi)$; and G $(\Diamond\Box\phi \to \Box\Diamond\phi)$. While the foregoing approaches are consistent with realism about mathematical objects, they are nevertheless not direct arguments thereof. The aim of this essay is to redress the foregoing lacuna, and thus to avail of the resources of modal ontology and epistemology in order to argue for the reality of mathematical entities and truth.

In Section **2**, I outline the elements of the abstractionist foundations of mathematics. In Section **3**, I examine Hale and Wright (2009)'s objections to the merits and need, in the defense of mathematical platonism and its epistemology, of the thesis of Necessitism, underlying the thought that whatever can exist actually does so. The Necessitist thesis is codified by the Barcan formula (see Barcan, 1946; 1947), and states that possibly if there is something which satisfies a condition, then there is something such that it possibly satisfies that condition: $\Diamond\exists x\phi x \to \exists x\Diamond\phi x$. I argue that Hale and Wright's objections to Necessitism as a requirement on admissible abstraction can be answered; and I examine both the role of the higher-order Necessitist proposal in their endorsement of an abundant conception of properties, as well as

---

[3]Compare Reinhardt (1974a) on the imaginative exercises taking the form of counterfactuals concerning supercompact cardinals and the truth of undecidable formulas. See Maddy (1988b) for discussion.

[4]The epistemic interpretation of two-dimensional intensional semantics is first advanced in Chalmers (1996; 2004).

[5]See Section **4**, for further discussion. Gödel (1951/1995: 11-12) anticipates a similar distinction between epistemic and metaphysical readings of the determinacy of mathematical truths, by distinguishing between mathematics in its subjective and objective senses.

cardinality issues that arise depending on whether Necessitism is accepted at first- and higher-order. In Section **4**, I provide an account of the role of epistemic and metaphysical modality in explaining the prima facie justification to believe the truth of admissible abstraction principles, and demonstrate how it converges with both Hale and Wright (op. cit.)'s and Wright (2012; 2014)'s preferred theory of default entitlement rationally to trust the truth of admissible abstraction. Section **5** provides concluding remarks.

9.2 The Abstractionist Foundations of Mathematics

The abstractionist foundations of mathematics are inspired by Frege (1884/1980; 1893/2013)'s proposal that cardinal numbers can be explained by specifying an equivalence relation, expressible in the signature of second-order logic, on first- or higher-order entities. At first-order, in Frege (1884/1980: 64), the direction of the line, a, is identical to the direction of the line, b, if and only if lines a and b are parallel. At second-order, in Frege (op. cit.: 68) and Wright (1983: 104-105), the cardinal number of the concept, **A**, is identical to the cardinal number of the concept, **B**, if and only if there is a one-to-one correspondence between **A** and **B**, i.e., there is an injective and surjective (bijective) mapping, R, from **A** to **B**. With Nx: a numerical term-forming operator,

- $\forall \mathbf{A} \forall \mathbf{B}[\text{Nx}: \mathbf{A} = \text{Nx}: \mathbf{B} \equiv \exists R[\forall x[\mathbf{A}x \to \exists y(\mathbf{B}y \land Rxy \land \forall z(\mathbf{B}z \land Rxz \to y = z))] \land \forall y[\mathbf{B}y \to \exists x(\mathbf{A}x \land Rxy \land \forall z(\mathbf{A}z \land Rzy \to x = z))]]]$.

The foregoing is referred to as 'Hume's Principle'.[6] Frege's Theorem states

---

[6]Frege (1884/1980: 68) writes: 'the Number which belongs to the concept F is the extension of the concept '[equinumerous] to the concept F' (op. cit.: 72-73). Boolos (1987/1998: 186) coins the name, 'Hume's Principle', for Frege's abstraction principle for cardinals, because Frege (op. cit.: 63) attributes equinumerosity as a condition on the concept of number to Hume (1739-1740/2007: Book 1, Part 3, Sec. 1, SB71), who writes: 'When two numbers are so combin'd, as that the one has always an unite answering to every unite of the other, we pronounce them equal ...'. Frege notes that identity of number via bijections is anticipated by the mathematicians, Ernst Schröder and Ernst Kossak, as well Cantor (1883/1996: Sec. 1), the last of whom writes: '[E]very well-defined set has a determinate power; two sets have the same power if they can be, element for element, correlated with one another reciprocally and one-to-one', where the power of a set corresponds to its cardinality (see Cantor, 1895/2007: 481).

that the Dedekind-Peano axioms for the language of arithmetic can be derived from Hume's Principle, as augmented to the signature of second-order logic.[7] Abstraction principles have further been specified both for the real numbers (see Hale, 2000a; Shapiro, 2000; and Wright, 2000), and for sets (see Wright, 1997; Shapiro and Weir, 1999; Hale, 2000b; and Walsh, 2016).

The philosophical significance of the abstractionist program consists primarily in its provision of a neo-logicist foundation for classical mathematics, and in its further providing a setting in which to examine constraints on the identity-conditions of mathematical concepts.[8] The philosophical significance of the abstractionist program consists, furthermore, in its circumvention of Benacerraf (1973)'s challenge to the effect that our knowledge of mathematical truths is in potential jeopardy, because of the absence of naturalistic, in particular causal, conditions thereon. Both Wright (1983: 13-15) and Hale (1987: 10-15) argue that the abstraction principles are epistemically tractable, only if (i) the surface syntax of the principles – e.g., the term-forming operators referring to objects – are a perspicuous guide to their logical form; and (ii) the principles satisfy Frege (1884/1980: X)'s context principle, such that the truth of the principles is secured prior to the reference of the terms figuring therein.

9.3 Abstraction and Necessitism

9.3.1 Hale and Wright's Arguments against Necessitism

One crucial objection to the abstractionist program is that – while abstraction principles might provide necessary and sufficient truth-conditions for the concepts of mathematical objects – an explanation of the actual truth of the principles has yet to be advanced (see Eklund, 2006; 2016). In response, Hale and Wright (2009: 197-198) proffer a tentative endorsement of an 'abundant' conception of properties, according to which fixing the sense

---

[7]See Dedekind (1888/1996) and Peano (1889/1967). See Wright (1983: 154-169) for a proof sketch of Frege's theorem; Boolos (1987) for the formal proof thereof; and Parsons (1964) for an initial conjecture of the theorem's validity.

[8]Thanks here to Chris Peacocke, for comments. Shapiro and Linnebo (2015) prove that Heyting arithmetic can be recovered from Frege's Theorem. Criteria for consistent abstraction principles are examined in, inter alia, Hodes (1984); Hazen (1985); Boolos (1990/1998); Heck (1992); Fine (2002); Weir (2003); Cook and Ebert (2005); Linnebo and Uzquiano (2009); Linnebo (2010); and Walsh (op. cit.).

of a predicate will be sufficient for predicate reference.[9] Eklund (2006: 102) suggests, by contrast, that one way for the truth of the abstraction principles to be explained is by presupposing what he refers to as a 'Maximalist' position concerning the target ontology.[10] According to the ontological Max-

---

[9]Hale and Wright (2009) and Wright (2012a) extend the abundant conception of properties to objects, although this extension is orthogonal to the discussion in this chapter. The aim of this and the following section is to examine the Necessitist commitments of the abundant conception of properties, especially as exploited by Hale (2013a,b). For the sake of completeness, however, the abundant conception of objects can be characterized as follows. Hale and Wright argue that, in the case of objects, the senses of singular terms are not sufficient for reference, but rather the following must be satisfied: the truth of the context, viz. the right-hand-side of abstraction principles, by way of which singular terms for the objects on the left-hand-sides can be defined. This figures as an Aristotelian constraint to the effect that those contexts are objective truths occurring on the side of the World given that sense alone is not sufficient for reference. Hale and Wright claim: 'As with the abundant conception of properties, there is no additional gap to cross which requires "hitting off" something on the other side by virtue of its fit with relevant specified conditions, as the property of being composed of the element with atomic number 79 is hit off (or so let's suppose) by the combination of conditions that control our unsophisticated use of "gold". But nor is it the case that reference is bestowed by the possession of sense alone' (207). And they continue: 'The abstractionist conception of the truth of the right-hand sides of instances of good abstractions as conceptually sufficient for the truth of the left-hand sides precisely takes the terms in question out of the market for "hitting off" reference to things whose metaphysical nature is broadly comparable to that of sparse properties, and assigns to them instead a referential role relevantly comparable to that of predicates as viewed by the abundant Aristotelian' (208). For similar comments, see Wright (2012a: 132): 'In contrast with any Meinongian view, we need the truth of the right-hand side kind of context before we can claim existence. It is not enough that the abstract terms have a sense. Appropriate (atomic) statements containing them have to be true. But those truths can be objective. And the truth of the left-hand sides of instances of abstraction principles will be an objective matter just if that of their right-hand side counterparts is, because that is given as a necessary and sufficient condition. Thus where it is objectively so that a pair of properties are one-one correspondent, it will correspondingly be objectively so that some one number is the number of them both. But there will be no metaphysical hostage, no "fishing", in drawing this conclusion about their number. The reason is that numbers, like all abstracts, are to be compared to abundant Aristotelian properties: entities knowledge of which is fully grounded in knowledge of the truth of atomic predications and identity statements, respectively, and embodies no further conjecture about the nature of the World'.

[10]For further discussion of ontological Maximalism, see Hawley (2007) and Sider (2007: IV). Chalmers (2009: 89, 121) endorses a type of 'lightweight' Maximalism, with regard to 'ordinary existence assertions' about numbers, by contrast to 'ontological existence assertions' about numbers. Chalmers writes: 'An *ordinary* existence assertion, to a first approximation, is an existence assertion of the sort typically made in ordinary first-order

imalist position, if it is possible that a term has an extension, then actually the term does have the extension.

Hale and Wright (op. cit.) raise two issues for the ontological Maximalist proposal. The first is that ontological Maximalism is committed to a proposal that they take to be independently objectionable, namely ontological Necessitism (185). They write: "Most obviously, maximalism denies the possibility of contingent non-existence, to which there are obvious objections" (op. cit.) Hale and Wright (op. cit.) raise a similar contention to the effect that actual, and not *merely possible*, reference is what the abstractionist program intends to target; and that Maximalism and Necessitism, so construed, are purportedly silent on the status of ascertaining when the possibilities at

---

discussion of the relevant subject matter. For example, a typical mathematician's assertion of "There are four prime numbers less than ten" is an ordinary existence assertion, as is a typical drinker's assertion of "There are three glasses on the table." An *ontological* existence assertion, to a first approximation, is an existence assertion of the sort typically made in broadly philosophical discussion where ontological considerations are paramount' (op. cit.: 81). The distinction corresponds to Carnap's distinction between internal and external questions (Carnap, 1950: §2). Chalmers (2009: 78) writes: 'An intermediate sort of *lightweight realism* has also developed, holding that while there are objective answers to ontological questions, these answers are somehow shallow or trivial, perhaps grounded in conceptual truths'. 'Other forms of lightweight realism include the *neo-Fregean realism* of Hale and Wright (2001[; 2009,] this volume), the *lightweight sortalism* of Thomasson ([2009,] this volume), and the *lightweight maximalism* of Eklund ([2009,] this volume). These closely related views are all liberal about the existence of objects, and hold that existence sentences can be conceptually analyzed in such a way that they can be analytically entailed by sentences without corresponding existence assertions. On the first view, ampliative conditions such as "If the the [sic - D.E.] Fs can be mapped one-to-one onto the Gs, there there [sic - D.E.] is a number that is the number of Fs and the number of Gs" are taken to be conceptual truths. On the second view, all existential assertions have the underlying form "There exists an F ... " for some sortal concept F, and these sortal concepts have associated application conditions such that the existential assertions can be analytic consequences of qualitative characterizations of the world. On the third view, it is taken to be a conceptual truth about existence that if the existence of an F is consistent with certain basic truths, then Fs exist. On all of these views, the truth-value of unproblematic ontological assertions is objective and (usually or always) determinate, so all can be seen as versions of lightweight ontological realism' (98). 'In fact, when doing philosophy it is often sensible to assume a maximalist framework, on which any entity whose existence is consistent with the nature of this world can be taken to exist (see Eklund, [2009,] this volume). This makes for considerable convenience, not least in the present project [...] Again, nothing here entails maximalism as a heavyweight ontological view. It simply reflects the advantage of (lightweight) maximalism as a framework for conducting one's theorizing about the world' (121).

issue are actual.

The second issue that Hale and Wright find with Maximalism is that it misconstrues the demands that the abstractionist program is required to address. The abstractionist program is supposed to be committed to ontological Maximalism, because the possibility that a term has an extension will otherwise not be sufficient for the success of the term's reference. It is further thought that, without an appeal to Maximalism, and despite the actuality of successful mathematical reference, there are yet possible situations in which the mathematical terms still do not refer (193). In response, they note that no 'collateral metaphysical assistance' – such as ontological Maximalism would be intended to provide – is necessary in order to explain the truth of abstraction principles (op. cit.). Rather, there is prima facie, default entitlement rationally to trust that the abstraction principles are actually true, and such entitlement is sufficient to foreclose upon the risk that possibly the mathematical terms therein do not refer (192).

In the remainder of this section, I will argue that Hale and Wright's objections to Necessitism and the ontological Maximalist approach to admissible abstraction both can be answered, and the proposals are in any case implicit in their endorsement of the abundant conception of properties.

The principle of the necessary necessity of being (NNE) can be derived from the Barcan formula.[11] NNE states that necessarily all objects are such that necessarily there is something to which each is identical; $\Box \forall x \Box \exists y (x = y)$. Informally, necessarily everything has necessary being, i.e. necessarily everything is necessarily something, even if contingently non-concrete. Applied to entities at higher-order, NNE can be formalized as follows: $\Box \forall X \Box \exists Y \Box \forall x (Xx \iff Yx)$ (op. cit.: 264). Williamson (2013: 6.1-6.4) targets issues for the comprehension principle for identity properties of individuals, i.e. haecceities, if the negations of the Barcan formula and NNE are true at first-order, and thus for objects. With regard to properties and relations at higher-order, Williamson's arguments have targeted closure conditions, given a modalized interpretation of comprehension principles (op. cit.). The latter take the form, $\text{Comp}_M := \exists X \Box \forall x (Xx \iff A)$, with x an individual variable which may occur free in A and X a monadic first-order predicate variable which does not occur free in A (262). The Contingentist, by contrast, can countenance only 'intra-world' comprehension principles in which the modal operators and iterations thereof take scope over the entire formula; e.g. $\Diamond \exists X \forall x (Xx$

---

[11]See Williamson (2013: 38).

⟺ A) (see Sider, 2016: 686). Williamson targets, in particular, a higher-order modal completeness property for a quasi-reflexive [for all x,y∈R, Rxy → □(Rxx ∧ Ryy)], anti-symmetric, and transitive relation, ≤. The relation codifies upper bounds and least upper bounds, as well as modalized versions thereof, where '[t]o be an upper bound of a property is to be at least as great (in the sense of the ordering) as everything that has the property. To be a least upper bound of the property is to be an upper bound of the property that every upper bound of the property is at least as great as' (Williamson, 2013: 286). The claim that 'any possible property that can have a modal upper bound can have a modal least upper bound' is recorded by 'prefixing every universal quantifier with a necessity operator and the other quantifiers and the ordering symbol itself with a possibility operator. Formally: [i] □∀X[◇∃y□∀x(Xx → ◇x ≤ y) → ◇∃y[□∀x(Xx → ◇x ≤ y) ∧ □∀z[□∀x(Xx → ◇x ≤ z) → ◇y ≤ z]]]' (287). Williamson notes that to apply this formula, one replaces Xx with the formula A, where x can be free in A but neither y nor z can be, in order to obtain the following: [ii] '◇∃y□∀x(A → ◇x ≤ y) → ◇∃y[□∀x(A → ◇x ≤ y) ∧ □∀z[□∀x(A → ◇x ≤ z) → ◇y ≤ z]]]'. However, one needs $Comp_M$ in order to derive [ii] from [i], because $Comp_M$ 'provide[s] a property over which the second-order quantifier [in [i]] ranges necessarily coextensive with A' (op. cit.). By rejecting $Comp_M$, Contingentists cannot preclude cases in which the parameters in A might be incompossible, such that there would be no property which is necessarily coextensive with A (op. cit.). The foregoing provides prima facie abductive support for the requirement of Necessitism in mathematics. The constitutive role of the Necessitist modal comprehension scheme in characterizing the relation between modal upper and least upper bounds answers Hale and Wright's first contention against the Necessitist commitments of ontological Maximalism.

Williamson refers to the assignments for models in the metaphysical setting as universal interpretations (59). The analogue for logical truth occurs when a truth is metaphysically universal, i.e., if and only if its second-order universal generalization is true on the intended interpretation of the metalanguage (200). The connection between truth-in-a-model and truth simpliciter is then that – as Williamson puts it laconically – when 'the framework at least delivers a condition for a modal sentence to be true in a universal interpretation, we can derive the condition for it to be true in the intended universal interpretation, which is the condition for it to be true simpliciter' (op. cit.).

One of the crucial interests of the metaphysical universality of propositions is that the models in the class need not be pointed, in order to countenance the actuality of the possible propositions defined therein.[12] Rather, the class of true propositions generated by the metaphysically universal propositions is sufficient for the propositions actually to be true (268-269).[13] Williamson writes that 'since whatever is is, whatever is actually is: if there is something, then there actually is such a thing' (23). Thus, the foregoing characterization of actuality can explain why the metaphysically universal propositions which are true simpliciter are actual.

This account of actuality answers Hale and Wright's contention that the interaction between the possible and actual truth of sentences such as abstraction principles cannot be accounted for.[14]

9.3.2 Hale on the Necessary Being of Purely General Properties and Objects

Note, further, that the abundant conception of properties endorsed by Hale and Wright depends upon the Necessitist Thesis, and the truth of ontological Maximalism thereby. Hale writes: '[I]t is sufficient for the *actual* existence of a property or relation that there *could* be a predicate with appropriate satisfaction conditions ... *purely general* properties and relations exist as a matter of (absolute) *necessity*', where a property is purely general if and only if there could be a predicate for which, and it embeds no singular terms (Hale, 2013b: 133, 135; see also 2013a: 99-100).[15]

Hale argues for the necessary necessity of being for properties and propositions as follows (op. cit.: 135; 2013b: 167). Suppose that p refers to the proposition that a property exists, and that q refers to the proposition that a

---

[12]A model is pointed if it includes a designated element. That the models are unpointed is noted in Williamson (2013: 100).

[13]Thanks here to Bruno Jacinto for discussion.

[14]Cook (2016: 398) demonstrates how formally to define modal operators within Hume's Principle, i.e. the consistent abstraction principle for cardinal numbers. Necessitist Hume's Principle takes the form: $\Box \forall X, Y[\#(X) =_\Box \#(Y) \iff X \approx Y]$, where X and Y are second-order variables, $\#$ is a numerical term-forming operator, $\approx$ is a bijection, and for variables, x,y, of arbitrary type '$x =_\Box y \iff \exists z[z = x \wedge z = y \wedge \Box \exists w(w = z)]$'. See Cook (op. cit.) for further discussion.

[15]Cook (op. cit.: 388) notes the requirement of Necessitism in the abundant conception of properties, and discusses one point at which Williamson's and Hale's Necessitist proposals might be inconsistent. The points of divergence between the two variations on the proposal are examined in some detail below.

predicate for the property exists. Let the necessity operator be defined as a counterfactual with an unrestricted, universally quantified antecedent, such that, for all propositions, ψ: $[\Box\psi \iff \forall\phi(\phi\,\Box\!\!\rightarrow \psi)]$ (135).[16] On the abundant conception of properties, $\Box[p \iff \Diamond q]$. Intuitively: Necessarily, there is a property if and only if possibly there is a predicate for that property. Given the counterfactual analysis of the modal operator: For all propositions about a property, if there were a proposition specifying a predicate s.t. the property is in the predicate's extension, then there would be that property.

From '$\Box[p \iff \Diamond q]$', one can derive both 'p $\iff$ $\Diamond$q', and – by the rule, RK – the necessitation thereof, '$\Box$p $\iff$ $\Box\Diamond$q' (op. cit.). By the 5 axiom in S5, $\Diamond q \iff \Box\Diamond q$ (op. cit.). So, '$\Box\Diamond$q $\iff$ $\Diamond$q'; '$\Diamond$q $\iff$ p'; and '$\Box\Diamond$q $\iff$ $\Box$p'. Thus – by transitivity – 'p $\iff$ $\Box$p' (op. cit.); i.e., all propositions about properties are necessarily true, such that the corresponding properties have necessary being. By the 4 axiom in S5, $\Box$p $\iff$ $\Box\Box$p; so, the necessary being of properties and propositions is itself necessary. Given the endorsement of the abundant conception of properties – Hale and Wright are thus committed to higher-order Necessitism, i.e., the necessary necessity of being.

Hale (2013b) endeavors to block the ontological commitments of the Barcan formula and its converse by endorsing a negative free logic. Thus, in the derivation:

Assumption,
1. $\Box\forall x[F(x)]$.
By $\Box$-elimination,
2. $\forall x[F(x)]$.
By $\forall$-elimination,
3. $F(x)$.
By $\Box$-introduction,
4. $\Box[F(x)]$.
By $\forall$-introduction,
5. $\forall\Box[F(x)]$.
By $\rightarrow$-introduction,
6. $\Box\forall x[F(x)] \rightarrow \forall x\Box[F(x)]$,

Hale imposes an existence-entailing assumption in the inference from lines (2) to (3), i.e.

---

[16] Proponents of the translation from modal operators into counterfactual form include Stalnaker (1968/1975), McFetridge (1990: 138), and Williamson (2007).

'(Free∀-Elimination) From ∀x[A(x)], together with an existence-entailing premise F(t), we may infer A(t) where t can be any term' (op. cit.: 208-209).

Because the concept of, e.g., cardinal number is defined by abstraction principles which are purely general because they embed no singular terms, the properties of numbers are argued to have necessary being. The necessary being of the essential properties of number – i.e., higher-order Necessitism about purely general properties – along with the necessary existence of second-level functions in Hume's Principle are argued then to explain in virtue of what abstract objects such as numbers have themselves necessary being. As Hale writes: 'This enables the essentialist to give a simple and straightforward explanation of the necessary existence of cardinal numbers. There necessarily exist cardinal numbers because they are the values of the pure function Nu...u...for a certain range of arguments – pure first-level sortal properties – and both that function and those arguments to it exist necessarily. In short, certain objects – the cardinal numbers – exist necessarily because their existence is a consequence of the existence of a certain function and certain properties which themselves exist necessarily' (176-177).

By contrast, essential properties defined by theoretical identity statements, which if true are necessarily so, do embed singular terms and are thus not purely general. So, the essential nature of water, i.e., the property 'being comprised of one oxygen and two hydrogen molecules', has contingent being, explaining in virtue of what samples of water have contingent being (216-217).

One objection to the foregoing concerns the necessary being of different types of numbers. While an abstraction principle for cardinal numbers can be specified using only purely general predicates – i.e., Hume's Principle – abstraction principles for imaginary and complex numbers have yet to be specified. Shapiro (2000) provides an abstraction principle for the concepts of the reals by simulating Dedekind cuts, where abstraction principles are provided for the concepts of the cardinals, natural numbers, integers, and rational numbers, from which the reals are thence defined: Letting F,G, and R denote rational numbers, $\forall F,G[\mathbf{C}(F) = \mathbf{C}(G) \iff \forall R(F \leq R \iff G \leq R)]$. [17] Hale (2000/2001)'s own definition of the concept of the reals is provided relative to a domain of quantities. The quantities are themselves taken to be abstract, rather than physical, entities (409). The quantitative domain can thus be comprised of both rational numbers as well as the abstracts for

---

[17]See Dedekind (1872/1996: Sec. 4), for the cut method for the definition of the reals.

lengths, masses, and points.[18] The reals are then argued not to be numbers, but rather quantities defined via an abstraction principle which states that a set of rational numbers in one quantitative domain is identical to a set of rational numbers in a second quantitative domain if and only if the two domains are isomorphic (407).[19] Hale argues, then, that it is innocuous for the real abstraction principle to be conditional on the existence of at least one quantitative domain, because the rational numbers can be defined, similarly as on Shapiro's approach, via cut-abstractions and abstractions on the integers, naturals, and cardinals. Thus, the reals can be treated as abstracts derived from purely general abstraction principles, and are thus possessed of necessary being.

However, abstraction principles for imaginary numbers such as $i = \sqrt{-1}$, and complex numbers which are defined as the sum of a real number and a second real multiplied by i, have yet to be accounted for. The provision of an abstraction principle for complex numbers would, in any case, leave open a host of questions concerning the applicability of the numbers, violating what is referred to as Frege's constraint. Frege's constraint is satisfied when the application for a concept of number figures in its definition (see e.g., Wright, 2000). Such questions might include the inquiry into how, e.g., complex-valued wave functions might interact with physical ontology; e.g., how a lower-(3)-dimensional real-valued configuration space for particles might relate to the higher-($3n$)-dimensional, complex-valued wave function (see Simons, 2016; Ney, 2013; Maudlin, 2013).

The modality in the Barcan-induced Necessitist proposal at first- and higher-order is, as noted, interpreted metaphysically, and incurs no similar issues with regard to the interaction between purely general properties and Frege's constraint. Further, because true on its second-order universal generalization on its intended, metaphysical interpretation, the possible truth-in-a-model of the relevant class of propositions is, as discussed in Section **3.1**, thus sufficient for entraining the actual truth of the relevant propositions.

---

[18] An abstraction principle for lengths, based on the equivalence property of congruence relations on intervals of a line, or regions of a space, is defined in Shapiro and Hellman (2015: 5, 9). Shapiro and Hellman provide, further, an abstraction principle for points, defined as comprising, respectively, the left- and right-ends of intervals (op. cit.: 5, 10-12).

[19] See Hale (op. cit.: 406-407), for the further conditions that the domains are required to satisfy.

9.3.3 Cardinality and Intensionality

An interesting residual question concerns the status of the worlds, upon the translation of modal propositional logic into non-modal first-order languages.[20] Fritz (op. cit.) notes that a world can be represented by a predicate, in the latter. However, whether objects satisfy the predicate can vary from point to point, in the non-modal first-order class of points.[21] Another issue is that modal propositional logic is equivalent only to the bisimulation-invariant fragment of first-order logic, rather than to the full variant of the logic (see van Benthem, 1983; Janin and Walukiewicz, 1996). Thus, there cannot be a faithful translation from each modal operator in modal propositional logic into a predicate of full first-order logic.[22]

One way to mitigate the foregoing issues might be by arguing that the language satisfies real-world rather than general validity, such that necessarily the predicate will be satisfied only at a designated point in a model – intuitively, the analogue of the concrete rather than some merely possible world, simulating thereby the translation from possibilist to actualist discourse (see Fine, op. cit.: 211,135-136, 139-140, 154, 166-168, 170-171) – by contrast to holding of necessity as interpreted as satisfaction at every point in the model. The reply would be consistent with what Williamson refers to as 'chunky-style Necessitism' which validates the following theorems: where the predicate C(x) denotes the property of being grounded in the concrete and P(x) is an arbitrary predicate, (a) '$\forall x \Diamond C(x)$', yet (b) '$\Box \forall x_1, \ldots, x_n P(x_1, \ldots, x_n) \to (Cx_1, \ldots, Cx_n)]$' (325-332). Williamson (p. 33, fn. 5) argues, however, in favor of general, rather than real-world validity. A second issue for the reply is that principle (b), in the foregoing, is inconsistent with Williamson's protracted defense of the 'being constraint', according to which $\Box \forall x_1, \ldots, x_n \Box [P(x_1, \ldots, x_n) \to \exists y_1, \ldots, y_n (x_1 = y_1, \ldots, x_n = y_n)]$, i.e. if $x_1, \ldots, x_n$ satisfy a predicate, then $x_1, \ldots, x_n$ are each something, even if possibly non-concrete (148).

A related issue concerns the translation of modalized, variable-binding, generalized quantifiers of the form:

---

[20] Thanks here to Alessandro Rossi, for discussion.

[21] Suppose that the model is defined over the language of second-order arithmetic, such that the points in the model are the ordinals. A uniquely designated point might then be a cardinal number whose height is accordingly indexed by the ordinals.

[22] For further discussion of the standard translation between propositional modal and first-order non-modal logics, see Blackburn et al. (2001: 84).

'there are n objects such that ... ',
'there are countably infinite objects such that ... ',
'there are uncountably infinite objects such that ... ' (Fritz and Goodman, 2017).

The generalized quantifiers at issue are modalized and consistent with first-order Necessitism, because the quantifier domains include all possible – including contingently non-concrete – objects.[23] It might be argued that the translation is not of immediate pertinence to the ontology of mathematics, because the foregoing first-order quantifiers can be restricted such that they range over only uncountably infinite *necessarily* non-concrete objects – i.e. abstracta – by contrast to ranging unrestrictedly over all modal objects, including the contingently non-concrete entities induced via the Barcan formula – i.e., the 'mere possibilia' that are non-concrete as a matter of contingency. However, the Necessitist thesis can be valid even in the quantifier domain of a first-order language restricted to necessarily non-concrete entities. If, e.g., a mathematician takes, despite iterated applications of set-forming operations, the cumulative hierarchy of sets to have a fixed cardinal height, then the first-order Necessitist thesis will still be valid, because all possible objects will actually be still something.

The first-order Necessitist proposal engendered by taking the height of the cumulative hierarchy to be fixed is further consistent with the addition to the

---

[23]Fritz and Goodman (2017) note that if there uncountably infinite concrete objects, such that there are uncountably infinite conditions which are necessary for those objects to become concrete, then a guide to how many merely possible objects there are might be just that number of conditions (13). Thanks here to Jeremy Goodman. The suggestion is interesting, because constraints on the size of the cardinality of the domain of possible objects, or on how the cardinality of the domain of possible objects might relate to the domain of concrete objects, have not, in the literature, been examined in protracted detail. An interesting exception is Fritz (2016). Let the truth-in-a-model of a formula of modal logic be metaphysically universal, if and only if the formula is a true proposition, i.e. it is true on its second-order universal generalization (Williamson, 2013: 200). Fritz shows that cases in which the cardinality of the domain of individuals in the model is less than or equal to that of the domain of worlds are supposed to determine the weakest normal modal logic for metaphysical universality – i.e. K: necessitation ($\phi \vdash \Box\phi$), K [$\Box(\phi \to \psi) \to [\Box\phi \to \Box\psi]$], modus ponens, and uniform substitution; see Williamson (96-98) – and cases in which the cardinality of individuals is greater than or equal to that of the worlds are shown to determine the strongest modal logic for metaphysical universality (Fritz, op. cit.: 594). Williamson himself notes that, when the cardinality of individuals is greater than the cardinality of worlds, the contingentist formula that $\neg\forall x \exists y (Px \iff x = y)$ will be metaphysically universal (Williamson, op. cit.: 144-146).

first-order language of additional intensional operators – such as those introduced by Vlach (1973) – in order to characterize the indefinite extensibility of the concept of set; i.e., that despite unrestricted universal quantification over all of the entities in a domain, another entity can be defined with reference to, and yet beyond the scope of, that totality, over which the quantifier would have further to range. First-order Necessitism is further consistent with the relatively expanding domains induced by Bernays (1942)'s Theorem. Bernays' Theorem states that class-valued functions from classes to sub-classes are not onto, where classes are non-sets (see Uzquiano, 2015a: 186-187). So, the cardinality of a class will always be less than the cardinality of its sub-classes. Suppose that that there is a generalization of Bernays' theorem, such that the non-sets are interpreted as possible objects. Thus, the cardinality of the class of possible objects will always be less than the cardinality of the sub-classes in the image of its mapping. Given iterated applications of Bernays' theorem, the cardinality of a domain of non-sets is purported then not to have a fixed height.

In both cases, however, the addition of Vlach's intensional operators permits there to be multiple-indexing in the array of parameters relative to which a cardinal can be defined. So, both the intensional characterization of indefinite extensibility and the generalization of Bernays' Theorem to possible objects are consistent with the first-order Necessitist proposal that all possible objects are actual, and so the cardinality of the target universe is fixed.[24]

Fritz and Goodman suggest that a necessary condition on the equivalence of propositions is that they define the same class of models (op. cit.: 1.4). The proposed translation of the modalized generalized quantifiers would be Contingentist, by taking (NNE) to be invalid, such that the domain in the translated model would be comprised of only possible concrete objects, rather than the non-concrete objects as well (op. cit.).

Because of the existence of non-standard models, the generalized quanti-

---

[24]Note that the proposal that the cardinality of the cumulative hierarchy of sets is fixed, despite continued iterated applications of set-forming operations, is anticipated by Cantor (1883/1996: Endnote [1]). Cantor writes: 'The absolutely infinite sequence of numbers thus seems to me to be an appropriate symbol of the absolute; in contrast the infinity of the first number-class (I) [i.e., the countable infinity comprising the class of natural numbers, $\aleph_0$ - D.E.], which has hitherto sufficed, because I consider it to be a graspable idea (not a representation), seems to me to dwindle into nothingness by comparison' (op. cit.; see Cantor, 1899/1967).

fier that 'there are countably infinitely many possible ... ' cannot be defined in first-order logic. Fritz and Goodman note that generalized quantifiers ranging over countably infinite objects can yet be simulated by enriching one's first-order language with countably infinite conjunctions. On the latter approach, finitary existential and universal quantifiers can be defined as the countably infinite conjunction of formulas stating that, for all natural numbers n, 'there are n possible ...' (2.3).

Crucially, however, there are some modalized generalized quantifiers that cannot be similarly paraphrased – e.g., 'there are uncountably infinite possible objects s.t. ...' – and there are some modalized generalized quantifiers that cannot even be defined in first-order languages – e.g. '*most* objects s.t. ...' (2.4-2.5)

In non-modal first-order logic, it is possible to define generalized quantifiers which range over an uncountably infinite domain of objects, by augmenting finitary existential and universal quantifiers with an uncountably infinite stock of variables and an uncountably infinite stock of conjunctions of formulas (2.4). Fritz and Goodman note, however, that the foregoing would require that the quantifiers bind the uncountable variables 'at once', s.t. they must have the same scope. The issue with the proposal is that, in the setting of modalized existential quantification over an uncountably infinite domain, the Contingentist paraphrase requires that bound variables take different scopes, in order to countenance the different possible sets that can be defined in virtue of the indefinite extensibility of cardinal number (op. cit.).

In order to induce the Contingentist paraphrase, Fritz and Goodman suggest defining 'strings of infinitely many existential and universal quantifiers', such that a modalized, i.e. Necessitist, generalized quantifier of the form, 'there are uncountably infinite possible ...' can be redefined by an uncountably infinite sequence of finitary quantifiers with infinite variables and conjunction symbols of the form:

'Possibly for some x_1, possibly for some x_2, *etc*.: x_1, x_2, *etc*. are pairwise distinct and are each possibly ...',

where *etc*. denotes an uncountable sequence of, respectively, 'an uncountable string of interwoven possibility operators and existential quantifiers', and an 'uncountable string of variables' (op. cit.).

An argument against the proposed translation of the quantifier for there being uncountably infinite possible objects is that it is contentious whether an uncountable sequence of operators or quantifiers has a definite meaning [see

Williamson (2013: 7.7)]. Thus, e.g., while negation can have a determinate truth condition which specifies its meaning, a string of uncountably infinite negation operators will similarly have determinate truth conditions and yet not have an intuitive, definite meaning (357). One can also define a positive or negative integer, x, such that sx is interpreted as the successor function, x+1, and px is interpreted as the inverse function, x−1. However, an infinitary expression consisting in uncountable, alternating iterations of the successor and inverse functions – $spsps...x$ – will similarly not have a definite meaning (op. cit.). Finally, one can define an operator O_i mapping truth conditions for an arbitrary formula A to the truth condition, p, of the formula $\Diamond \exists x_i (Cx_i \wedge A)$, with Cx being the predicate for being concrete (258). Let the operators commute, such that $O_i O_j$ iff $O_j O_i$ (op. cit.). A total ordering of truth conditions defined by an infinite sequence of the operators can be defined, such that the relation is reflexive, anti-symmetric, transitive, and connected [$\forall x,y (x \leqslant y \vee y \leqslant x)$] (op. cit.). However, total orders need not have a least upper bound; and the sequence, $O_i O_i O_i ...(p)$, would thus not have a non-arbitrary, unique value (op. cit.). The foregoing might sufficiently adduce against Fritz and Goodman's Contingentist paraphrase of the uncountable infinitary modalized quantifier.

The philosophical significance of the barrier to a faithful translation from modal propositional to extensional full first-order languages, as well as a faithful translation from modalized, i.e. Necessitist, generalized quantifiers to Contingentist quantification, is arguably that the modal resources availed of in the abstractionist program might then be ineliminable.

9.4 Epistemic Hyperintensionality, Epistemic Utility, and Entitlement

In this section, I address, finally, Hale and Wright's second issue with regard to the role of modality in guaranteeing that the possible truth of abstraction principles provides warrant for the belief in their actual truth. While Necessitism is not immediately pertinent to the default entitlement to trust that abstraction principles are true, I will argue that epistemic modalities or hyperintensional states are yet relevant to Wright's application of the notion of entitlement and 'expected epistemic utility' to abstraction principles. As noted, Hale and Wright argue that there is non-evidential entitlement rationally to trust that acceptable abstraction principles are true, and thus that the terms defined therein actually refer. In response, I will proceed by targeting the explanation in virtue of which there is such epistemic, default entitlement. I will outline two proposals concerning the foregoing grounding claim – advanced, respectively, in chapter **8** and by Wright (2012a; 2014) – and I will argue that the approaches converge.

Wright's elaboration of the notion of rational trust, which is intended to subserve epistemic entitlement, appeals to a notions of expected epistemic utility in the setting of decision theory (2014: 226, 241). In order better to understand this notion of expected epistemic utility, we must be more precise.

There are two, major interpretations of (classical) expected utility. A model of decision theory is a tuple $\langle A, O, K, V \rangle$, where A is a set of acts; O is a set of outcomes; K encodes a set of counterfactual conditionals, where an act from A figures in the antecedent of the conditional and O figures in the conditional's consequent; and V is a function assigning a real number to each outcome. The real number is a representation of the value of the outcome. In evidential decision theory, the expected utility of an outcome is calculated as the product of the agent's credence, conditional on her action, by the utility of the outcome. In causal decision theory, the expected utility of an outcome is calculated as the product of the agent's credence, conditional on both her action and the causal efficacy thereof, by the utility of the outcome.

First, because the causal efficacy of one's choice of acts is presumably orthogonal to the non-evidential rational trust to believe that mathematical abstraction principles are true, I will assume that the notion of expected epistemic utility theory that Wright (op. cit.) avails of relies only on the subjective credence of the agent conditional on her action (where the action might be a mental action of trusting the truth of the relevant proposition), multiplied by the utility that she assigns to the outcome of the proposition in which she's placing her rational trust. Thus expected epistemic utility in the setting of decision-theory will be calculated within the (so-called) evidential, rather than causal, interpretation of the latter.

Second, there are two, major interpretations concerning how to measure the subjective credences of an agent. The philosophical significance of this choice point is that it bears directly on the very notion of the *epistemic utility* that an agent's beliefs will possess.

The epistemic utility associated with the pragmatic approach is, generally, utility maximization. By contrast to the pragmatic approach, the epistemic approach to measuring the accuracy of one's beliefs is grounded in the notion of dominance (see Joyce, 1998; 2009). According to the epistemic approach, there is an ideal, or vindicated, probability concerning a proposition's obtaining, and if an agent's subjective probability measure does not satisfy the Kolmogorov axioms, then one can prove that it will always be dominated by a distinct measure; i.e. it will always be the case that a distinct subjective probability measure will be closer to the vindicated world than one's own. The epistemic utility associated with the epistemic approach is thus the minimization of inaccuracy (see Pettigrew, 2014).[25]

Wright notes that the rational trust subserving epistemic entitlement will be pragmatic, and makes the intriguing point that 'pragmatic reasons are not a special genre of reason, to be contrasted with e.g. epistemic, prudential, and moral reasons' (2012: 484). He provides an example according to which one might be impelled to prefer the 'alleviation of Third world suffering' to one's own 'eternal bliss' (op. cit.); and so presumably has the pragmatic approach to expected utility in mind. The intriguing point to note, however, is that epistemic utility is variegated; one's epistemic utility might consist, e.g., in both the reduction of epistemic inaccuracy and in the satisfaction of one's preferences. Wright concludes that there is thus 'no good cause to deny certain kinds of pragmatic reason the title 'epistemic'. This will be the case where, in the slot in the structure of the reasons for an action that is to be filled by the desires of the agent, the relevant desires are focused on epistemic goods and goals' (op. cit.).

Third, and most crucially: The very idea of expected epistemic utility in the setting of decision theory makes implicit appeal to the notion of possible worlds. The full and partial beliefs of an agent will have to be defined on a probability distribution, i.e. a set of epistemically possible worlds. The philosophical significance of this point is that it demonstrates how Hale and

---

[25]The distinction between the epistemic (also referred to as the alethic) and the pragmatic approaches to epistemic utility is anticipated by Clifford (1877) and James (1896), with Clifford endorsing the epistemic approach, and James the pragmatic. The distance measures comprising the scoring rules for the minimization of inaccuracy are examined in, inter alia, Fitelson (2001) and Leitgeb and Pettigrew (2010). A generalization of Joyce's argument for probabilism to models of non-classical logic is examined in Paris (2001) and Williams (2012). A dominance-based approach to decision theory is examined in Easwaran (2014) and Helzner (ms), and a dominance-based approach to the notion of coherence – which can accommodate phenomena such as the preface paradox, and is thus weaker than the notion of consistency in an agent's belief set – is examined in Easwaran and Fitelson (2015).

Wright's appeal to default, rational entitlement to trust that abstraction principles are true converges with the modal approach to the epistemology of mathematics advanced in chapter **8**. The latter proceeds by examining large cardinal axioms and undecidable sentences via the epistemic interpretation of two-dimensional semantics. The latter can be understood as recording the thought that the semantic value of a proposition relative to a first parameter which ranges over epistemically possible worlds, will constrain the semantic value of the proposition relative to a second parameter which ranges over metaphysically possible worlds. The formal clauses for epistemic and metaphysical mathematical modalities are as follows:

- **Epistemic Mathematical Necessity**

  $[\![\blacksquare\phi]\!]^{c,w} = 1 \iff \forall c' [\![\phi]\!]^{c',c'} = 1$

  (ϕ is true at all points in epistemic modal space).

- **Epistemic Mathematical Possibility**

  $[\![\blacklozenge\phi]\!] \neq \varnothing \iff [\![\neg\blacksquare\neg\phi]\!] = 1$

  (ϕ might be true if and only if it is not epistemically necessary for ϕ to be false).

Epistemic mathematical modality is constrained by consistency, and the formal techniques of provability and forcing. A mathematical formula is metaphysically impossible, if it can be disproved or induces inconsistency in a model.

According, then, to the latter, the possibility of deciding mathematical propositions which are currently undecidable relative to a background mathematical language such as ZFC should be two-dimensional. The epistemic possibility of deciding Orey sentences can thus be a guide to the metaphysical possibility thereof.[26] Further, both the numerical term-forming operator, Nx, in abstraction principles, as well as entire abstraction principles themselves, can receive a two-dimensional treatment, such that the value of numerical terms relative to epistemic possibilities considered as actual can determine the value of numerical terms relative to metaphysical possibilities, and the

---

[26]See Kanamori (2008) and Woodin (2010), for further discussion of the mathematical properties at issue.

epistemic possibility of an abstraction principle's truth can determine the metaphysical possibility thereof.[27]

The convergence between Wright's and my approaches consists, then, in that – on both approaches – there is a set of epistemically possible worlds. In the former case, the epistemically possible worlds subserve the preference rankings for the definability of expected epistemic utility. Epistemic mathematical modality is thus constitutive of the notion of rational entitlement to which Hale and Wright appeal, and – in virtue of its convergence with the two-dimensional semantics here proffered – epistemically possible worlds can serve as a guide to the metaphysical mathematical possibility that mathematical propositions, such as abstraction principles for cardinals, reals, and sets, are true.

Novel epistemic abstractionist modalities can, further, be countenanced.

Linnebo writes: 'Let us add to our language the modal operators $\Box$ and $\Diamond$. We may think of '$\Box \phi$' as meaning "no matter what abstraction steps we carry out, it will remain the case that ϕ", and '$\Diamond \phi$' as "we can abstract so as to make it the case that ϕ". Obviously, this interpretation of the modal operators is different from the more familiar one in terms of metaphysical modality. In the useful terminology of (Fine, 2006), the present interpretation is "interpretational" rather than "circumstantial"; that is, it is concerned with how the language is interpreted, not with how reality is. In particular, every interpretational possibility is compatible with the metaphysically actual world' (2018: 61-62).

I propose to treat Linnebo's interpretational abstractionist modalities as epistemic modalities.

Both Linnebo's and my operators are 'transcendental', or 'extended' in the sense outlined by Fine (2005a: 324, 326; 2020). Transcendental modality is 'true regardless of the circumstances [i.e., not 'in', but 'at' or 'of' all possible worlds], for we can recognize it to be true on the basis of its logical form alone and without regard to the circumstances' (Fine, 2005a: 324, 326), by

---

[27] After writing this chapter in 2015, a remark about conceivability-based, though not two-dimensional, approaches to abstraction principles was published in 2020 by Bob Hale. Hale (2020: 270) writes: 'If Hume's Principle is true – that is, if conceivability implies possibility – then we should insist upon the more guarded description', with the description pertaining to whether conceiving that ϕ ought to be understood 'in terms of imagining (our) finding out or discovering [that ϕ] or in more guarded terms as imagining (our) having compelling evidence or good reason to believe' that ϕ. This remark is made in the context of a discussion of the counterconceivability of the essentiality of origins.

contrast to necessity which is an 'unextended' modality, and which can be interpreted as truth 'in' a possible world or truth 'in' all possible worlds (326-327).[28] A 'superextended' modality applies to hybrid sentences which use worldly and unworldly expressions (op. cit.). Worldly entities have necessary existence as depending on the kinds of 'existents', and unworldly entities have transcendental existence as expressed by logical terms, like quantifiers and identity, and as expressed by predicates which pick out essential properties like being rational (324, 350-351). Necessary existence is 'object-driven', and two types can be distinguished depending on whether the objects' existence is indexed to times. Transcendental existence is an 'ontic notion of existence', 'domain-driven', and tied to the notion of being (351). 'For something to exist in this sense is simply for there to be something that it is. This is the sense of existence that is tied to our understanding of the quantifier; where '∃y' is the unrestricted quantifier, x will exist in this sense if ∃y(x = y)' (350). Necessary existence is defined via worldly existents, i.e. 'the character of the object in question' (351) and 'the kind of thing that exists' (354), whereas transcendental existence is 'ontic' and tied to 'ontology' (351), the notion of being, and is expressed by unworldly, logical expressions like quantifiers and identity (350).

My epistemic abstractionist box operator is a transcendental modality, a weak modality, and satisfies the condition of 'general validity'. ϕ is weakly necessary if and only if it is not false in all possible worlds and converges with the notion of general validity, which is necessity interpreted as 'falsity at no world in a model' (Davies and Humberstone, 1979: 1). The epistemic

---

[28]The distinction is anticipated by Evans (1979/1985: 188, fn. 17). Davies (2004: 3.1) writes of Evans' distinction, between truth with respect to a world and truth in a world, that: 'If we think of a sentence's being *made true by* a state of affairs along the lines of the sentence's being *true with respect to* a possible world, then this additional constraint is bound to seem puzzling. In general, s and '[Actually]s' are true with respect to different possible worlds. That is why it may be that '☐s' is false even though '☐(As)' is true. So how could s and 'As' be made true by the same states of affairs? The way out of this apparent puzzle is to observe that Evans insists that we distinguish between truth with respect to a world and truth in a world (p. 188, note 17). Truth *with respect to* possible worlds is relevant to the evaluation of '☐'-modalisations and so it belongs with the notions of superficial contingency and necessity. But the notions of deep contingency and necessity go along with truth *in* possible worlds. A sentence is deeply necessary just in case it is true *in* every possible world. Truth in a world w is glossed as: if w were to obtain, or were to be actual, then would be true (p. 207). And it is subject to the constraint that s and 'As' are true in the same worlds'.

abstractionist box operator contrasts with the epistemic abstractionist diamond operator, where the latter records necessity interpreted as real world validity. ϕ is real world valid 'if for no model is it false at the actual world of the model' (op. cit.).[29] The relation between the interpretation of the diamond operator as an abstraction yielding ϕ and real world validity might be captured by analogy to forcing, i.e. a possible extension of a ground model, where forcing can, too, be interpreted as validity, as, for example, in the semantics for modal logic.[30]

9.4.1 Topic-Sensitive Epistemic Two-dimensional Truth-maker Semantics

If one prefers hyperintensional semantics to possible worlds semantics – in order e.g. to avoid the situation in intensional semantics according to which all necessary formulas express the same proposition because they are true at all possible worlds – one can avail of the topic-sensitive epistemic two-dimensional truthmaker semantics outlined in chapters 2 and 4.

The epistemic abstractionist box operator can be defined as an epistemically necessary truthmaker, and the epistemic abstractionist diamond operator can be defined as an epistemic truthmaker. The foregoing yields the first account of transcendental hyperintensionality in the literature.

The application of truthmaker semantics to abstraction principles might coincide with Cameron (2008)'s suggestion that truthmaker theory be appealed to in order to account for the truth of abstraction principles rather than the prior existence of objects to which the quantifiers in the principles are ontologically committed. Hale and Wright (2009: p. 186, fn. 19) object to this maneuver that the target conception of ontological commitment is necessary for understanding how truth-conditions are fixed, and so ought not to be eschewed. In response, the role of ontological commitment in satisfying the truth-conditions of abstraction principles appears to be consistent with a truthmaker conception of hyperintensional states which verify the principles (both epistemically and metaphysically on the view proffered in this chapter). Cameron (op. cit.: 11) notes that: 'Whether or not we

---

[29]Kuhn (2020) develops a semantics for transcendental modalities which makes them necessity operators applied to unworldly truths, although does not distinguish between general and real world validity in his semantics.

[30]See Avigad (2004), for historical discussion of the interpretation of validity as forcing.

are ontologically committed to numbers depends solely on whether we need them as truthmakers', so truthmaker theory itself does not entirely adduce against the requirement that the prior existence of objects in a quantifier domain is necessary in order to fix truth-conditions for a target sentence. The truthmaker approach is also consistent with a predicative conception of abstraction principles, as advanced by Linnebo (2018a), according to which objects are introduced via the principle and iterations thereof rather than there being a totality of objects prior to the stipulation of the principle.

9.5 Concluding Remarks

In this essay, I have endeavored to provide an account of the modal foundations of mathematical platonism. Hale and Wright's objections to the idea that Necessitism cannot account for how possibility and actuality might converge were shown to be readily answered. In response, further, to Hale and Wright's objections to the role of modalities in countenancing the truth of abstraction principles and the success of mathematical predicate reference, I demonstrated how my two-dimensional intensional and hyperintensional approaches to the epistemology of mathematics are consistent with Hale and Wright's conception of the epistemic entitlement rationally to trust that abstraction principles are true. Epistemic and metaphysical states and possibilities may thus be shown to play a constitutive role in vindicating the reality of mathematical objects and truth, and in explaining our possible knowledge thereof.

Chapter 10

Hyperintensional Ω-Logic

10.1 Introduction

This essay examines the philosophical significance of Ω-logic in Zermelo-Fraenkel set theory with choice (ZFC). I argue that the philosophical significance of the foregoing is two-fold. First, because the epistemic and modal and hyperintensional profiles of Ω-logical validity correspond to those of second-order logical consequence, Ω-logical validity is genuinely logical. Second, the foregoing provides a hyperintensional account of the interpretation of mathematical and metamathematical vocabulary.

10.2 Definitions

In this section, I define the axioms of Zermelo-Fraenkel set theory with choice. I define the mathematical properties of the large cardinal axioms which can be adjoined to ZFC, and I provide a detailed characterization of the properties of Ω-logic for ZFC. Coalgebras are dual to Boolean-valued algebraic models of Ω-logic. Modal and hyperintensional coalgebras are then argued to provide a precise characterization of the modal and hyperintensional profiles of Ω-logical validity.

10.2.1 Axioms[1]

- Extensionality
  $\forall x,y.(\forall z.z \in x \iff z \in y) \to x = y$

- Empty Set
  $\exists x.\forall y.y \notin x$

- Pairing
  $\forall x,y.\exists z.\forall w.w \in z \iff w = x \lor w = y$

- Union
  $\forall x.\exists y.\forall z.z \in y \iff \exists w.w \in x \land z \in w$

- Power Set
  $\forall x.\exists y.\forall z.z \in y \iff z \subseteq x$

- Separation (with $\vec{x}$ a parameter)
  $\forall \vec{x},y.\exists z.\forall w.w \in z \iff w \in y \land A(w,\vec{x})$

- Infinity
  $\exists x.\emptyset \in x \land \forall y.y \in x \to y \cup \{y\} \in x$

- Foundation
  $\forall x.(\exists y.y \in x) \to \exists y \in x.\forall z \in x.z \notin y$

- Replacement
  $\forall x, \vec{y}.[\forall z \in x.\exists!w.A(z,w,\vec{y})] \to \exists u.\forall w.w \in u \iff \exists z \in x.A(z,w,\vec{y})$

- Choice
  $\forall x.\emptyset \notin x \to \exists f \in (x \to \cup x).\forall y \in x.f(y) \in y$

---
[1] For a standard presentation, see Jech (2003). The presentation here follows Avigad (2021). For detailed, historical discussion, see Maddy (1988a).

10.2.2 Large Cardinals

Borel sets of reals are subsets of ω^ω or $\mathbb{R}$, closed under countable intersections and unions.[2] For all ordinals, a, such that $0 < a < \omega_1$, and $b < a$, Σ^0_a denotes the open subsets of ω^ω formed under countable unions of sets in Π^0_b, and Π^0_a denotes the closed subsets of ω^ω formed under countable intersections of Σ^0_b.

Projective sets of reals are subsets of ω^ω, formed by complementations ($\omega^\omega - u$, for $u \subseteq \omega^\omega$) and projections [$p(u) = \{\langle x_1, \ldots, x_n \rangle \in \omega^\omega \mid \exists y \langle x_1, \ldots, x_n, y \rangle \in u\}$]. For all ordinals a, such that $0 < a < \omega$, Π^1_0 denotes closed subsets of ω^ω; Π^1_a is formed by taking complements of the open subsets of ω^ω, Σ^1_a; and Σ^1_{a+1} is formed by taking projections of sets in Π^1_a.

The full power set operation defines the cumulative hierarchy of sets, V, such that $V_0 = \emptyset$; $V_{a+1} = \wp(V_0)$; and $V_\lambda = \bigcup_{a<\lambda} V_a$.

In the inner model program (see Woodin, 2001a, 2010, 2011a; Kanamori, 2012a,b), the definable power set operation defines the constructible universe, $L(\mathbb{R})$, in the universe of sets, V, where the sets are transitive such that $a \in C \iff a \subseteq C$; $L(\mathbb{R}) = V_{\omega+1}$; $L_{a+1}(\mathbb{R}) = \mathrm{Def}(L_a(\mathbb{R}))$; and $L_\lambda(\mathbb{R}) = \bigcup_{a<\lambda}(L_a(\mathbb{R}))$.

Via inner models, Gödel (1940/1990) proves the consistency of the Generalized Continuum Hypothesis, $\aleph_a^{\aleph_a} = \aleph_{a+1}$, as well as the axiom of choice, relative to the axioms of ZFC. However, for a countable transitive set of ordinals, M, in a model of ZF without choice, one can define a generic set, G, such that, for all formulas, ϕ, either ϕ or $\neg\phi$ is forced by a condition, f, in G. Let $M[G] = \bigcup_{a<\kappa} M_a[G]$, such that $M_0[G] = \{G\}$; with $\lambda < \kappa$, $M_\lambda[G] = \bigcup_{a<\lambda} M_a[G]$; and $M_{a+1}[G] = V_a \cap M_a[G]$.[3] G is a Cohen real over M, and comprises a set-forcing extension of M. The relation of set-forcing, $\Vdash$, can then be defined in the ground model, M, such that the forcing condition, f, is a function from a finite subset of ω into $\{0, 1\}$, and $f \Vdash u \in G$ if $f(u) = 1$ and $f \Vdash u \notin G$ if $f(u) = 0$. The cardinalities of an open dense ground model, M, and a generic extension, G, are identical, only if the countable chain condition (c.c.c.) is satisfied, such that, given a chain – i.e., a linearly ordered subset of a partially ordered (reflexive, anti-symmetric, transitive) set – there is a countable, maximal antichain consisting of pairwise incompatible forcing conditions. Via set-forcing extensions, Cohen (1963, 1964) constructs a model of ZF which negates the Generalized Continuum Hypothesis, and thus

---
[2] See Koellner (2013), for the presentation, and for further discussion, of the definitions in this and the subsequent paragraph.
[3] See Kanamori (2012a: 2.1; 2012b: 4.1), for further discussion.

proves the independence thereof relative to the axioms of ZF.[4]

Gödel (1946/1990: 1-2) proposes that the value of Orey sentences such as the GCH might yet be decidable, if one avails of stronger theories to which new axioms of infinity – i.e., large cardinal axioms – are adjoined.[5] He writes that: 'In set theory, e.g., the successive extensions can be represented by stronger and stronger axioms of infinity. It is certainly impossible to give a combinatorial and decidable characterization of what an axiom of infinity is; but there might exist, e.g., a characterization of the following sort: An axiom of infinity is a proposition which has a certain (decidable) formal structure and which in addition is true. Such a concept of demonstrability might have the required closure property, i.e. the following could be true: Any proof for a set-theoretic theorem in the next higher system above set theory ... is replaceable by a proof from such an axiom of infinity. It is not impossible that for such a concept of demonstrability some completeness theorem would hold which would say that every proposition expressible in set theory is decidable from present axioms plus some true assertion about the largeness of the universe of sets'.

For cardinals, κ, a, C, C $\subseteq \kappa$ is closed unbounded in κ, if it is closed [if a $< \kappa$ and $\bigcup(C \cap a) = a$, then a$\in$C] and unbounded ($\bigcup C = \kappa$) (Kanamori, op. cit.: 360). A cardinal, S $\subseteq \kappa$, is stationary in κ, if, for any closed unbounded C $\subseteq \kappa$, C $\cap$ S $\neq \emptyset$ (op. cit.). An ideal is a subset of a set closed under countable unions, whereas filters are subsets closed under countable intersections (361). A cardinal κ is regular if the cofinality of κ is identical to κ. Uncountable regular limit cardinals are weakly inaccessible (op. cit.). A strongly inaccessible cardinal is regular and has a strong limit, such that if $\lambda < \kappa$, then $2^\lambda < \kappa$ (op. cit.).

Large large cardinal axioms are defined by elementary embeddings.[6] Elementary embeddings can be defined thus. For models A,B, and conditions ϕ, j: A $\rightarrow$ B, $\phi\langle a_1, \ldots, a_n\rangle$ in A if and only if $\phi\langle j(a_1), \ldots, j(a_n)\rangle$ in B (363). A measurable cardinal is defined as the ordinal denoted by the critical point of j, crit(j) (Koellner and Woodin, 2010: 7). Measurable cardinals are

---

[4]See Kanamori (2008), for further discussion.

[5]See Kanamori (2007), for further discussion. Kanamori (op. cit.: 154) notes that Gödel (1931/1986: fn. 48a) makes a similar appeal to higher-order languages, in his proofs of the incompleteness theorems. The incompleteness theorems are examined in further detail, in Section **6.4**, above.

[6]The definitions in the remainder of this subsection follow the presentations in Koellner and Woodin (2010) and Woodin (2010, 2011a).

inaccessible (Kanamori, op. cit.).

Let κ be a cardinal, and $\eta > \kappa$ an ordinal. κ is then η-strong, if there is a transitive class M and an elementary embedding, j: V → M, such that crit(j) = κ, j(κ) > η, and $V_\eta \subseteq$ M (Koellner and Woodin, op. cit.).

κ is strong if and only if, for all η, it is η-strong (op. cit.).

If A is a class, κ is η-A-strong, if there is a j: V → M, such that κ is η-strong and j(A ∩ V_κ) ∩ V_η = A ∩ V_η (op. cit.).

κ is a Woodin cardinal, if κ is strongly inaccessible, and for all A ⊆ V_κ, there is a cardinal $\kappa_A < \kappa$, such that κ_A is η-A-strong, for all η such that κ_A < η < κ (Koellner and Woodin, op. cit.: 8).

κ is superstrong, if j: V → M, such that crit(j) = κ and $V_{j(\kappa)} \subseteq$ M, which entails that there are arbitrarily large Woodin cardinals below κ (op. cit.).

Large large cardinal axioms can then be defined as follows.

∃xΦ is a large large cardinal axiom, because:

(i) Φx is a Σ_2-formula, where 'a sentence ϕ is a Σ_2-sentence if it is of the form: There exists an ordinal α such that $V_\alpha \Vdash \psi$, for some sentence ψ' (Woodin, 2019);

(ii) if κ is a cardinal, such that V ⊨ $\Phi(\kappa)$, then κ is strongly inaccessible, where a cardinal κ is regular if the cofinality of κ – comprised of the unions of sets with cardinality less than κ – is identical to κ, and a strongly inaccessible cardinal is regular and has a strong limit, such that if $\lambda < \kappa$, then $2^\lambda < \kappa$ (see Kanamori, 2012: 360); and

(iii) for all generic partial orders $\mathbb{P} \in V_\kappa$, and all V-generics G ⊆ $\mathbb{P}$, V[G] ⊨ Φx (Koellner, 2006: 180).

For all generic partial orders $\mathbb{P} \in V_\kappa$, $V^\mathbb{P}$ ⊨ $\Phi(\kappa)$; I_{NS} is a non-stationary ideal – 'I_{NS} is the σ-ideal of all sets A ⊆ ω_1 such that ω_1\A contains a closed unbounded set' (Woodin, 2001b: 686); A^G is the canonical representation of reals in L($\mathbb{R}$), i.e. the interpretation of A in M[G]; H(κ) is comprised of all of the sets whose transitive closure is < κ (see Woodin, 2001a: 569); and L($\mathbb{R}$)$^{\mathbb{P}max}$ ⊨ $\langle H(\omega_2), \in, I_{NS}, A^G \rangle$ ⊨ 'ϕ'. $\mathbb{P}$ is a homogeneous partial order in L($\mathbb{R}$), such that the generic extension of L($\mathbb{R}$)$^\mathbb{P}$ inherits the absoluteness, i.e. the generic invariance, of L($\mathbb{R}$). Thus, L($\mathbb{R}$)$^{\mathbb{P}max}$ is (i) effectively complete, i.e. invariant under set-forcing extensions; and (ii) maximal, i.e. satisfies all Π_2-sentences and is thus consistent by set-forcing over ground models (Woodin, ms: 28).

Assume ZFC and that there is a proper class of Woodin cardinals; A∈$\mathbb{P}$($\mathbb{R}$) ∩ L($\mathbb{R}$); ϕ is a Π_2-sentence; and V(G), s.t. $\langle H(\omega_2), \in, I_{NS}, A^G \rangle$ ⊨ 'ϕ': Then, it can be proven that L($\mathbb{R}$)$^{\mathbb{P}max}$ ⊨ $\langle H(\omega_2), \in, I_{NS}, A^G \rangle$ ⊨ 'ϕ', where 'ϕ' :=

$\exists A \in \Gamma^{\infty} \langle H(\omega_1), \in, A \rangle \models \psi$.

The axiom of determinacy (AD) states that every set of reals, $a \subseteq \omega^{\omega}$ is determined.

Woodin (1999)'s Axiom (*) can be thus countenanced:
$AD^{L(\mathbb{R})}$ and $L[(\mathbb{P}\omega_1)]$ is a $\mathbb{P}$max-generic extension of $L(\mathbb{R})$,
from which it can be derived that $2^{\aleph_0} = \aleph_2$. Thus, ¬CH; and so CH is absolutely decidable.

In more recent work, Woodin (2019) provides evidence that CH might, by contrast, be true. The truth of CH would follow from the truth of Woodin's Ultimate-L conjecture. The following definitions are from Woodin (op. cit.): 'A transitive class is an inner model if[, for the class of ordinals, Ord, - D.E.] Ord $\subset$ M, and M ⊩ ZFC'. L, the constructible reals, and HOD, the hereditarily ordinal definable sets, are inner models. 'Suppose N is an inner model and that δ is an uncountable (regular) cardinal of V. N has the δ-cover property if for all $\sigma \subset N$, if $|\sigma| < \delta$ then there exists $\tau \in N$ such that: $\sigma \subset \tau$ and $|\tau| < \delta$. N has the δ-approximation property if for all sets $X \subset N$, the following are equivalent: (i) $X \in N$ and (ii) For all $\sigma \in N$, if $|\sigma| < \delta$, then $\sigma \cap X \in N$. Suppose N is an inner model and that $\sigma \subset N$. Then $N[\sigma]$ denotes the smallest inner model M such that $N \subseteq M$ and $\sigma \in M$. Suppose that N is an inner model and δ is strongly inaccessible. Then N has the δ-genericity property if for all $\sigma \subseteq \delta$, if $|\sigma| < \delta$ then $N[\sigma] \cap V_{\delta}$ is a Cohen extension of N $\cap V_{\delta}$. The axiom for V = Ultimate-L states then that '(i) There is a proper class of Woodin cardinals, and (ii) For each Σ_2-sentence ϕ, if ϕ holds in V then there is a universally Baire set $A \subseteq \mathbb{R}$ such that $HOD^{L(A,\mathbb{R})} \Vdash \phi$, where a set is universally Baire if for all topological spaces Ω and for all continuous functions $\pi : \Omega \to \mathbb{R}^n$, the preimage of A by π has the property of Baire in the space Ω'. The property of Baire holds if, for a subset of a topological space $A \subseteq X$, there is an open set $U \subset X$ such that $A \, \Xi \, U$ is a meagre subset, where Ξ is the symmetric difference, i.e. the union of relative complements, and a subset of a topological space is meagre if it is a countable union of nowhere dense sets, where nowhere dense subsets of the topology hold if their union with an open set is not dense.[7] The Ultimate-L Conjecture is then as follows: 'Suppose that δ is an extendible cardinal. δ is an extendible cardinal if for each $\lambda > \delta$ there exists an elementary embedding $j : V_{\lambda+1} \to V_{j(\lambda)+1}$ such that $CRT(j) = \delta$ and $j(\delta) > \lambda$. Then provably there is an inner model N

---

[7]https://en.wikipedia.org/wiki/PropertyofBaire, https://en.wikipedia.org/wiki/Symmetricdifference, https://en.wikipedia.org/wiki/Meagreset.

such that: 1. N has the δ-cover and δ-approximation properties. 2. N has the δ-genericity property. 3. N ⊩ 'V = Ultimate-L' (Woodin, op. cit.).

10.2.3 Ω-Logic

For partial orders, $\mathbb{P}$, let $V^{\mathbb{P}} = V^{\mathbb{B}}$, where $\mathbb{B}$ is the regular open completion of $(\mathbb{P})$.[8] $M_a = (V_a)^M$ and $M_a^{\mathbb{B}} = (V_a^{\mathbb{B}})^M = (V_a^{M^{\mathbb{B}}})$. Sent denotes a set of sentences in a first-order language of set theory. $T \cup \{\phi\}$ is a set of sentences extending ZFC. c.t.m abbreviates the notion of a countable transitive ∈-model. c.B.a. abbreviates the notion of a complete Boolean algebra.

Define a c.B.a. in V, such that $V^{\mathbb{B}}$. Let $V_0^{\mathbb{B}} = \emptyset$; $V_\lambda^{\mathbb{B}} = \bigcup_{b<\lambda} V_b^{\mathbb{B}}$, with λ a limit ordinal; $V_{a+1}^{\mathbb{B}} = \{f: X \to \mathbb{B} \mid X \subseteq V_a^{\mathbb{B}}\}$; and $V^{\mathbb{B}} = \bigcup_{a \in On} V_a^{\mathbb{B}}$.

ϕ is true in $V^{\mathbb{B}}$, if its Boolean-value is $1^{\mathbb{B}}$, if and only if
$V^{\mathbb{B}} \models \phi$ iff $[\![\phi]\!]^{\mathbb{B}} = 1^{\mathbb{B}}$.

Thus, for all ordinals, a, and every c.B.a. $\mathbb{B}$, $V_a^{\mathbb{B}} \equiv (V_a)^{V^{\mathbb{B}}}$ iff for all $x \in V^{\mathbb{B}}$, $\exists y \in V^{\mathbb{B}} [\![x = y]\!]^{\mathbb{B}} = 1^{\mathbb{B}}$ iff $[\![x \in V^{\mathbb{B}}]\!]^{\mathbb{B}} = 1^{\mathbb{B}}$.

Then, $V_a^{\mathbb{B}} \models \phi$ iff $V^{\mathbb{B}} \models$ '$V_a \models \phi$'.

Ω-logical validity can then be defined as follows:
For $T \cup \{\phi\} \subseteq$ Sent,
$T \models_\Omega \phi$, if for all ordinals, a, and c.B.a. $\mathbb{B}$, if $V_a^{\mathbb{B}} \models T$, then $V_a^{\mathbb{B}} \models \phi$.

Supposing that there exists a proper class of Woodin cardinals and if $T \cup \{\phi\} \subseteq$ Sent, then for all set-forcing conditions, $\mathbb{P}$:
$T \models_\Omega \phi$ iff $V^T \models$ '$T \models_\Omega \phi$',
where $T \models_\Omega \phi \equiv \emptyset \models$ '$T \models_\Omega \phi$'.

The Ω-Conjecture states that $V \models_\Omega \phi$ iff $V^{\mathbb{B}} \models_\Omega \phi$ (Woodin, ms). Thus, Ω-logical validity is invariant in all set-forcing extensions of ground models in the set-theoretic universe.

The soundness of Ω-Logic is defined by universally Baire sets of reals. For a cardinal, e, let a set A be e-universally Baire, if for all partial orders $\mathbb{P}$ of cardinality e, there exist trees, S and T on $\omega \times \lambda$, such that $A = p[T]$ and if $G \subseteq \mathbb{P}$ is generic, then $p[T]^G = \mathbb{R}^G - p[S]^G$ (Koellner, 2013). A is universally Baire, if it is e-universally Baire for all e (op. cit.).

Ω-Logic is sound, such that $V \vdash_\Omega \phi \to V \models_\Omega \phi$. However, the completeness of Ω-Logic has yet to be resolved.

A **E**-coalgebra is a pair $\mathbb{A} = (A, \mu)$, with A an object of C referred to as the carrier of $\mathbb{A}$, and $\mu: A \to \mathbf{E}(A)$ is an arrow in C, referred to as the

---

[8] The definitions in this section follow the presentation in Bagaria et al. (2006).

transition map of $\mathbb{A}$ (390).

$\mathbb{A} = \langle A, \mu: A \to \mathbf{E}(A) \rangle$ is dual to the category of algebras over the functor μ (417-418). If μ is a functor on categories of sets, then coalgebraic models are dual to Boolean-algebraic models of Ω-logical validity.

Leach-Krouse (ms) defines the modal logic of Ω-consequence as satisfying the following axioms:

For a theory $\mathbf{T}$ and with $\Box\phi := \mathbf{T}^\mathbb{B}_\alpha \Vdash \mathrm{ZFC} \Rightarrow \mathbf{T}^\mathbb{B}_\alpha \Vdash \phi$,

$\mathrm{ZFC} \vdash \phi \Rightarrow \mathrm{ZFC} \vdash \Box\phi$

$\mathrm{ZFC} \vdash \Box(\phi \to \psi) \to (\Box\phi \to \Box\psi)$

$\mathrm{ZFC} \vdash \Box\phi \to \phi \Rightarrow \mathrm{ZFC} \vdash \phi$

$\mathrm{ZFC} \vdash \Box\phi \to \Box\Box\phi$

$\mathrm{ZFC} \vdash \Box(\Box\phi \to \phi) \to \Box\phi$

$\Box(\Box\phi \to \psi) \vee \Box(\Box\psi \wedge \psi \to \phi)$), where this clause added to GL is the logic of 'true in all V_κ for all κ strongly inaccessible' in ZFC.

10.2.4 Two-dimensional Hyperintensionality and Ω-logic

Finally, the axioms of the modal logic of Ω-consequence can be rendered hyperintensional as follows. As with the two-dimensional hyperintensional profile of the Epistemic Church-Turing Thesis and the axioms of epistemic set theory, the two-dimensional hyperintensional profile of Ω-logical consequence can be countenanced by adding a topic-sensitive truthmaker from a metaphysical state space and making its value dependent on the value of the epistemically necessary truthmaker, $A(\phi)$.

For a theory $\mathbf{T}$ and with $A(\Box\phi) :=$ for all $s \in P$ there is a $s' \in P$ such that $s' \sqcup s \in P$ and $s' \vdash \text{'}\mathbf{T}^\mathbb{B}_\alpha \Vdash \mathrm{ZFC} \Rightarrow \mathbf{T}^\mathbb{B}_\alpha \Vdash \phi\text{'}$, where $\Box$ is interpreted as $\mathbf{T}^\mathbb{B}_\alpha \Vdash \mathrm{ZFC} \Rightarrow \mathbf{T}^\mathbb{B}_\alpha \Vdash \phi$,

$\mathrm{ZFC} \vdash \phi \Rightarrow \mathrm{ZFC} \vdash A^{(s \cap t)}(\Box\phi)$

$\mathrm{ZFC} \vdash A^{(s \cap t)}[\Box(\phi \to \psi) \to (\Box\phi \to \Box\psi)]$

$\mathrm{ZFC} \vdash A^{(s \cap t)}(\Box\phi) \to \phi \Rightarrow \mathrm{ZFC} \vdash \phi$

$\mathrm{ZFC} \vdash A^{(s \cap t)}(\Box\phi) \to A^{(s \cap t)}(\Box\Box\phi)$

$\mathrm{ZFC} \vdash A^{(s \cap t)}[\Box(\Box\phi \to \phi)] \to A^{(s \cap t)}(\Box\phi)$

$A^{(s \cap t)}[\Box(\Box\phi \to \psi) \vee \Box(\Box\psi \wedge \psi \to \phi)]$.

10.3 Discussion

10.3.1 Ω-Logical Validity is Genuinely Logical

Frege (1884/1980; 1893/2013)'s proposal – that cardinal numbers can be explained by specifying a biconditional between the identity of, and an equivalence relation on, concepts, expressible in the signature of second-order logic – is the first attempt to provide a foundation for mathematics on the basis of logical axioms rather than rational or empirical intuition. In Frege (1884/1980. cit.: 68) and Wright (1983: 104-105), the number of the concept, **A**, is argued to be identical to the number of the concept, **B**, if and only if there is a one-to-one correspondence between **A** and **B**, i.e., there is a bijective mapping, R, from **A** to **B**. With Nx: a numerical term-forming operator,

- ∀**A**∀**B**[Nx: **A** = Nx: **B** ≡ ∃R[∀x[**A**x → ∃y(**B**y ∧ Rxy ∧ ∀z(**B**z ∧ Rxz → y = z))] ∧ ∀y[**B**y → ∃x(**A**x ∧ Rxy ∧ ∀z(**A**z ∧ Rzy → x = z))]]].

Frege's Theorem states that the Dedekind-Peano axioms for the language of arithmetic can be derived from the foregoing abstraction principle, as augmented to the signature of second-order logic and identity.[9] Thus, if second-order logic may be counted as pure logic, despite that domains of second-order models are definable via power set operations, then one aspect of the philosophical significance of the abstractionist program consists in its provision of a foundation for classical mathematics on the basis of pure logic as augmented with non-logical implicit definitions expressed by abstraction principles.

There are at least three reasons for which a logic defined in ZFC might not undermine the status of its consequence relation as being logical. The first reason for which the mathematical entanglement of Ω-logical validity might be innocuous is that, as Shapiro (1991: 5.1.4) notes, many mathematical properties cannot be defined within first-order logic, and instead require the expressive resources of second-order logic.[10] For example, the notion of well-foundedness cannot be expressed in a first-order framework, as evinced by

---

[9] See Dedekind (1888/1963) and Peano (1889/1967). See Wright (1983: 154-169) for a proof sketch of Frege's theorem; Boolos (1987) for the formal proof thereof; and Parsons (1964) for an incipient conjecture of the theorem's validity.

[10] Koellner cites Charles Parsons, with regard to the phrase, 'mathematical entanglement' (2010: 2).

considerations of compactness. Let E be a binary relation. Let m be a well-founded model, if there is no infinite sequence, $a_0, \ldots, a_i$, such that $Ea_0, \ldots, Ea_{i+1}$ are all true. If m is well-founded, then there are no infinite-descending E-chains. Suppose that T is a first-order theory containing m, and that, for all natural numbers, n, there is a T with $n+1$ elements, $a_0, \ldots, a_n$, such that $\langle a_0, a_1 \rangle, \ldots, \langle a_n, a_{n-1} \rangle$ are in the extension of E. By compactness, there is an infinite sequence such that that $a_0 \ldots a_i$, s.t. $Ea_0, \ldots, Ea_{i+1}$ are all true. So, m is not well-founded.

By contrast, however, well-foundedness can be expressed in a second-order framework:

$\forall X[\exists x Xx \rightarrow \exists x[Xx \wedge \forall y(Xy \rightarrow \neg Eyx)]]$, such that m is well-founded iff every non-empty subset X has an element x, s.t. nothing in X bears E to x.

One aspect of the philosophical significance of well-foundedness is that it provides a distinctively second-order constraint on when the membership relation in a given model is intended. This contrasts with Putnam (1980)'s claim, that first-order models *mod* can be intended, if every set s of reals in *mod* is such that an ω-model in *mod* contains s and is constructible, such that – given the Downward Löwenheim-Skolem theorem[11] – if *mod* is non-constructible but has a submodel satisfying 's is constructible', then the model is non-well-founded and yet must be intended. The claim depends on the assumption that general understanding-conditions and conditions on intendedness must be co-extensive, to which I will return in Section **10.3.2**.

A second reason for which Ω-logic's mathematical entanglement might not be pernicious, such that the consequence relation specified in the Ω-logic might be genuinely logical, may again be appreciated by its comparison with second-order logic. Shapiro (1998) defines the model-theoretic characterization of logical consequence as follows:

'(10) Φ is a logical consequence of [a model] Γ if Φ holds in all possibilities under every interpretation of the nonlogical terminology which holds in Γ' (148).

A condition on the foregoing is referred to as the 'isomorphism property', according to which 'if two models M, M' are isomorphic vis-a-vis the non-logical items in a formula Φ, then M satisfies Φ if and only if M' satisfies Φ'

---

[11]The Downward Löwenheim-Skolem theorem claims that for any first-order model M, M has a submodel M' whose domain is at most denumerably infinite, s.t. for all assignments s on, and formulas $\phi(x)$ in M', $M, s \Vdash \phi(x) \iff M', s \Vdash \phi(x)$.

(151).

Shapiro argues, then, that the consequence relation specified using second-order resources is logical, because of its modal and epistemic profiles. The epistemic tractability of second-order validity consists in 'typical soundness theorems, where one shows that a given deductive system is truth-preserving' (154). He writes that: '[I]f we know that a model is a good mathematical model of logical consequence (10), then we know that we won't go wrong using a sound deductive system. Also, we can know that an argument is a logical consequence . . . via a set-theoretic proof in the metatheory' (154-155).

The modal profile of second-order validity provides a second means of accounting for the property's epistemic tractability. Shapiro argues, e.g., that: 'If the isomorphism property holds, then in evaluating sentences and arguments, the only 'possibility' we need to 'vary' is the size of the universe. If enough sizes are represented in the universe of models, then the modal nature of logical consequence will be registered . . . [T]he only 'modality' we keep is 'possible size', which is relegated to the set-theoretic metatheory' (152).

Shapiro's remarks about the considerations adducing in favor of the logicality of non-effective, second-order validity generalize to Ω-logical validity. As with Shapiro's definition of logical consequence, where Φ holds in all possibilities in the universe of models and the possibilities concern the 'possible size' in the set-theoretic metatheory, the Ω-Conjecture states that $V \models_\Omega \phi$ iff $V^\mathbb{B} \models_\Omega \phi$, such that Ω-logical validity is invariant in all set-forcing extensions of ground models in the set-theoretic universe.

10.3.2 Hyperintensionality and the Concept of Set

In this section, I argue, finally, that the hyperintensional profile of Ω-logic can be availed of in order to account for the understanding-conditions of the concept of set.

Putnam (op. cit.: 473-474) argues that defining models of first-order theories is sufficient for both understanding and specifying an intended interpretation of the latter. Wright (1985: 124-125) argues, by contrast, that understanding-conditions for mathematical concepts cannot be exhausted by the axioms for the theories thereof, even on the intended interpretations of the theories. He suggests, e.g., that:

'[I]f there really were uncountable sets, their existence would surely have to flow from the concept of set, as intuitively satisfactorily explained. Here, there is, as it seems to me, no assumption that the content of the ZF-axioms cannot exceed what is invariant under all their classical models. [Benacerraf] writes, e.g., that: 'It is granted that they are to have their 'intended interpretation': '∈' is to mean set-membership. Even so, and conceived as encoding the intuitive concept of set, they fail to entail the existence of uncountable sets. So how can it be true that there are such sets? Benacerraf's reply is that the ZF-axioms are indeed faithful to the relevant informal notions only if, in addition to ensuring that '∈' means set-membership, we interpret them so as to observe the constraint that 'the universal quantifier has to mean all or at least all sets' (p. 103). It follows, of course, that if the concept of set does determine a background against which cantor's theorem, under its intended interpretation, is sound, there is more to the concept of set that can be explained by communication of the intended sense of '∈' and the stipulation that the ZF-axioms are to hold. And the residue is contained, presumably, in the informal explanations to which, Benacerraf reminds us, Zermelo intended his formalization to answer. At least, this must be so if the 'intuitive concept of set' is capable of being explained at all. Yet it is notable that Benacerraf nowhere ventures to supply the missing informal explanation – the story which will pack enough into the extension of 'all sets' to yield cantor's theorem, under its intended interpretation, as a highly non-trivial corollary' (op. cit).

In order to provide the foregoing type of explanation of the concept of set,[12] I will argue that there are several points in the model theory and epistemology of set-theoretic languages at which the interpretation of the concept of set constitutively involves hyperintensional notions. The hyperintensionality at issue is consistent with realist positions with regard to both truth values and the ontology of abstracta.[13]

One point is in the coding of the signature of the theory, T, in which Gödel's incompleteness theorems are proved (see Halbach and Visser, 2014). The choice of coding bridges the numerals in the language with the properties of the target numbers. The choice of coding is therefore intensional, and has been marshaled in order to argue that the very notion of syntactic computability – via the equivalence of partial recursive functions, λ-definable terms, and the transition functions of discrete-state automata such as Tur-

---

[12]Thanks here to Branden Fitelson.

[13]For the hyperintensional commitments of the abstractionist foundations of mathematics, see chapter **9**.

ing machines – is constitutively semantic (see Rescorla, 2015). The choice of coding can be hyperintensional if the coding is topic-sensitive. Further points at which hyperintensionality can be witnessed in the phenomenon of self-reference in arithmetic are introduced by Reinhardt (1986). Reinhardt (op. cit.: 470-472) argues that the provability predicate can be defined relative to the minds of particular agents – similarly to Quine (1968)'s and Lewis (1979)'s suggestion that possible worlds can be centered by defining them relative to parameters ranging over tuples of spacetime coordinates or agents and locations – and that a theoretical identity statement can be established for the concept of the foregoing minds and the concept of a computable system. A hyperintensional semantics for provability logic is suggested in §**8.1.2**.

A second point at which understanding-conditions may be shown to be constitutively hyperintensional can be witnessed by the conditions on the epistemic entitlement to assume that the theory in which Gödel's second incompleteness theorem is proved is consistent (see Dummett, 1963/1978; Wright, 1995). Wright suggests that '[T]o treat [...] Gentzen[(1936/1969)'s] proof [of the consistency of Peano Arithmetic] as establishing consistency is implicitly to exclude any antecedent doubt about the *coherence* of the concept of natural number' [op. cit.: p. 91, fn. 9].[14] Wright's elaboration of

---

[14]'Any permissible antecedent doubt about the consistency of first-order number theory would have, therefore, to concern whether the (fully coherent) concept of natural number is faithfully reflected by the standard first-order axioms – specifically, by all admissible instances of the induction schema. Such a doubt would have to concern whether, even when restricted to first-order arithmetical vocabulary, "tricksy" predicates might not somehow be formulated whose use in inductions could lead to contradiction' [op. cit.; see Waxman (ms$_2$) for discussion]. Wright notes: 'Prescinding from the intuitionistic demonstration which we are about to consider, I suspect the real reason why so many of us tend to regard[, the universally quantified undecidable sentence,] U as informally demonstrated by Gödel's construction is actually rather unflattering, *viz.* we succumb to a simple conflation, confusing the discovery of a *commitment* with the discovery of a *truth*. The proof of Putnam's conditional (3)[If a Turing machine, T, is consistent, U is true (p. 90)] for an intuitively correct[, arithmetical system,] S is a deeply impressive result which teaches us, on pain of accepting the inconsistency of our arithmetical thought, that we are *committed* to regarding each instance of U as computationally verifiable. Since we want to believe that our arithmetical thought is consistent – since, indeed, it is doubtful if agnosticism on the matter could be a practical option – there can therefore be no sitting on the fence as far as U is concerned. But it is quite another thing to view this commitment to U as something we incur on the basis of a *demonstration* of its truth. We are, as a rough parallel, similarly committed by everything we ordinarily think and do to the existence of the material

the notion of epistemic entitlement, appeals to a notion of rational 'trust', which he argues is recorded by the calculation of 'expected epistemic utility' in the setting of decision theory (2004; 2014: 226, 241). Wright notes that the rational trust subserving epistemic entitlement will be pragmatic, and makes the intriguing point that 'pragmatic reasons are not a special genre of reason, to be contrasted with e.g. epistemic, prudential, and moral reasons' (2012: 484). Crucially, however, the very idea of expected epistemic utility in the setting of decision theory makes implicit appeal to epistemically possibly worlds or hyperintensional epistemic states.

A third consideration adducing in favor of the thought that grasp of the concept of set might constitutively possess a hyperintensional profile is that the concept can have a hyperintension – i.e., a function from states to extensions. The modal similarity types in the coalgebraic modal logic may be interpreted as dynamic-interpretational modalities, where the dynamic-interpretational modal operator has been argued to entrain the possible reinterpretations both of the domains of the theory's quantifiers (see Fine, 2005b, 2006), as well as of the intensions of non-logical concepts, such as the membership relation (see Uzquiano, 2015). See chapter 2 for a hyperintensional semantics for dynamic-interpretational modalities.

The fourth consideration avails directly of the hyperintensional profile of Ω-logical consequence. While the above dynamic-interpretational modality will suffice for possible reinterpretations of mathematical terms, the absoluteness of the consequence relation is such that, if the Ω-conjecture is true, then Ω-logical validity is invariant in all possible set-forcing extensions of ground models in the set-theoretic universe. The truth of the Ω-conjecture would thereby place an indefeasible necessary condition on a formal understanding of the hyperintension for the concept of set.

---

world; no agnosticism on the point is practical. It would be very much easier than it is to dispose of material-world scepticism if this commitment could immediately be taken as the reflection of a cognitive achievement. If this somewhat deflating account is correct, we are obliged to conclude that no informal demonstration of the undecidable sentence attends Gödel's proof, and nothing takes place worthy of dignification as "recognition" of its truth. We are merely brought to see that our standing commitment to the consistency of our arithmetical thought embraces a plethora of unsuspected, specifically arithmetical commitments, each associated with a Gödelian undecidable sentence' (pp. 91-92, fn. 10). See Peacocke (2000), for further discussion.

10.4 Concluding Remarks

In this essay I have examined the philosophical significance of the duality between coalgebras and Boolean-valued algebraic models of Ω-logic. I argued that – as with the property of validity in second-order logic – Ω-logical validity is genuinely logical. I argued, then, that modal and hyperintensional coalgebras, which characterize the hyperintensional profile of Ω-logical consequence, are constitutive of the interpretation of mathematical concepts such as the membership relation.

Chapter 11

A Modal Logic and Hyperintensional Semantics for Gödelian Intuition

11.1 Introduction

'The incompleteness results do not rule out the possibility that there is a theorem-proving computer which is in fact equivalent to mathematical intuition' – Gödel, quoted in Wang (1986: 186).[1]

In his remarks on the epistemology of mathematics, Gödel avails of a notion of non-sensory intuition – alternatively, 'consciousness', or 'phenomenology' (see Gödel, 1961/1995: 383) – as a fundamental, epistemic conduit into mathematical truths.[2] According to Gödel, the defining properties of mathematical intuition include (i) that it either is, or is analogous to, a type

---

[1] Note however that, in the next subsequent sentence, Gödel records scepticism about the foregoing. He remarks: 'But they imply that, in such a – highly unlikely for other reasons – case, either we do not know the exact specification of the computer or we do not know that it works correctly' [Gödel, quoted in Wang (op. cit.)].

[2] Another topic that Gödel suggests as being of epistemological significance is the notion of 'formalism freeness', according to which the concepts of computability, demonstrability, and (ordinal) definability can be specified independently of a background formal language [see Gödel (1946/1990), and Kennedy (2013) for further discussion]. Kennedy notes however that, in his characterizations of demonstrability and definability, Gödel assumes ZFC as his metatheory (op. cit.: 383). Further examination of the foregoing is beyond the scope of the present essay.

of perception (1951/1995: 323; 1953/9-V/1995: 359; 1964/1990: 268); (ii) that it enables subjects to alight upon new axioms which are possibly true (1953/9-III/1995: p. 353, fn. 43; 1953/9-V/1995: p. 361; 1961/1995: pp. 383, 385; 1964/1990: p. 268); (iii) that it is associated with modal properties, such as provability and necessity (1933/1986: 301; 1964/1990: 261); and (iv) that the non-sensory intuition of abstracta such as concepts entrains greater conceptual 'clarification' (1953/9-III/1995: p. 353, fn. 43; 1961/1995: p. 383). Such intuitions are purported to be both *of* abstracta and formulas, as well as to the effect *that* the formulas are true. The distinction between 'intuition-of' and 'intuition-that' is explicitly delineated in Parsons (1980: 145-146), and will be further discussed in Section **2**.[3]

In this paper, I aim to outline the logical foundations for rational intuition, by examining the nature of property (iii).[4] The primary objection to Gödel's approach to mathematical knowledge is that the very idea of ra-

---

[3]For differences between Gödel's conception of intuition, and that of, respectively, Kant, Brouwer, and Hilbert, see Parsons (2008: ch. 5). See Copeland and Shagrir (2013), for differences between Gödel's and Turing's conceptions of intuition.

[4]Parsons (1980) suggests that intuition, which he refers to, further, as the imagination, can be captured by a type of mathematical modality, although he does not provide a modal logic for the modality. Parsons writes: 'We can call the possibility in question *mathematical possibility*; this expresses the fact that we are not thinking of the capabilities of the human organism, and it may even be extraneous to think of this "construction" as an act of the *mind*. The latter construal agrees with the viewpoint of Kant and Brouwer. It is very tempting if we want to say that any string of strokes is *perceptible* or *imaginable*. (It is preferable to reserve these words for tokens, but then one can speak of the *intuitability* of the type.) The idea is that no matter how many times the operation of constructing one more stroke in imagination has been repeated, "we" can still construct one more. However, I think there is really a hidden assumption that there is no constraint on what "we" can perceive beyond the open temporality of these experiences, and some very gross aspects of spatial structure. Kant and Brouwer thought these were contributions of our minds to the way we experience the world. Kant of course thought that we could not know these things *a priori* unless our minds had contributed them. I am not persuaded by this, and in any case I do not want my argument to rest on the notion of *a priori* knowledge' (158-159). Parsons suggests that intuition can be of structures, along with formulas and objects (152). Parsons (2008) is, however, sceptical about the interpretation of mathematical modality as an epistemic modality. Parsons interprets informal provability as an epistemic modality (82), and writes that formal provability is 'prima facie epistemic' and epistemic modalities might 'be related to the fact that mathematical knowledge is characteristically obtained by proof' (81-82). Formulas and the modality in modal-structuralism are argued, however, not to be epistemic, because their truth and the existence of mathematical structures are independent of epistemic states (80-83).

tional intuition is insufficiently constrained.⁵ Subsequent research has thus endeavored to expand upon the notion, and to elaborate on intuition's roles. Chudnoff (2013) suggests, e.g., that intuitions are non-sensory experiences which represent non-sensory entities, and that the justificatory role of intuition is that it enables subjects to be aware of the truth-makers for propositions (p. 3; ch. 7). He argues, further, that intuitions both provide evidence for beliefs as well as serve to guide actions (145).⁶ Bengson (2015: 718-723) suggests that rational intuition can be identified with the 'presentational', i.e., phenomenal, properties of representational mental states – namely, cognitions – where the phenomenal properties at issue are similarly non-sensory; are not the product of a subject's mental acts, and so are 'non-voluntary'; are gradational; and they both 'dispose or incline assent to their contents' and further 'rationalize' assent thereof.⁷ Boghossian countenances intuitions as 'pre-judgmental and pre-doxastic' (2020: 201). He defines intuition as follows: 'An intuition, as I understand it (following many others), is an intellectual *seeming*. An intellectual seeming is similar to a sensory seeming in being a presentation of a proposition's being true; yet dissimilar to it in not having a sensory phenomenology' (200). Boghossian writes: '[T]he idea is that we are equipped with a special capacity ['rational insight' (Boghssian and Williamson, 2020: 31) - D.E.] for non-empirical observation, a capacity whose exercise is capable of yielding insights into necessary truths' [op. cit.; see Williamson (2023: 385)]. He suggests that 'intuitive judgments appear to instantiate a type of *three-step process*: you consider a scenario and a question about it; after sufficient reflection, a particular answer to that

---

⁵See, e.g., Hale and Wright (2002). Wright (2004) provides a vivid articulation of the issue: 'A major — but not the only - problem is that, venerable as the tradition of postulating intuitive knowledge of first principles may be, no-one working within it has succeeded at producing even a moderately plausible account of how the claimed faculty of rational intuition is supposed to work — how exactly it might be constituted so as to be reliably responsive to basic logical validity as, under normal circumstances, vision, say, is reliably responsive to the configuration of middle-sized objects in the nearby environment of a normal human perceiver' (op. cit.: 158).

⁶A similar proposal concerning the justificatory import of cognitive phenomenology – i.e., the properties of consciousness unique to non-sensory mental states such as belief – can be found in Smithies (2013a,b). Smithies prescinds, however, from generalizing his approach to the epistemology of mathematics.

⁷Compare Kriegel (2015: 68), who stipulates that 'making a judgment that p involves a feeling of involuntariness' and 'making a judgment always involves the feeling of mobilizing a concept'.

question comes to seem true to you, either because, as we saw earlier, you work out that it is true, or because, without working it out, it just comes to strike you as true; finally, you endorse this proposition' (201). Nagel (2013) examines an approach to intuitions which construes the latter as a type of cognition. She distinguishes, e.g., between intuition and reflection, on the basis of experimental results which corroborate that they are distinct types of cognitive processing (op. cit.: 226-228). Intuitive and reflective cognitive processing are argued to interact differently with the phenomenal information comprising subjects' working memory stores. Nagel notes that – by contrast to intuitive cognition – reflective cognition 'requires the sequential use of a progression of conscious contents to generate an attitude, as in deliberation' (231). We will here follow Chudnoff (op. cit.), Bengson (op. cit.), and Boghossian (op. cit.) in taking rational intuition to be a non-sensory phenomenal property of mental states such as judgment, and a non-sensory intellectual seeming which presents a proposition as being true.

Rather than target objections to the foregoing essays, the present discussion aims to rebut the primary objection to mathematical intuition alluded to above, by providing a logic for its defining properties. The significance of the proposal is thus that it will make the notion of intuition formally tractable, and might thus serve to redress the contention that the notion is mysterious and ad hoc.

In his (1933/1986) and (1964/1990), Gödel suggests that intuition has a constitutively modal profile. Constructive intuitionistic logic is shown to be translatable into the modal logic, S4, with the rule of necessitation, while the modal operator is interpreted as concerning provability. Mathematical intuition of set-theoretic axioms is, further, purported both to entrain 'intrinsic' justification, and to illuminate the 'intrinsic necessity' thereof. Gödel (1964/1990: 260-261) suggests that intrinsic necessity is a property of axioms which are 'implied' by mathematical concepts, such as that of set. Gödel (1964/1990) writes: 'First of all the axioms of set theory by no means form a system closed in itself, but, quite on the contrary, the very concept of set on which they are based suggests their extension by new axioms which assert the existence of still further iterations of the operation 'set of' ... These axioms show clearly, not only that the axiomatic system of set theory as used today is incomplete, but also that it can be supplemented without arbitrariness by new axioms which only unfold the content of the concept of set explained above' (260-261).[8] Extrinsic justifications are associated, by contrast, with both the evidential probability of propositions, and the 'fruitful' consequences of a mathematical theory subsequent to adopting new axioms (op. cit.: 261, 269).

Following Gödel's line of thought, I aim, in this paper, to provide a modal logic for the notion of 'intuition-that', i.e. intuitions that an axiom might be true.[9] Via correspondence results between modal propositional logic and the bisimulation-invariant fragment of first-order logic, and fixed point modal propositional logic and the bisimulation-invariant fragment of monadic second-order logic [see van Benthem (1983; 1984/2003), Janin and Walukiewicz (1996); and Venema (2014, ms)], a precise translation can be provided between the notion of 'intuition-of', i.e., intuitions of objects and properties, and the modal operators regimenting the notion of 'intuition-that'. I argue that intuition-that can thus be codified by an operator in fixed point modal propositional logic, where the logic is given a dynamic-interpretational interpretation. There is thus a formal correspondence between the operators codifying the notion of 'intuition-that' and the predicates of the second-order logic in which the predicates are interpreted so as to concern the properties of 'intuition-of'. This provides a precise answer to the inquiry advanced by Parsons (1993: 233) with regard to how 'intuition-that' relates to 'intuition-of'.

This result is of relevance to Williamson (2023). Williamson argues that a

---

[8]Note that intrinsic necessity and Gödel's notion of analyticity as true in virtue of the concepts involved might, for Gödel, be convergent notions. With regard to analyticity, Gödel (1972b/1990) writes: 'It may be doubted whether evident axioms in such great numbers (or of such great complexity) can exist at all, and therefore the theorem mentioned might be taken as an indication for the existence of mathematical yes or no questions undecidable for the human mind. But what weighs against this interpretation is the fact that there *do* exist unexplored series of axioms which are analytic in the sense that they only explicate the content of the concepts occurring in them, e.g., the axioms of infinity in set theory, which assert the existence of sets of greater and greater cardinality or of higher and higher transfinite types and which only explicate the content of the general concept of set' (305-306). The convergence in the notions of intrinsic necessity and analyticity for Gödel might have been inspired by his interactions with the logical positivists of the Vienna Circle, who similarly identified analyticity with necessity [See Feferman in Gödel (1986: p. 30, fn. 19), who attributes the foregoing to Parsons; and Gödel (1953/9-III/1995: §§39-40, fns. 41-45)]. Parsons (2014: 146) argues for the identification of the two notions for Gödel, although doesn't draw on the similarity between the definitions of the two notions in the 1964/1990 and 1972b/1990 works quoted above as evidence.

[9]See Parsons (1979-1980; 1983: p. 25, chs.10-11; 2008: 176).

knowledge-first account of the phenomenology of pre-doxastic states is preferable to one which appeals to intuition. My modal logic for rational intuition might provide a counterexample to Williamson's claim that the 'feeling of knowing' (Williamson, 2023: 386) as an account of the phenomenology of pre-doxastic states is preferable to one which appeals to intuition, because epistemic logics don't capture all of the properties which my modal logic and hyperintensional semantics for rational intuition can capture, and such properties are relevant to the phenomenology of pre-doxastic states, namely, the relation between intuition-that and intuition-of. I believe, too, that my modal logic for rational intuition is preferable to Artemov (1995)'s Justification Logic, where justification terms can be applied to mathematical axioms, for similar reasons.[10]

I argue, then, that intuition-that can further be shown to entrain property (iv), i.e. conceptual elucidation, by way of figuring as an interpretational modality which induces the reinterpretation of domains of quantification (see Fine, 2005; 2006) and the reinterpretation of the intensions and hyperintensions of mathematical vocabulary (see Uzquiano, 2015).[11] Fine (op. cit.) has

---

[10] See Artemov and Fitting (2019), for further discussion.

[11] Gödel (1951: §30) writes: 'Namely, it is correct that a mathematical proposition says nothing about the physical or psychical reality existing in space and time, because it is true already owing to the meaning of the terms occurring in it, irrespectively of the world of real things. What is wrong, however, is that the meaning of the terms (that is, the concepts they denote) is asserted to be something man-made and consisting merely in semantical conventions. The truth, I believe, is that these concepts form an objective reality of their own, which we cannot create or change, but only perceive and describe'.

Gödel (1961/?: 8) writes: 'In what manner, however, is it possible to extend our knowledge of these abstract concepts, i.e., to make these concepts themselves precise and to gain comprehensive and secure insight into the fundamental relations that subsist among them, i.e., [into] the axioms that hold for them? Obviously, not, or in any case not exclusively, by trying to give explicit definitions for concepts and proofs for axioms, since for that one obviously needs other undefinable abstract concepts and axioms holding for them. Otherwise one would have nothing from which one could define or prove. The procedure must thus consist, at least to a large extent, in a clarification of meaning that does not consist in giving definitions.

'Now in fact, there exists today the beginning of a science which claims to possess a systematic method for such a clarification of meaning, and that is the phenomenology founded by Husserl. Here clarification of meaning consists in focusing more sharply on the concepts concerned by directing our attention in a certain way, namely, onto our own acts in the use of these concepts, onto our powers in carrying out our acts, etc. But one must keep clearly in mind that this phenomenology is not a science in the same sense as the other sciences. Rather it is [or in any case should be] a procedure or technique that

countenanced both postulational interpretational and postulational dynamic modalities, where the latter are imperatival. I propose to combine interpretational and dynamic modalities. Modalized rational intuition is therefore expressively equivalent to – and can crucially serve as a guide to the interpretation of – mathematical concepts which are formalizable in monadic first- and second-order formal languages.

Gödel writes: 'I said that in math[ematical] reasoning the non-comput[ational] (*i.e.* intuitive) element consists in intuitions of higher & higher infinities. This is quite true but it [sic - D.E.] this situation can be further analysed & then it turns out that they result (as becomes perfectly clear when these things are carried out in detail) from a deeper & deeper self knowledge of reason [to be more precise from a more & more complete rational knowledge of the <u>essence</u> of reason (of which essence the fac[ulty] of self knowledge is itself a constituent part)] [I believe that comput[ational] reason also results from self knowledge of reason but not from essential but factual knowledge] It seems to me that this is a verification (in the field of math[ematics]) of some tenets of idealistic philosophy' (Gödel, 1963, cited in van Atten, 2015: 160).[12] See chapter **5**, for the relation between rational

---

should produce in us a new state of consciousness in which we describe in detail the basic concepts we use in our thought, or grasp other basic concepts hitherto unknown to us'. See Husserl (1921/2001: Investigation VI) and Tieszen (1989; 2005; 2013).

[12] Gödel (1951/1995: pp. 311-312) writes of computational reason resulting from factual self knowledge of reason that: 'Corresponding to [Gödel's Disjunction; see §**2.3** - D.E.] about the incompletability of mathematics, the philosophical implications *prima facie* will be disjunctive too; however, under either alternative they are very decidedly opposed to materialistic philosophy. Namely, if the first alternative holds, this seems to imply that the working of the human mind cannot be reduced to the working of the brain, which to all appearances is a finite machine with a finite number of parts, namely, the neurons and their connections. So apparently one is driven to take some vitalistic viewpoint. On the other hand, the second alternative, where there exist absolutely undecidable mathematical propositions, seems to disprove the view that mathematics is only our own creation; for | the creator necessarily knows all properties of his creatures, because they can't have any others except those he has given to them. So this alternative seems to imply that mathematical objects and facts (or at least *something* in them) exist objectively and independently of our mental acts and decisions, that is to say, [it seems to imply] some form or other of Platonism or "realism" as to the mathematical objects.[Footnote17] For, the empirical interpretation of mathematics,[Footnote18] that is, the view that mathematical facts are a special kind of physical or psychological facts, is too absurd to be seriously maintained (see below). | It is not known whether the first alternative holds, but at any rate it is in good agreement with the opinions of some of the leading men in brain and nerve physiology, who very decidedly deny the possibility | of a purely mechanistic explanation

intuition and self-knowledge, countenanced via the fixed points of the epistemic hyperintensional μ-calculus. I advance a novel account of the iteration of epistemic states based on modal and hyperintensional fixed points in the hyperintensional μ-calculus, instead of modal axiom 4 interpreted epistemically i.e. the KK principle. The modal logic and hyperintensional semantics for Gödelian intuition treated as a mathematical modality relates to epistemic operators interpreted as concerning knowledge, because intuition has dynamic-interpretational modalities and hyperintensional verifiers which effect conceptual elucidation, and to self-knowledge because of fixed points of epistemic states. See §§**8.1** and **9.4**.

In Section **2**, I countenance and motivate a modal logic, which embeds dynamic logic within the modal μ-calculus (see Carreiro and Venema, 2014). I argue that the dynamic interpretational properties of modalized rational intuition provide a precise means of accounting for the manner by which intuition can yield the reinterpretation of quantifier domains and mathematical vocabulary; and thus explain the role of rational intuition in entraining conceptual elucidation. In Section **3**, I examine remaining objections to the viability of rational intuition and provide concluding remarks.

11.2 Modalized Rational Intuition and Conceptual Elucidation

In this section, I will outline the logic for Gödelian intuition. The motivation for providing a logic for rational intuition will perhaps be familiar from treatments of the property of knowledge in formal epistemology. The analogy between rational intuition and the property of knowledge is striking: Just as

---

of psychical and nervous processes.

'As far as the second alternative is concerned, one might object that the constructor need not necessarily know *every* property of what he constructs. For example, we build machines and still cannot predict their behaviour in every detail. But this objection is very poor. For we don't create the machines out of nothing, but build them out of some given material. If the situation were similar in mathematics, then this material or basis for our constructions would be something objective and would force some realistic viewpoint upon us even if certain other ingredients of mathematics were our own creation. *The same would be true if in our creations we were to use some instrument in us but different from our ego (such as "reason" interpreted as something like a thinking machine). For mathematical facts would then (at least in part) express properties of this instrument, which would have an objective existence'.* [My emphasis added - D.E.]

knowledge has been argued to be a mental state (Williamson, 2001; Nagel, 2013b); to be propositional (Stanley and Williamson, 2001); to be factive; and to possess modal properties (Hintikka, 1962; Nozick, 1981; Fagin et al., 1995; Meyer and van der Hoek, 1995), so rational intuition can be argued to be a property of mental states; to be propositional, as recorded by the notion of intuition-that; and to possess modal properties amenable to rigorous treatment in systems of modal logic.

I should like to suggest that the modal logic of Gödelian intuition is the modal μ-calculus (see Carreiro and Venema, 2014).

Suppose that there is a language, L, with the following operations: $\neg$ (negation), $\wedge$ (conjunction), $\vee$ (disjunction), $\Diamond$, $\Box$, $\mu x.$ (least fixed point), $\upsilon x.$ (greatest fixed point); and the following grammar:

$$\phi := \top \mid \bot \mid \phi \wedge \phi \mid \phi \vee \phi \mid \phi \rightarrow \psi \mid \Diamond\phi \mid \Box\phi \mid \mu x.\phi \mid \upsilon x.\phi$$

Let M be a model over the Kripke frame, $\langle W, R \rangle$; so, M = $\langle W, R, V \rangle$. W is a non-empty set of possible worlds. R is a binary relation on W. R[w] denotes the set $\{v \in W \mid (w, v) \in R\}$. V is a function assigning proposition letters, ϕ, to subsets of W.

$\langle M, w \rangle \Vdash \phi$ if and only if $w \in V(\phi)$.
$\langle M, w \rangle \Vdash \neg\phi$ iff it is not the case that $\langle M, w \rangle \Vdash \phi$
$\langle M, w \rangle \Vdash \phi \wedge \psi$ iff $\langle M, w \rangle \Vdash \phi$ and $\langle M, w \rangle \Vdash \psi$
$\langle M, w \rangle \Vdash \phi \vee \psi$ iff $\langle M, w \rangle \Vdash \phi$ or $\langle M, w \rangle \Vdash \psi$
$\langle M, w \rangle \Vdash \phi \rightarrow \psi$ iff, if $\langle M, w \rangle \Vdash \phi$, then $\langle M, w \rangle \Vdash \psi$
$\langle M, w \rangle \Vdash \phi \iff \psi$ iff $[\langle M, w \rangle \Vdash \phi$ iff $\langle M, w \rangle \Vdash \psi]$
$\langle M, w \rangle \Vdash \Diamond\phi$ iff $\langle R_w \rangle(\phi)$
$\langle M, w \rangle \Vdash \Box\phi$ iff $[R_w](\phi)$, with
$\langle R \rangle(\phi) := \{w \in W \mid R[w] \cap \phi \neq \emptyset\}$
$[R](\phi) := \{w \in W \mid R[w] \subseteq \phi\}$
$\langle M, w \rangle \Vdash \mu x.\phi$ iff $\bigcap\{U \subseteq W \mid \phi \subseteq U\}$ (Fontaine, 2010: 18)
$\langle M, w \rangle \Vdash \upsilon x.\phi$ iff $\bigcup\{U \subseteq W \mid U \subseteq \phi\}$ (op. cit.; Fontaine and Place, 2010),
$R_A := \bigcap_{a \in A'} R_a$.

This last clause characterizes the intersection of accessibility relations in the modal logic for rational intuition, such that the pooled intuition can be thought of as a type of distributive property among a set of agents. Interpreting the property as knowledge, Baltag and Smets (2020: 3) write: 'One can now introduce, for each group [A'] ⊆ A, a distributed knowledge operator $K_{A'}\phi$ as the Kripke modality $[R_A]$ ... The logic of distributed knowledge LD has as language the set of all formulas built recursively from atomic formulas p ∈ Prop by using negation ¬ϕ, conjunction $\phi \wedge \psi$, and distributed knowledge operators $K_{A'}\phi$ (for all groups [A'] ⊆ A). The logic of distributed knowledge and common knowledge LDC is obtained by extending the language of LD with common knowledge modalities $[O]_{A'}$. These logics are known to be decidable and have the finite model property. [The following comprise] complete proof systems LDC and LD for these logics:

(I) Axioms and rules of classical propositional logic
(II) S5 axioms and rules for distributed knowledge
(K-Necessitation) From ϕ, infer $K_{A'}\phi$
(K-Distribution) $K_{A'}(\phi \to \psi) \to (K_{A'}\phi \to K_{A'}\psi)$
(Veracity) $K_{A'}\phi \to \phi$
(Pos. Introspection) $K_{A'}\phi \to K_{A'}K_{A'}\phi$
(Neg. Introspection) ¬$K_{A'}\phi \to K_{A'}\neg K_{A'}\phi$
(III) Special axiom for distributed knowledge:
(Monotonicity) $K_{A'}\phi \to K_Q\phi$, for all A' ⊆ Q ⊆ A.'

The notion of pooled rational intuition among a set of agents, as formalized by the logic of distributive knowledge, might be one way formally to account for one aspect of the communitarian conditions on practices, techniques, uses, customs, and institutions which subserve the metasemantic determination and normative status of linguistic contents in the work of the later Wittgenstein (2009).

With regard to the axioms which rational intuition satisfies, K states that $\Box(\phi \to \psi) \to (\Box\phi \to \Box\psi)$; i.e., if one has an intuition that ϕ entails ψ, then if one has the intuition that ϕ then one has the intuition that ψ. 4 states that $\Box\phi \to \Box\Box\phi$; i.e., if one has the intuition that ϕ, then one intuits that one has the intuition that ϕ. Necessitation states that $\vdash\phi \to \vdash\Box\phi$. Because intuition-that is non-factive, we eschew in our modal system of axiom T, which states that $\Box\phi \to \phi$; i.e., one has the intuition that ϕ only if ϕ is the case [see BonJour (1998: 4.4); Parsons (2008: 141)].

In order to account for the role of rational intuition in entraining conceptual elucidation (see Gödel, 1961/1995: 383), I propose to follow Fine (2006) and Uzquiano (2015) in suggesting that there are interpretational modalities associated with the possibility of reinterpreting both domains of quantification (Fine, op. cit.) and the non-logical vocabulary of mathematical languages, such as the membership relation in ZF set theory (Uzquiano, op. cit.).[13]

Fine (2005) has advanced modalities which are postulational, and prescriptive. He (op. cit.) suggests, further, that the postulational modality might be characterized as a program in dynamic logic, whose operations can take the form of 'simple' and 'complex' postulates which enjoin subjects to reinterpret the domains. Uzquiano (op. cit.)'s generalization of the interpretational modality, in order to target the reinterpretation of the intensions of terms such as the membership relation, can similarly be treated.

In propositional dynamic logic (PDL), there are an infinite number of diamonds, with the form $\langle \pi \rangle$.[14] π denotes a non-deterministic program, which in the present setting will correspond to Fine's postulates adumbrated in the following. $\langle \pi \rangle \phi$ abbreviates 'some execution of π from the present state entrains a state bearing information ϕ'. The dual operator is $[\pi]\phi$, which abbreviates 'all executions of π from the present state entrain a state bearing information ϕ'. π^* is a program that executes a distinct program, π, a number of times ≥ 0. This is known as the iteration principle. PDL is similarly closed under finite and infinite unions. This is referred to as the 'choice' principle: If π_1 and π_2 are programs, then so is $\pi_1 \cup \pi_2$. The forth condition is codified by the 'composition' principle: If π_1 and π_2 are programs, then $\pi_1;\pi_2$ is a program (intuitively: the composed program first executes π_1 then π_2). The back condition is codified by Segerberg's induction axiom (Blackburn et al., op. cit: p. 13). Formally, $[\pi^*](\phi \rightarrow [\pi]\phi) \rightarrow (\phi \rightarrow [\pi^*]\phi)$.

Crucially, the iteration principle permits π^* to be interpreted as a fixed point for the equation: x $\iff \phi \vee \Diamond$x. The smallest solution to the equation will be the least fixed point, μx.$\phi \vee \Diamond$x, while the largest solution to the equation, when $\pi^* \vee \infty_\Diamond$, will be the greatest fixed point, νx.$\phi \vee \Diamond$x. Janin and Walukiewicz (op. cit.) have proven that the modal μ-calculus is equivalent to the bisimulation-invariant fragment of monadic second-order

---

[13] A variant strategy is pursued by Eagle (2008). Eagle suggests that the relation between rational intuition and conceptual elucidation might be witnessed via associating the fundamental properties of the entities at issue with their Ramsey sentences; i.e., existentially generalized formulas, where the theoretical terms therein are replaced by second-order variables bound by the quantifiers. However, the proposal would have to be expanded upon, if it were to accommodate Gödel's claim that mathematical intuitions possess a modal profile.

[14] See Blackburn et al., op. cit.: 12-14. Semantics and proof-theory for PDL are outlined in Hoare (1969); Pratt (1976); Goldblatt (1987: ch. 10; 1993: ch. 7) and van Benthem (2010: 158).

logic.

Fine's simple postulational dynamic modality takes, then, the form:

'(i) Introduction. !x.C(x)', which states the imperative to: '[I]ntroduce an object x [to the domain] conforming to the condition C(x)'.

Fine's complex dynamic-postulational modalities are the following:

(ii) 'Composition. Where β and γ are postulates, then so is $\beta;\gamma$. We may read $\beta;\gamma$ as: do β and then do γ; and $\beta;\gamma$ is to be executed by first executing β and then executing γ.
(iii) Conditional. Where β is a postulate and A an indicative sentence, then A $\to$ β is a postulate. We may read A $\to$ β as: if A then do β. How A $\to$ β is executed depends upon whether or not A is true: if A is true, A $\to$ β is executed by executing β; if A is false, then A $\to$ β is executed by doing nothing.
(iv) Universal. Where $\beta(x)$ is a postulate, then so is $\forall x \beta(x)$. We may read $\forall x \beta(x)$ as: do $\beta(x)$ for each x; and $\forall x \beta(x)$ is executed by simultaneously executing each of $\beta(x1)$, $\beta(x2)$, $\beta(x3)$, ..., where x1, x2, x3, ... are the values of x (within the current domain). Similarly for the postulate $\forall F \beta(F)$, where F is a second-order variable.
(v) Iterative Postulates. Where β is a postulate, then so is β^*. We may read β^* as: iterate β; and β^* is executed by executing β, then executing β again, and so on ad infinitum' (op. cit.: 91-92).

Whereas Fine avails of postulational interpretational modalities in order both to account for the notion of indefinite extensiblity and to demonstrate how relatively unrestricted quantification can be innocuous without foundering upon Russell's paradox (op. cit.: 26-30), the primary interest in adopting modal μ-logic with modal operators interpreted as dynamic interpretational modalities as the logic of rational intuition is its capacity to account for dynamic reinterpretations of mathematical vocabulary and quantifier domains; and thus to illuminate how the precise mechanisms codifying modalized rational intuition might be able to entrain advances in conceptual elucidation.

The computational profile of modalized rational intuition can be outlined as follows. In category theory, a category C is comprised of a class Ob(C) of objects and a family of arrows for each pair of objects C(A,B) (Venema, 2007: 421). An **E-coalgebra** is a pair $\mathbb{A} = (A, \mu)$, with A an object of C referred to as the carrier of $\mathbb{A}$, and μ: A → $\mathbf{E}$(A) is an arrow in C, referred to as the transition map of $\mathbb{A}$ (390). A coalgebraic model of deterministic automata can be thus defined (391). An automaton is a tuple, $\mathbb{A} = \langle$A, a_I, C, Ξ, F$\rangle$, such that A is the state space of the automaton $\mathbb{A}$; $a_I \in$A is the automaton's initial state; C is the coding for the automaton's alphabet, mapping numerals to the natural numbers; Ξ: A X C → A is a transition function, and F $\subseteq$ A is the collection of admissible states, where F maps A to $\{1, 0\}$, such that F: A → 1 if a$\in$F and A → 0 if a$\notin$F (op. cit.). The modal profile of coalgebraic automata can be witnessed both by construing the transition function as a counterfactual conditional (see Stalnaker, 1968; Williamson, 2007), and in virtue of the convergence of coalgebraic automata with coalgebraic models of modal logic (407). Let

$\Diamond \phi \equiv \nabla \{\phi, \top\}$,
$\Box \phi \equiv \nabla \varnothing \vee \nabla \phi$ (op. cit.)

$[\![\nabla \Phi]\!] = \{w \in W \mid R[w] \subseteq \bigcup \{[\![\phi]\!] \mid \phi \in \Phi\}$ and $\forall \phi \in \Phi, [\![\phi]\!] \cap R[w] \neq \varnothing\}$ (Fontaine, 2010: 17).

Let an $\mathbf{E}$-coalgebraic modal model, $\mathbb{A} = \langle$S, λ, R[.]$\rangle$, where λ(s) is 'the collection of proposition letters true at s in S, and R[s] is the successor set of s in S', such that $\mathbb{S}, s \Vdash \nabla \Phi$ if and only if, for all (some) successors σ of s$\in$S, $[\Phi, \sigma(s) \in \mathbf{E}(\Vdash_A)]$ (Venema, 2007: 399, 407), with $\mathbf{E}(\Vdash_A)$ a relation lifting of the satisfaction relation $\Vdash_A \subseteq$ S x Φ. Let a functor, $\mathbf{K}$, be such that there is a relation $\overline{\mathbf{K}} \subseteq \mathbf{K}$(A) x $\mathbf{K}$(A') (Venema, 2012: 17)). Let Z be a binary relation s.t. Z $\subseteq$ A x A' and $\wp \overline{Z} \subseteq \wp$(A) x $\wp$(A'), with

$\wp \overline{Z} := \{(X, X') \mid \forall x \in X \exists x' \in X'$ with $(x, x') \in Z \wedge \forall x' \in X' \exists x \in X$ with $(x, x') \in Z\}$ (op. cit.). Then, we can define the relation lifting, $\overline{\mathbf{K}}$, as follows:

$\overline{\mathbf{K}} := \{[(\pi, X), (\pi', X')] \mid \pi = \pi'$ and $(X, X') \in \wp \overline{Z}\}$ (op. cit.), with π a projection mapping of $\overline{\mathbf{K}}$.[15]

Modal automata are defined over a modal one-step language (Venema, 2020: 7.2). With A being a set of propositional variables, the set, Latt(X), of lattice terms over X has the following grammar:

$$\phi ::= \bot \mid \top \mid x \mid \phi \wedge \phi \mid \phi \vee \phi,$$

with x$\in$X and $\phi \in$Latt(A) (op. cit.).

---

[15] The projections of a relation R, with R a relation between two sets X and Y such that R $\subseteq$ X x Y, are

X ←—(π_1) R (π_2)—→ Y such that $\pi_1((x, y)) = x$, and $\pi_2((x, y)) = y$. See Rutten (2019: 240).

The set, 1ML(A), of modal one-step formulas over A has the following grammar:

$$\alpha \in A ::= \bot \mid \top \mid \Diamond\phi \mid \Box\phi \mid \alpha \wedge \alpha \mid \alpha \vee \alpha \text{ (op. cit.)}.$$

A modal P-automaton $\mathbb{A}$ is a quadruple (A, Θ, a_I), with A a non-empty finite set of states, $a_I \in A$ an initial state, and the transition map
Θ: A x $\wp$P → 1ML(A)
maps states to modal one-step formulas, with $\wp$P the power set of the set of proposition letters, P (op. cit.: 7.3).

The philosophical significance of the foregoing is that the modal logic of rational intuition can be interpreted by modal coalgebraic automata. The foregoing accounts for the distinctively computational nature of the modal profile of rational intuition, while the modalities are interpreted as dynamic-interpretational modalities which effect reinterpretations of (hyper-)intensions and quantifier domains and thus effect conceptual elucidation.

11.3 Hyperintensionality

The hyperintensionality of rational intuition can be countenanced in two ways. The first way is via truthmaker semantics, which we here present in a two-dimensional guise. The truthmakers are interpreted as states of intuition. The second way is via topic-sensitivity, which countenances 'two-component' contents comprised of worlds and a mereology of topics i.e. subject matters. In this paper, I will render two-dimensional truthmakers topic-sensitive.

We here propose a topic-sensitive truthmaker semantics for dynamic epistemic logic and dynamic interpretational modalities.

The language of public announcement logic has the following syntax (see Baltag and Renne, 2016):

$$\phi := p \mid \phi \wedge \phi \mid \neg\phi \mid [a]\phi \mid [\phi!]\psi$$

$[a]\phi$ is interpreted as the 'the agent knows ϕ'. $[\phi!]\psi$ is an announcement formula, and is intuitively interpreted as "whenever ϕ is true, ψ is true after we eliminate all not-ϕ possibilities (and all arrows to and from these possibilities)".

Semantics for public announcement logic is as follows:
M, w $\Vdash \phi$ if and only if w$\in$V(ϕ)

M, w ⊩ φ ∧ ψ if and only if M, w ⊩ φ and M, w ⊩ ψ
M, w ⊩ ¬φ if and only if M, w ⊮ φ
M, w ⊩ [a]φ if and only if M, v ⊩ φ for each v satisfying wR$_a$v
M, w ⊩ [φ!]ψ if and only if M, w ⊮ φ or M[φ!], w ⊩ ψ,
where M[φ!] = (W[φ!], R[φ!], V[φ!]) is defined by
W[φ!] := (v∈W | M, v ⊩ φ) (intuitively, "retain only the worlds where φ is true") (op. cit.),
xR[φ!]$_a$y if and only if xR$_a$y (intuitively, "leave arrows between remaining words unchanged"), and
v∈V[φ!](p) if and only if v∈V(p) (intuitively, "leave the valuation the same at remaining worlds").

My proposal is that both announcement formulas, [φ!]ψ, and Fine and Uzquiano's dynamic modalities ought to be rendered hyperintensional, such that the box operators are interpreted as topic-sensitive necessary truthmakers. The dynamic interpretational modalities can just take the clause for A(φ) as above. For announcement formulas, [φ!]ψ if and only if either (i) for all s∈P there is no s'∈P such that s' ⊔ s ∈P and s' ⊢ φ or (ii) M[φ!], s ⊢ ψ, where M[φ!] = ⟨S[φ!], ≤[φ!], v[φ!]⟩ is defined by
S[φ!] := s'∈S | M, s' ⊢ φ (intuitively, retain only states which verify φ),
≤[φ!] if and only if s ≤ s' (intuitively, leave relations between remaining states unchanged), and
v[φ!] if and only if v: Prop → (2^S x 2^S) which assigns a bilateral proposition ⟨φ$^+$, φ$^-$⟩ to φ∈Prop (intuitively, leave the valuation the same at remaining states). States are topic-sensitive such that s in the foregoing abbreviates s ∩ t. Thus topic-sensitive truthmakers, conceived as states of intuition, can receive an interpretation on which they induce reinterpretations of (hyper-)intensions and quantifier domains, and thus effect conceptual elucidation.

11.4 Concluding Remarks

In this note, I have endeavored to outline the modal logic of Gödel's conception of intuition, in order both (i) to provide a formally tractable foundation thereof, and thus (ii) to answer the primary objection to the notion as a viable approach to the epistemology of mathematics. I have been less concerned with providing a defense of the general approach from the array of objections that have been proffered in the literature. Rather, I have sought to demonstrate how the mechanisms of rational intuition can be formally codified and thereby placed on a secure basis.

Among, e.g., the most notable remaining objections, Koellner (2009) has argued that the best candidates for satisfying Gödel's conditions on being intrinsically justified are reflection principles, which state that the height of the hierarchy of sets, V, cannot be constructed 'from below', because, for all true formulas in V, the formulas will be true in a proper initial segment of the hierarchy. Koellner's limitative results are, then, to the effect that reflection principles cannot exceed the use of second-order parameters without entraining inconsistency or triviality (op. cit.). Another crucial objection is that the properties of rational intuition, as a species of cognitive phenomenology, lack clear and principled criteria of individuation. Burgess (2014) notes, e.g., that the role of rational intuition in alighting upon mathematical truths might be distinct from the functions belonging to what he terms a 'heuristic' type of intuition. The constitutive role of the latter might be to guide a mathematician's non-algorithmic insight as she pursues an informal proof. A similar objection is advanced in Cappelen (2012: 3.2-3.3), who argues that – by contrast to the properties picked out by theoretical terms such as 'utility function' – terms purporting to designate cognitive phenomenal properties both lack paradigmatic criteria of individuation and must thereby be a topic of disagreement, in virtue of the breadth of variation in the roles that the notion has been intended to satisfy. Williamson (2020) advances an argument against whether intuitions – as understood by Boghossian (2020) and described above – are fit for purpose in internalist epistemologies, as well as an argument against intuition's theoretical significance in general. As explicated by Boghossian (2020: 222), Williamson's argument against whether intuitions are fit for purpose for the internalist runs as follows:

'1. To have reason to believe in a certain type of mental state, it must either be consciously available or reasonably postulated.

2. Intuitions are not consciously available (mere introspection does not reveal them).

3. Hence, to have reason to believe in intuitions, they must be reasonably postulated.

4. If a type of mental state is postulated, then it is not consciously available.

5. If a type of mental state is to serve an internalist purpose, it must be consciously available.

6. Hence, even if we had reason to believe in intuitions, they would not

be able to serve an internalist purpose' (op. cit.).

Our proposal above would reject premise 2 of Williamson's argument, because intuitions are defined as real phenomenal properties, and thus as being consciously available.

Williamson's argument against the theoretical significance of intuitions in general is that they are 'in danger of justifying bigoted beliefs' (Williamson, 2020: 237). He mentions the possibility of there being e.g. Nazis with consistent belief sets (238), and asks of intuitions: 'Why should they be impervious to all the usual distortions from ignorance and error, bigotry and bias' (213)? Arguably, there are not sufficient constraints on intuition to rule out that intuitions provide prima facie justification even for the most unethical belief sets.

The foregoing issues notwithstanding, I have endeavored to demonstrate that – as with the property of knowledge – an approach to the notion of intuition-that which construes the notion as a modal operator, and the provision thereof with a philosophically defensible logic, might be sufficient to counter the objection that the very idea of rational intuition is mysterious and constitutively unconstrained.

Chapter 12

Hyperintensional Category Theory and Indefinite Extensibility

12.1 Introduction

This essay endeavors to provide a characterization of the defining properties of indefinite extensibility for set-theoretic truths in category theory, i.e. generation and definiteness. The concept of indefinite extensibility is introduced by Russell (1905: 36), and reintroduced by Dummett (1963/1978) in the setting of a discussion of the philosophical significance of Gödel (1931/1986)'s first incompleteness theorem. Gödel's first incompleteness theorem can be characterized as stating that – relative to a coding defined over the signature of first-order arithmetic, a predicate expressing the property of provability, and a fixed point construction – the formula can be defined as not satisfying the provability predicate. Dummett's concern is with the conditions on our grasp of the concept of natural number, given that the latter figures in a formula whose truth appears to be satisfied despite the unprovability – and thus non-constructivist profile – thereof (186). His conclusion is that the concept of natural number 'exhibits a particular variety of inherent vagueness, namely indefinite extensibility', where a 'concept is indefinitely extensible if, for any definite characterisation of it, there is a natural extension of this characterisation, which yields a more inclusive concept; this extension will be made according to some general principle for generating such extensions, and, typically, the extended characterisation will be formulated by reference to the previous, unextended, characterisation' (195-196). Elaborating on the notion of indefinite extensibility, Dummett (1996: 441) redefines the concept as follows: an 'indefinitely extensible concept is one such that, if we can form a definite conception of a totality all of whose members fall under the concept, we can, by reference to that totality, characterize a larger totality all of whose members fall under it'. Subsequent approaches to the notion have endeavored to provide a more precise elucidation thereof, either by providing an explanation of the property which generalizes to an array of examples in number theory and set theory (see Wright and Shapiro, 2006), or by availing of modal notions in order to capture the properties of definiteness and extendability which are constitutive of the concept (see Fine, 2006; Linnebo, 2013; Uzquiano, 2015). However, the foregoing modal characterizations of indefinite extensibility have similarly been restricted to set-theoretic languages. Furthermore, the modal notions that the approaches avail of are taken to belong to a proprietary type which is irreducible to either the metaphysical or the logical interpretations of the operator.

The aim of this essay is to redress the foregoing, by providing a modal characterization of indefinite extensibility in the setting of category theory, rather than number or set theory. One virtue of the category-theoretic, modal definition of indefinite extensibility is that it provides for a robust account of the epistemological foundations of modal approaches to the ontology of mathematics. A second aspect of the philosophical significance of the examination is that it can serve to redress the lacuna noted in the appeal to an irreducible type of mathematical modality, which is argued (i) to be representational, (ii) still to bear on the ontological expansion of domains of sets, and yet (iii) not to range over metaphysical possibilities. By contrast to the latter approach, the category-theoretic characterization of indefinite extensibility is able to identify the functors of coalgebras with elementary embeddings and the modal and hyperintensional properties of set-theoretic, Ω-logical consequence. Functors receive their values relative to two parameters, the first ranging over epistemically possible worlds and states and the second ranging over objective, though not metaphysical, possible worlds and states. Functors thus receive their values in an epistemic two-dimensional semantics.

In Section **2**, I examine the extant approaches to explaining both the property and the understanding-conditions on the concept of indefinite extensibility. In Section **3**, hyperintensional coalgebras are availed of to model

Grothendieck Universes, and I define the notion of indefinite extensibility in the category-theoretic setting. I argue that the category-theoretic definition of indefinite extensibility yields an explanation of the generative property of indefinite extensibility for set-theoretic truths as well as of the notion of definiteness which figures in the definition. I argue that the generative property of indefinite extensibility for set-theoretic truths in category theory is identifiable with the Grothendieck Universe Axiom and the elementary embeddings in Vopenka's principle. I argue, then, that the notion of definiteness for set-theoretic truths can be captured by the role of modal coalgebras in characterizing the modal profile of Ω-logical consequence, where the latter accounts for the absoluteness of mathematical truths throughout the set-theoretic universe. The category-theoretic definition is shown to circumvent the issues faced by rival attempts to define indefinite extensibility via extensional and intensional notions within the setting of set theory. Section 4 provides concluding remarks.

12.2 Indefinite Extensibility in Set Theory

Characterizations of indefinite extensibility have so far occurred in the language of set theory, and have availed of both extensional and intensional resources. In an attempt to define the notion of definiteness, Wright and Shapiro (op. cit.) argue, for example, that indefinite extensibility may be characterized as follows (266).

Formally, let Π be a higher-order concept of type τ. Let P be a first-order concept falling under Π of type τ. Let f be a function from entities to entities of the same type as P. Finally, let X be a sub-concept of P. The notion of definiteness is defined as the limitless preservation of 'Π-hood' by sub-concepts thereof 'under iteration of the relevant operation', f (269).

The foregoing impresses as a necessary condition on the property of indefinite extensibility. Wright and Shapiro note, e.g., that the above formalization generalizes to an array of concepts countenanced in first-order number theory and analysis, including concepts of the finite ordinals (defined by iterations of the successor function); of countable ordinals (defined by countable order-types of well-orderings); of regular cardinals (defined as occurring when the cofinality of a cardinal, κ, is identical to κ); of large cardinals (defined by elementary embeddings from the universe of sets into proper subsets thereof, which specify critical points measured by the ordinals); of real numbers (defined as cuts of sets of rational numbers); and of Gödel numbers (defined as codings by natural numbers of symbols and formulas) (266-267).

As it stands, however, the definition might not be sufficient for the definition of indefinite extensibility, by being laconic about the reasons for which new sub-concepts – comprised as the union of preceding sub-concepts with a target operation defined thereon – are presumed interminably to generate. In response to the above desideratum, concerning the reasons for which indefinite extensibility might be engendered, philosophers have recently appealed to modal properties of the formation of sets. In his (2018a), Linnebo countenances both interpretational and metaphysical modalities, and he argues that the former also satisfy S4.2 i.e. K $[\Box(\phi \to \psi) \to (\Box\phi \to \Box\psi)]$, T $(\Box\phi \to \phi)$, 4 $((\Box\phi \to \Box\Box\phi)$, and G $(\Diamond\Box\phi \to \Box\Diamond\phi)$. Fine (2006) argues, e.g., that – in order to avoid the Russell property when quantifying over all sets – there are postulational interpretational modalities which induce a reinterpretation of quantifier domains, and serve as a mechanism for tracking the ontological inflation of the hierarchy of sets via, e.g., the power-set operation (2007). Reinhardt (1974a) and Williamson (2007) argue that modalities are inter-definable with counterfactuals. While Williamson (2016) argues both that imaginative exercises take the form of counterfactual presuppositions and that it is metaphysically possible to decide propositions which are undecidable relative to the current axioms of extensional mathematical languages such as ZF, Reinhardt (op. cit.) argues that large cardinal axioms and undecidable sentences in extensional ZF can similarly be imagined as obtaining via counterfactual presupposition.

In an examination of the iterative hierarchy of sets, Parsons (1977/1983) notes that the notion of potential infinity, as anticipated in Book 3, ch. 6 of Aristotle's *Physics*, may be codified in a modal set theory by both a principle which is an instance of the Barcan formula (namely, for predicates P and rigidifying predicates Q, $\forall x(Px \iff Qx) \land \Box\{\forall x(\Box Qx \lor \Box\neg Qx) \land \forall R[\forall x\Box(Qx \to Rx) \to \Box\forall x(Qx \to Rx)]\}$, as well as a principle for definable set-forming operations (e.g., unions) for Borel sets of reals $\Box(\forall x)\Diamond(\exists y)[y = x \cup \{x\}]$ (pp. 520, fn. 24; 528). The modal extension is argued to be a property of the imagination, or intuition, and to apply further to iterations of the successor function in an intensional variant of arithmetic (1979-1980).

Hellman (1990) develops the program intimated in Putnam (1967), and thus argues for an eliminativist, modal approach to mathematical structuralism as applied to second-order plural ZF. The possibilities at issue are taken to be logical – concerning both the consistency of a set of formulas as well as the possible satisfaction of existential formulas – and he specifies, further, an 'extendability principle', according to which 'every natural model [of ZF] has a proper extension' (421).

Extending Parsons' project, Linnebo (2009, 2013) avails of a second-order, plural modal set theory in order to account for both the notion of potential infinity as well as the notion of definiteness. Similarly to Parsons' use of the Barcan formula (i.e., $\forall \Box \phi \rightarrow \Box \forall \phi$), Linnebo's principle for the foregoing is as follows: $\forall u(u < xx \rightarrow \Box \phi) \rightarrow \Box \forall u(u < xx \rightarrow \phi)$ (2013: 211). He argues, further, that the logic for the modal operator is S4.2. Studd (2013) examines the notion of indefinite extensibility by availing of a bimodal temporal logic. Uzquiano (2015)'s approach to defining the concept of indefinite extensibility argues that the height of the cumulative hierarchy is in fact fixed, and that indefinite extensibility can similarly be captured via the use of modal operators in second-order plural modal set theory. The modalities are taken to concern the possible reinterpretations of the intensions of the non-logical vocabulary – e.g., the set-membership relation – which figures in the augmentation of the theory with new axioms and the subsequent climb up the fixed hierarchy of sets (see Gödel, 1964/1990).

Chapter **8** proffers a novel epistemology of mathematics, based on an application of the epistemic interpretation of two-dimensional semantics in set-theoretic languages to the values of large cardinal axioms and undecidable sentences. Modulo logical constraints such as consistency and absoluteness in the extensions of ground models of the set-theoretic universe, the epistemic possibility that an undecidable proposition receives a value may serve, then, as a guide to the metaphysical possibility thereof. Finally, chapter **10** argues that the modal and hyperintensional profiles of validity, in the Ω-logic defined in Boolean-valued models of set-theory, can be captured by coalgebras, and provides a necessary condition on the formal grasp of the concept of 'set'.

The foregoing accounts of the metaphysics and epistemology of indefinite extensibility are each defined in the languages of number and set theory. In the following section, I examine the nature of indefinite extensibility in the setting of category theory, instead. One aspect of the philosophical significance of the examination is that it can serve to provide an analysis of the mathematical modality at issue. By contrast to Hellman's approach, which takes the mathematical modality at issue to be logical (see Field, 1989: 37; Rayo, 2013), and Fine (op. cit.)'s approach, which takes the mathematical modality to be either interpretational or dynamic, I argue in the following sections that the mathematical modality can be captured by mappings in coalgebras, relative to two parameters, the first ranging over epistemically possible worlds and the second ranging over objective possible worlds.

If one prefers hyperintensional semantics to possible worlds semantics – in order e.g. to avoid the situation in intensional semantics according to which all necessary formulas express the same proposition because they are true at all possible worlds – one can avail of the foregoing epistemic two-dimensional truthmaker semantics, which specifies a notion of exact verification in a state space and where states are parts of whole worlds (Fine 2017a,b; Hawke and Özgün, 2023).

In the section that follows, I examine the properties of indefinite extensibility in the category-theoretic setting.

12.3 Hyperintensional Coalgebras and Indefinite Extensibility

This section examines the reasons for which category theory provides a more theoretically adequate setting in which to define indefinite extensibility than do competing approaches such as the Neo-Fregean epistemology of mathematics. According, e.g., to the Neo-Fregean program, concepts of number in arithmetic and analysis are definable via implicit definitions which take the form of abstraction principles. Abstraction principles specify biconditionals in which – on the left-hand side of the formula – an identity is taken to hold between numerical term-forming operators from entities of a type to abstract objects, and – on the right-hand side of the formula – an equivalence relation on such entities is assumed to hold.

In the case of cardinal numbers, the relevant abstraction principle is referred to as Hume's principle, and states that, for all x and y, the number of the x's is identical to the number of the y's if and only if the x's and the y's can be put into a one-to-one correspondence, i.e., there is a bijection from the x's onto the y's.

- $\forall \mathbf{A} \forall \mathbf{B}[\text{Nx: }\mathbf{A} = \text{Nx: }\mathbf{B} \equiv \exists R[\forall x[\mathbf{A}x \to \exists y(\mathbf{B}y \wedge Rxy \wedge \forall z(\mathbf{B}z \wedge Rxz \to y = z))] \wedge \forall y[\mathbf{B}y \to \exists x(\mathbf{A}x \wedge Rxy \wedge \forall z(\mathbf{A}z \wedge Rzy \to x = z))]]]$.

Abstraction principles for the concepts of other numbers have further been specified.

The abstractionist program faces several challenges, including whether conditions can be delineated for the abstraction principles, in order for the principles to avoid entraining inconsistency;[1] whether unions of abstraction principles can avoid the problem of generating more abstracts than concepts (Fine, 2002); and whether abstraction principles can be specified for mathematical entities in branches of mathematics beyond first and second-order arithmetic (see Boolos, 1997; Hale, 2000; Shapiro, op. cit.; and Wright, 2000). I will argue that the last issue – i.e., being able to countenance definitions for the entities and structures in branches of mathematics beyond first and second-order arithmetic – is a crucial desideratum, the satisfaction of which remains elusive for the Neo-Fregean program while yet being satisfiable and thus adducing in favor of the hyperintensional platonist approach that is outlined in what follows.

One issue for the attempt, along abstractionist lines, to provide an implicit definition for the concept of set is that doing so with an unrestricted comprehension principle yields a principle identical to Frege (1893/2013)'s Basic Law V; and thus – in virtue of Russell's paradox – entrains inconsistency. However, two alternative formulas can be defined, in order to provide a suitable restriction to the inconsistent abstraction principle. The first, conditional principle states that $\forall F,G[[Good(F) \vee Good(G)] \rightarrow [\{x \mid Fx\} = \{Gx\} \iff \forall x(Fx \iff Gx)]]$. The second principle is an unconditional version of the foregoing, and states that $\forall F,G[\{x \mid Fx\} = \{Gx\} \iff [Good(F) \vee Good(G) \rightarrow \forall x(Fx \iff Gx)]]$. Following von Neumann (1925/1967: 401-402)'s suggestion that Russell's paradox can be avoided with a restriction of the set comprehension principle to one which satisfies a constraint on the limitation of its size, Boolos (1997) suggests that the 'Good' predicate in the above principles is intensionally isomorphic to the notion of smallness in set size, and refers to the principle as New V. However, New V is insufficient for deriving all of the axioms of ZF set theory, precluding, in particular, both the axioms of infinity and the power-set axiom (see Wright and Hale, 2005: 193). Further, there are other branches of number theory for which it is unclear whether acceptable abstraction principles can be specified. Wiles' proof of Fermat's Last Theorem (i.e., that, save for when one of the variables is 0, the Diophantine equation, $x^n = y^n = z^n$, has no solutions when n > 2; see Hardy and Wright, 1979: 190) relies, e.g., on both invariants and

---

[1] See Hodes (1984); Hazen (1985); Boolos (1990); Heck (1992); Fine (2002); Weir (2003); Cook and Ebert (2005); Linnebo and Uzquiano (2009); Linnebo (2010); and Walsh (2016).

Grothendieck Universes in cohomological number theory (see McLarty, 2009: 4).

The foregoing issues with regard to the definability of abstracta in number theory, algebraic geometry (McLarty, op. cit.: 6-8), set theory, et al., can be circumvented in the category-theoretic setting; and in particular by coalgebras. In the remainder of this section, I endeavor to demonstrate how modal coalgebras are able to countenance two of the fundamental properties of indefinite extensibility. The first concerns the property of generation. The second includes the properties of intensional and extensional definiteness.

A labeled transition system is a tuple, LTS, comprised of a set of worlds, M; a valuation, V, from M to its power set, $\wp(M)$; and a family of accessibility relations, R. So LTS = $\langle M, V, R \rangle$ (see Venema, 2012: 7). Coalgebras can be thus characterized. Let a category C be comprised of a class Ob(C) of objects and a family of arrows for each pair of objects C(A, B) (Venema, 2007: 421). A functor from a category C to a category D, **E**: C → D, is an operation mapping objects and arrows of C to objects and arrows of D (422). An endofunctor on C is a functor, **E**: C → C (op. cit.).

A **E**-coalgebra is a pair $\mathbb{A} = (A, \mu)$, with A an object of C referred to as the carrier of $\mathbb{A}$, and μ: A → **E**(A) is an arrow in C, referred to as the transition map of $\mathbb{A}$ (390). A Kripke coalgebra combines V and R into a Kripke functor, σ_s; i.e. the set of binary morphisms from M to $\wp(M)$ (op. cit.: 7-8). Thus, for an s∈M, $\sigma(s) := [\sigma_V(s), \sigma_R(s)]$ (op. cit.). $\sigma(s)$ can be interpreted both epistemically and metaphysically. Thus, $\sigma(s)^{sol,wnl}$. Satisfaction for the system is defined inductively as follows: For a formula ϕ defined at a state, s, in M,

$[\![\phi]\!]^M = V(s)$ [2]
$[\![\neg\phi]\!]^M = S - V(s)$
$[\![\bot]\!]^M = \varnothing$
$[\![T]\!]^M = M$
$[\![\phi \vee \psi]\!]^M = [\![\phi]\!]^M \cup [\![\psi]\!]^M$
$[\![\phi \wedge \psi]\!]^M = [\![\phi]\!]^M \cap [\![\psi]\!]^M$
$[\![\Diamond\phi]\!]^M = \langle R \rangle [\![\phi]\!]^M$
$[\![\Box\phi]\!]^M = [R][\![\phi]\!]^M$, with
$\langle R \rangle(\phi) := \{s \in S \mid R[s] \cap \phi \neq \varnothing\}$ and
$[R](\phi) := \{s \in S \mid R[s] \subseteq \phi\}$ (9).[3]

---
[2] Equivalently, M,s ⊩ ϕ if s∈V(ϕ) (9).
[3] Hamkins and Linnebo (2022) argue that the modal logic of Grothendieck potentialism

Kripke coalgebras are the dual representations of Boolean-valued models of the Ω-logic of set theory (see Elohim, 2019). Modal coalgebras are able, then, to countenance the constitutive conditions of indefinite extensibility. Modal coalgebras are capable, e.g., of defining both the generative property of indefinite extensibility, as well as the notion of definiteness which figures therein. Further, the category-theoretic definition of indefinite extensibility is arguably preferable to those advanced in the set-theoretic setting, because modal and hyperintensional coalgebras can account for both the modal and hyperintensional profile and the epistemic tractability of Ω-logical consequence.

The *generative* property of indefinite extensibility for set-theoretic truths is captured by the Grothendieck Universe Axiom and the elementary embeddings in Vopenka's principle, j: A → B, $\phi\langle a_1, \ldots, a_n\rangle$ in A if and only if $\phi\langle j(a_1), \ldots, j(a_n)\rangle$ in B. The large cardinals countenanced by Grothendieck Universes are restricted to strongly inaccessible cardinals. A cardinal κ is regular if the cofinality of κ is identical to κ. Uncountable regular limit cardinals are weakly inaccessible (op. cit.). A strongly inaccessible cardinal is regular and has a strong limit, such that if $\lambda < \kappa$, then $2^\lambda < \kappa$ (Kanamori, 2012: 361). Indefinite extensibility follows from the Universe Axiom which states that for each set, the set belongs to a Grothendieck Universe such that the cardinality of inaccessible cardinals is unbounded. However, functors interpreted as elementary embeddings in category theory are such that Vopenka's principle can be satisfied, yielding Woodin cardinals. Vopenka's principle is secured via elementary embeddings between first-order structures interpreted as categories.

The notion of *definiteness* for set-theoretic truths is captured by the role of modal and hyperintensional coalgebras in characterizing the modal and hyperintensional profiles of Ω-logical validity. (See chapter **10**.)

The absoluteness of set-theoretic truths in virtue of Ω-logical validity corresponds to a type of objective – perhaps maximal, and thus metaphysical – necessity. This characterization of definiteness for set-theoretic truths would thus satisfy Linnebo (2018b)'s characterizations of both intensional and extensional definiteness. According to Linnebo (op. cit.: 203), a concept is intensionally definite if it 'has a sharp application condition', and extensionally definite if 'it has a fixed extension in all the circumstances in which the

---

has a lower bound of S4.3 [.3: $(\Diamond\phi \wedge \Diamond\psi) \rightarrow \Diamond[(\phi \wedge \Diamond\psi) \vee (\psi \wedge \Diamond\phi)]$ and an upper bound of S5 (KTE; E: $\neg\Box\phi \rightarrow \Box\neg\Box\phi$).

concept is available'. Linnebo (2018a: 210) argues that extensional definiteness is satisfied if set-membership is rigid: $\exists xx \Box \forall u[u \prec xx \iff \phi(u)]$, with $\prec$ a plurality membership relation. With regard to extensional definiteness as secured via an absoluteness condition, Koellner (2010) writes of the invariance property of Ω-logical consequence: '[T]he logical consequence relation is not perturbed by passing to a generic extension. If we think of the models $V^{\mathbb{B}}$ as possible worlds, then this is tantamount to saying that the logical consequence relation is invariant across the possible worlds'.

Whereas the Neo-Fregean approach to comprehension for the concept of set relies on an unprincipled restriction of the size of the universe in order to avoid inconsistency, and one according to which the axioms of ZF still cannot all be recovered, modal coalgebras provide a natural means for defining the minimal conditions necessary for formal grasp of the concept set. The category-theoretic definition of indefinite extensibility is sufficient for uniquely capturing both the generative property as well as the notion of definiteness which are constitutive of the concept. The category-theoretic definition of indefinite extensibility avails of a notion of mathematical modality which captures both the epistemic property of possible interpretations of quantifiers, as well as the objective, circumstantial property of set-theoretic ontological expansion.

One objection to the two-dimensional characterization of indefinite extensibility might be that modality is itself indefinitely extensible, such that it would be circular to define indefinite extensibility via modal notions.[4] In response, the modal characterization of indefinite extensibility can be illuminating, even if modality is itself indefinitely extensible. The explanations required for each phenomenon are distinct. Thus, indefinite extensibility can have interpretational and objective modalities, expressing epistemically possible reinterpretations of quantifier domains and objectively possible ontological expansion. However, the explanation for modality being itself indefinitely extensible can proceed via, e.g., Fritz (2017)'s puzzle. Fritz proffers two modal principles which are inconsistent. The first states that $\Box \uparrow_1 \Diamond \exists x (Ax \land \downarrow_1 \neg Ax)$ (op. cit.: 551). The second principle states that $\Diamond \uparrow_1 \Box \forall x (Ax \to \downarrow_1 Ax)$ (550). The first principle claims that possibilia are indefinitely extensible (557; see also Rayo, 2020).[5] So the indefinite extensibility of modality would not have to be explained by further modalities, but is rather explained

---

[4]Thanks to Justin Clarke-Doane for the objection.

[5]Fritz suggests that the notion of metaphysical possibility might not be in good standing, in light of the inconsistency of the two foregoing principles. By contrast, I take the

by a principle which shows that it is always possible to expand the domain of possibilia.

12.4 Concluding Remarks

In this essay, I outlined a number of approaches to defining the notion of indefinite extensibility, each of which restricts the scope of their characterization to set-theoretic languages. I endeavored, then, to define indefinite extensibility in the setting of category-theoretic languages, and examined the benefits accruing to the approach, by contrast to the extensional and modal approaches pursued in ZF.

The extensional definition of indefinite extensibility in ZF was shown to be insufficient for characterizing the generative property in virtue of which number-theoretic concepts are indefinitely extensible. The generative property of indefinite extensibility for set-theoretic truths in the category-theoretic setting was argued, by contrast, to be identifiable with the elementary embeddings by which large cardinal axioms can be specified. The modal definitions of indefinite extensibility in ZF were argued to be independently problematic, in virtue of endeavoring simultaneously to account for the epistemic properties of indefinite extensibility – e.g., possible reinterpretations of quantifier domains and mathematical vocabulary – as well as the objective properties of indefinite extensibility – i.e., the ontological expansion of the target domains, without providing an account of how this might be achieved. Coalgebraic functors can secure these two dimensions, by having both epistemic and objective interpretations. The mappings are interpreted both epistemically and objectively, such that the mappings are defined relative to two parameters, the first ranging over epistemically possible worlds and the second ranging over objective possible worlds. The mappings thus receive their values in a hyperintensional epistemic two-dimensional semantics.

Finally, against the Neo-Fregean approach to defining concepts of number, and the limits thereof in the attempt to define concepts of mathematical objects in other branches of mathematics beyond arithmetic, I demonstrated how – by characterizing the modal profile of Ω-logical validity and thus the absoluteness of mathematical truths concerning large cardinals throughout the set-theoretic universe – modal coalgebras are capable of capturing the

modalities at issue to be objective, though not maximal and thus not metaphysical, where non-maximal objective modalities can be indefinitely extensible.

notion of definiteness within the concept of indefinite extensibility for set-theoretic truths in category theory.

Chapter 13

Truth, Modality, and Paradox

13.1 Introduction

This essay targets a series of potential issues for the discussion of, and resolution to, the alethic paradoxes advanced by Scharp, in his *Replacing Truth* (2013).

In Section 2, Scharp's replacement strategy is outlined, and his semantic model is described in detail.

In Section 3, novel extensions of Scharp's theory to the preface paradox; to the property version of Russell's paradox in the setting of unrestricted quantification; to probabilistic self-reference; and to the sorites paradox are examined.

Section 4 examines six crucial issues for the approach and the semantic model that Scharp proffers. The six issues target the following points of contention:

(i) The status of revenge paradoxes in Scharp's theory;

(ii) Whether a positive theory of validity might be forthcoming on Scharp's approach, given that Scharp expresses sympathy with treatments on which – in virtue of Curry (1942)'s paradox – validity is not identical with necessary truth-preservation;

(iii) The failure of compositionality in Scharp's Theory of Ascending and Descending Truth (ADT) and whether the theory is not, then, in tension with natural language semantics. The foregoing might be pernicious, given Scharp's use of consistency with natural language semantics as a condition for the success of approaches to the paradoxes. A related issue concerns whether it is sufficient to redress the failure of compositionality by availing of hybrid conditions which satisfy both Ascending and Descending Truth;

(iv) Whether ADT can generalize, in order to account for other philosophical issues that concern indeterminacy;

(v) Whether Descending Truth and Ascending Truth can countenance the manner in which truth interacts with objectivity. It is unclear, e.g., how the theorems unique to each of Descending Truth and Ascending Truth – respectively, T-Elimination and T-Introduction – can capture distinctions between the reality of the propositions mapping to 1 in mathematical inquiry, by contrast to propositions – about humor, e.g. – whose mapping to 1 might be satisfied by more deflationary conditions; and

(vi) Whether the replacement strategy in general and ADT in particular can be circumvented, in virtue of approaches to the alethic paradoxes which endeavor to resolve them by targeting constraints on the contents of propositions and the values that they signify.

Section 5 provides concluding remarks.

13.2 Scharp's Replacement Theory

Scharp avers that two main alethic principles target the use of the predicate as a device of endorsement and as a device of rejection. When the truth predicate is governed by (T-Out), then it can be deployed in the guise of a device of endorsement. When the truth predicate is governed by (T-In), then it can be deployed in the guise of a device of rejection.

Scharp's theory aims to replace truth with two distinct concepts. His explicit maneuver is to delineate the two, smallest inconsistent subsets of alethic principles; and then to pair one of the subsets with one of the replacement concepts, and the other subset with the second replacement concept.

Thus, one replacement concept will be governed by (T-In) and not by (T-Out); and the second replacement concept will be governed by (T-Out) and not by (T-In).

Scharp refers to one of his two, preferred replacement concepts as *Descending Truth* (henceforth DT). DT is governed by (T-Out).

Scharp refers to the second of his two, preferred replacement concepts as *Ascending Truth* (henceforth AT). AT is governed by (T-In).

In his "Syntactical Treatments of Modality, with Corollaries on Reflexion Principles and Finite Axiomatizability" (1963), Montague proved that, for any predicate $H(x)$, the following conditions on the predicate are inconsistent.

Montague (1963)'s Lemma 3:
(i) All instances of $H(\phi) \to \phi$ are theorems.
(ii) All instances of $H[H(\phi) \to \phi]$ are theorems.
(iii) All instances of $H(\phi)$, where ϕ is a logical axiom, are theorems.
(iv) All instances of $H(\phi \to \psi) \to [H(\phi) \to H(\psi)]$ are theorems.
(v) Q – i.e., Robinson Arithmetic – is a subtheory.[1]

Scharp notes that Montague's conditions target only Predicate-Elimination, and are thus apt for governing DT.

Scharp argues that (v) is necessary, in order for languages that express the theory to refer to their own sentences. Condition (i) is necessary, because it captures (T-Out). Condition (ii) is necessary, because denying iterations of DT entrains a version of the revenge paradox.

Thus, either Condition (iii) or Condition (iv) must be rejected. Condition (iii) states that all tautologies are Descending True. Condition (iv) is an instance of closure. In virtue of considerations pertaining to validity (see Section 3), Scharp is impelled to reject (iv), s.t. DT cannot satisfy closure (151).

13.2.1 Properties of DT and AT

Scharp argues that the alethic principles, DT and AT, ought to include the following.
DT ought to satisfy:
[¬-Exc: $D(\neg\phi) \to \neg D(\phi)$];
[∧-Exc: $D(\phi \land \psi) \to D(\phi) \land D(\psi)$; and
[∨-Imb: $D(\phi) \lor D(\psi) \to D(\phi \lor \psi)$.
However, DT is not governed by:
[¬-Imb: $\neg T(\phi) \to T(\neg\phi)$; nor by
[∨-Exc: $T(\phi \lor \psi) \to T(\phi) \lor T(\psi)$.
AT ought to satisfy:
[¬-Imb: $\neg A(\phi) \to A(\neg\phi)$;
[∧-Exc: $A(\phi \land \psi) \to A(\phi) \land A(\psi)$; and

---

[1] For contemporary developments of predicate treatments of modality and knowledge, see Halbach et al (2003); Halbach and Welch (2009); Stern (2016); and Koellner (2016; 2018a,b).

[∨-Imb: $A(\phi) \vee A(\psi) \rightarrow A(\phi \vee \psi)$].
However, AT is not governed by:
[¬-Exc: $T(\neg\phi) \rightarrow \neg T(\phi)$]; nor by
[∧-Imb: $T(\phi) \wedge T(\psi) \rightarrow T(\phi \wedge \psi)$].
Scharp argues, further:
–that classical tautologies are Descending True;
–that the axioms governing the syntax of the theory are Descending True;
–that the axioms of PA are Descending True, in order to induce self-reference via Gödel-numbering; and
–that the axioms of the theories for both AT and DT are themselves Descending True (152).

DT takes classical values, and, in Scharp's theory, there are no restrictions on the language's expressive resources. This is problematic, because 'a' := '¬A(x)' and 'd' := '¬D(x)' can be countenanced in the language, and thereby yield contradictions:

Because $A(x)$ is governed by (T-In), 'a' entails that $A(a)$, although a states of itself that $\neg A(x)$. Contradiction.

Because $D(x)$ is governed by (T-Out), 'd' entails that replacing 'd' for x in 'd' is not a descending truth, i.e., $\neg D(d)$]. So – by condition (ii) – 'D[D(x)] → x' entails that it is not a descending truth that replacing 'd' for x in 'd' is not a descending truth [i.e., $\neg D(\neg d)$].

Thus, Scharp concedes that there must be problematic sentences in the language for his theory, s.t. both the sentences and their negations are Ascending True, and s.t. the sentences and their negations are not Descending True (op. cit.).

Scharp endeavors to block the foregoing, by suggesting that DT can be governed by both unrestricted (T-Out), as well as a restricted version of (T-In). Similarly, AT can be governed by both unrestricted (T-In), as well as a restricted version of (T-Out).

To induce the foregoing, Scharp introduces a 'Safety' predicate, $S(x)$. A sentence ϕ is *safe* if and only if ϕ is either (DT and AT) or not AT.
Thus,
$S(\phi) \wedge \phi \rightarrow D(\phi)$; and
$S(\phi) \wedge A(\phi) \rightarrow \phi$.
A sentence ϕ is *unsafe* if and only if ϕ is AT and not DT:
$S(\phi) \iff D(\phi) \vee \neg A(\phi)$.
From which it follows that:
$\neg S(\phi) \iff \neg D(\phi) \wedge A(\phi)$, s.t.

A(ϕ) → ¬D(ϕ);
D(ϕ) → A(ϕ);
¬A(ϕ) → ¬D(ϕ);
¬∃ϕ[D(ϕ) ∧ ¬A(ϕ)] (153).
Scharp avers too that AT and DT are duals. Thus,
D(ϕ) ⟺ ¬A¬(ϕ); and
A(ϕ) ⟺ ¬D¬(ϕ) (152).

13.2.2 Scharp's Theory: ADT

Scharp's Theory is referred to as ADT. The necessary principles comprising ADT are as follows (see p. 154):

- Descending Truth

(D1): D(ϕ) → ϕ
(D2): D(¬ϕ) → ¬D(ϕ)
(D3): D(ϕ ∧ ψ) → [D(ϕ) ∧ D(ψ)]
(D4): [D(ϕ) ∨ D(ψ)] → D(ϕ ∨ ψ)
(D5): If ϕ is a classical tautology, then D(ϕ)
(D6): If ϕ is a theorem of PA, then D(ϕ)
(D7): If ϕ is an axiom of ADT, then D(ϕ).

- Ascending Truth

(A1): ϕ → A(ϕ)
(A2): ¬A(ϕ) → A(¬ϕ)
(A3): [A(ϕ) ∨ A(ψ)] → A(ϕ ∨ ψ)
(A4): A(ϕ ∧ ψ) → [A(ϕ) ∧ A(ψ)]
(A5): If ϕ maps to the falsum constant, then ¬A(ϕ)
(A6): If ϕ negates an axiom of PA, then ¬A(ϕ)

- Transformation Rules

(M1): D(ϕ) ⟺ ¬A¬(ϕ)
(M2): S(ϕ) ⟺ D(ϕ) ∨ ¬A(ϕ)
(M3): S(ϕ) ∧ ϕ → D(ϕ)
(M4): A(ϕ) ∧ S(ϕ) → ϕ

(E1): If s and t are terms; s = t; and replacing s with t in a sentence p yields a sentence q; then D(p) $\iff$ D(q)

(E2): If s and t are terms; s = t; and replacing s with t in a sentence p yields a sentence q; then A(p) $\iff$ A(q)

(E3): If s and t are terms; s = t; and replacing s with t in a sentence p yields a sentence q; then S(p) $\iff$ S(q).

13.2.3 Semantics for ADT

Scharp advances a combination of relational semantics for a non-normal modal logic, as augmented by a neighborhood semantics. (A modal logic is normal if and only if it includes axiom K and the rule of Necessitation; respectively '$\Box[\phi \to \psi] \to [\Box\phi \to \Box\psi]$' and '$\vdash \phi \to \vdash \Box\phi$'.) He refers to this as *xeno semantics*.

A model, M, of ADT is a tuple, $\langle D, W, R, I \rangle$, where D is a non-empty domain of entities constant across worlds, W denotes the space of worlds, R denotes a relation of accessibility on W, and I is an interpretation-function mapping subsets of D to W. The clauses for defining truth in a world in the model are familiar:

$\langle M,w \rangle \Vdash \phi$ iff $w \in V(\phi)$

$\langle M,w \rangle \Vdash \neg\phi$ iff it is not the case that $\langle M,w \rangle \Vdash \phi$

$\langle M,w \rangle \Vdash \phi \wedge \psi$ iff $\langle M,w \rangle \Vdash \phi$ and $\langle M,w \rangle \Vdash \psi$

$\langle M,w \rangle \Vdash \phi \vee \psi$ iff $\langle M,w \rangle \Vdash \phi$ or $\langle M,w \rangle \Vdash \psi$

$\langle M,w \rangle \Vdash \phi \to \psi$ iff, if $\langle M,w \rangle \Vdash \phi$, then $\langle M,w \rangle \Vdash \psi$

$\langle M,w \rangle \Vdash \phi \iff \psi$ iff $[\langle M,w \rangle \Vdash \phi$ iff $\langle M,w \rangle \Vdash \psi]$

$\langle M,w \rangle \Vdash \Box\phi$ iff $\forall w'[$If $R(w,w')$, then $\langle M,w' \rangle \Vdash \phi]$

$\langle M,w \rangle \Vdash \Diamond\phi$ iff $\exists w'[R(w,w')$ and $\langle M,w' \rangle \Vdash \phi]$

Scharp augments his relational semantics with a neighborhood semantics. $M = \langle D, W, R, I \rangle$ is thus enriched with a neighborhood function, N, which maps sets of subsets of W to each world in W.

Necessity takes then the revised clause:

$\langle M,w \rangle \Vdash \Box\phi$ iff $\exists X \in N(w) \forall w'[\langle M,w' \rangle \Vdash \phi \iff w' \in X]$

Possibility takes the revised clause:

$\langle M,w \rangle \Vdash \Diamond\phi$ iff $\neg[\exists X \in N(w) \forall w'[\langle M,w' \rangle \Vdash \neg\phi \iff w' \in X]]$

Let L be a language with Boolean connectives, and the operators $\Box$, $\Diamond$, and Σ. $\Box$ is the Descending Truth operator. $\Diamond$ is the Ascending Truth operator. Σ is the Safety operator. A *xeno model* $M = \langle F, R, N, V \rangle$ where F denotes a xeno frame, R is an accessibility relation on wff in L, N is a function from W to 2^{2^w}, and V is an assignment-function from wff in L to the values $\{0, 1\}$.

Truth in a world is defined inductively as above.
The operators take the following clauses:
Descending Truth:
$\langle M,w \rangle \Vdash \Box\phi$ iff $\forall w' \in W[R_\phi(w,w') \rightarrow \exists X \in N(w') \forall v \in W[\langle M,v \rangle \Vdash \phi \iff v \in X]$

Ascending Truth:
$P(\phi)$ denotes the neighborhood structure – i.e., the set of subsets of worlds – at which ϕ is true.
$\langle M,w \rangle \Vdash \Diamond\phi$ iff $\neg[\forall w' \in W[R_{\neg\phi}(w,w') \rightarrow P(\neg\phi) \in N(w')]]$
Safety:
$\langle M,w \rangle \Vdash \Sigma\phi$ iff $\forall w' \in W[R_\phi(w,w') \rightarrow P(\phi) \in N(w')] \vee \exists w' \in W[R_{\neg\phi}(w,w') \wedge P(\neg\phi) \neg\in N(w')]$

A reflexive and co-reflexive xeno frame is equivalent to a neighborhood frame:
(Reflexivity) $\forall\phi \forall w \in W[R_\phi(w,w)] \wedge$
(Co-reflexivity) $\forall\phi \forall w \in W \forall w' \in W[R_\phi(w,w') \rightarrow w = w']$
A sentential xeno frame is *acceptable* iff
(i) $\forall w \in W\ N(w) \neq \emptyset$
(ii) $\forall w \in W\ \forall X \in N(w)\ X \neq \emptyset$
(iii) $\forall w \in W\ \forall X \in N(w)\ w \in X$
(iv) $\forall\phi \in L \forall w \in W[R_\phi(w,w)]$
(v) if ϕ and ψ are of the same syntactic type, then $R_\phi = R_\psi$

A constant-domain xeno frame is a tuple, $F = \langle W, N, R_f, D \rangle$. A constant-domain xeno model adds an interpretation-function I to F, s.t. I maps pairs of F and worlds w to subsets of D, s.t. $M = \langle F, R_M, I \rangle$.

A substitution is a function from a set of variables to elements of D. A substitution v' is *x-variant* of v, if $v(y) = v'(y)$ for all variables y.

Thus,
$\langle M,w \rangle \Vdash_v F[(a_1), \ldots, F(a_m)]$, where

a_i is either an individual constant or variable iff $\langle f(a_1), \ldots, f(a_m) \rangle \in I(F, w)$, s.t. if a_i is a variable x_i, then $f(a_i) = v(x_i)$, and if a_i is an individual constant c_i, then $f(a_i) = I(c_i)$

$\langle M,w \rangle \Vdash_v \neg\phi$ iff it is not the case that $\langle M,w \rangle \Vdash_v \phi$
$\langle M,w \rangle \Vdash_v \phi \wedge \psi$ iff $\langle M,w \rangle \Vdash_v \phi$ and $\langle M,w \rangle \Vdash_v \psi$
$\langle M,w \rangle \Vdash_v \phi \vee \psi$ iff $\langle M,w \rangle \Vdash_v \phi$ or $\langle M,w \rangle \Vdash_v \psi$
$\langle M,w \rangle \Vdash_v \phi \rightarrow \psi$ iff, if $\langle M,w \rangle \Vdash_v \phi$ then $\langle M,w \rangle \Vdash_v \psi$

⟨M,w⟩ ⊩$_v$φ ⟺ ψ iff [⟨M,w⟩ ⊩$_v$φ iff ⟨M,w⟩ ⊩$_v$ψ]
⟨M,w⟩ ⊩$_v$ ∀x[φ(x)] iff for all x-variant v' ⟨M,w⟩ ⊩$_{v'}$ φ(x)
⟨M,w⟩ ⊩$_v$ ∃x[φ(x)] iff for some x-variant v' ⟨M,w⟩ ⊩$_{v'}$ φ(x)
⟨M,w⟩ ⊩ □φ iff ∀w'∈W[R$_φ$(w,w') → ∃X∈N(w')∀v∈W[⟨M,w⟩ ⊩ [φ ⟺ v∈X]]]
⟨M,w⟩ ⊩ ◊φ iff ∃w'∈W[R$_{¬φ}$(w,w') ∧ P(¬φ) is not in N(w')].

13.3 New Extensions of ADT

In his discussion of Priest (2006)'s inclosure schema, Scharp disavows of a unified solution to the gamut of paradoxical phenomena (Scharp, 2013: 288). Despite the foregoing, I believe that there are at least four positive extensions of Scharp's theory of Ascending and Descending Truth that he does not discuss, and yet that might merit examination.

13.3.1 First Extension: The Preface Paradox

The first extension of the theory of ADT might be to the preface paradox. A set of credence functions is *Easwaran-Fitelson-coherent* if and only if (i) the credences are governed by the Kolmogorov axioms; and it is not the case both (ii) that one's credence is dominated by a distinct credence, s.t. the distinct credence is closer to the ideal, vindicated world, while (iii) one's credence is assigned the same value as the remaining credences, s.t. they are tied for closeness (see Easwaran and Fitelson, 2015).[2] Rather than eschew consistency in favor of a weaker epistemic norm such as Easwaran-Fitelson coherence, the ADT theorist might argue that consistency can be preserved, because the preface sentence, 'All of the beliefs in my belief set are true, and one of them is false' might be Ascending True rather than Descending True. Because the models in Scharp's replacement theory can preserve consistency in response to the Preface, ADT might, then, provide a compelling alternative to the Easwaran-Fitelson proposal.

---

[2] A credence function is here assumed to be a real variable, interpreted as a subjective probability density. The real variable is a function to the [0,1] interval, and is further governed by the Kolmogorov axioms: normality, 'Cr(T) = 1'; non-negativity, 'Cr(φ) ⩾ 0'; finite additivity, 'for disjoint φ and ψ, Cr(φ ∪ ψ) = Cr(φ) + Cr(ψ)'; and conditionalization, 'Cr(φ | ψ) = Cr(φ ∩ ψ) / Cr(ψ)'.

13.3.2 Second Extension: Absolute Generality

A second extension of Scharp's ADT theory might be to a central issue in the philosophy of mathematics; namely unrestricted quantification. A response to the latter might further enable the development of the property versions of AT and DT: i.e., being Ascending-True-of and being Descending-True-of. For example, Fine (2005c) and Linnebo (2006) advance a distinction between sets and interpretations, where the latter are properties; and suggest that inconsistency might be avoided via a suitable restriction of the property comprehension scheme.[3] A proponent of Scharp's ADT theory might be able: (i) to adopt the distinction between extensional and intensional groups (sets and properties, respectively); yet (ii) circumvent restriction of the property comprehension scheme, if they argue that the Russell property, R, is Ascending True-of yet not Descending True-of. The foregoing maneuver would parallel Scharp's treatment of the derivation, within ADT, of the Ascending and Descending Liars and their revenge analogues.

13.3.3 Third Extension: Probabilistic Self-reference

A third extension of ADT might be to a self-referential paradox in the probabilistic setting. Caie (2013) outlines a puzzle, according to which:

(1) '*' := $\neg \text{CrT}(*) \geq .5$

that is, (*) says of itself that it is not the case that an agent has credence in the truth of (*) greater than or equal to .5. As an instance of the T-scheme, (1) yields: 'T(*) $\iff$ $\neg$CrT(*) $\geq$ .5'. However, CrT(*) ought to map to the interval between .5 and 1. Then, 'Cr(ϕ) + Cr($\neg\phi$) $\neq$ 1', violating the normality condition which states that one's credences ought to sum to 1.

In ADT, the probabilist self-referential paradox might be blocked as follows. Axiom (A2) states that $\neg A(\phi) \to A(\neg\phi)$; so if it is not an Ascending Truth that ϕ, then it is an Ascending Truth that not ϕ. However, (A2) does not hold for Descending Truth. Thus, in the instance of the T-scheme which states that 'T(*) $\iff$ $\neg$CrT(*) $\geq$ .5', the move from '$\neg$CrT(*) $\geq$ .5' to 'CrT$\neg$(*) $\geq$ .5' is Ascending True, but not Descending True. So, if the move from '$\neg$CrT(*)' to 'CrT$\neg$(*)' is not Descending True, then the transition

---

[3] See Field (2004; 2008) for a derivation of the Russell property, R, given the 'naive comprehension scheme: $\forall u_1 \ldots u_n \exists y [\text{Property}(y) \land \forall x (x \text{ instantiates } y \iff \Theta(x, u_1 \ldots u_n)]$' (2008: 294). R denotes 'does not instantiate itself', i.e. $\forall x[x \in R \iff \neg(x \in x)]$, s.t. R$\in$R $\iff \neg$(R$\in$R) (2004: 78).

from 'Cr(ϕ) + ¬Cr(ϕ) = 1' to 'Cr(ϕ) + Cr(¬ϕ) = 1' is not Descending True. Similarly, then, to the status of the Descending Liar in ADT, the derivation of probabilistic incoherence from the probabilist self-referential sentence, (1), is Ascending True, but not Descending True.[4]

13.3.4 Fourth Extension: The Sorites Paradox

A fourth extension of ADT might, finally, be to the sorites paradox. Scharp's xeno semanics is non-normal, such that the accessibility relation is governed by the axioms T (reflexivity) and 4 (transitivity), although not by axiom K. Suppose that there is a bounded, phenomenal continuum from orange to red, beginning with a color hue, c_i, and such that – by transitivity – if c_i is orange, then c_{i+1} is orange. The terminal color hue, in the continuum, would thereby be orange and not red. The transitivity of xeno semantics explains the generation of the sorites paradox. However, xeno semantics appears to be perfectly designed in order to block the paradox, as well: The neighborhood function in Scharp's xeno semantics for ADT is such that one can construct a model according to which transitivity does not hold. Let M_k be a neighborhood model, s.t. W_k = {a, b, c}; N_k(a) = {a, b}; N_k(b) ={a, b, c}; N_k(c) = {b, c}; $V_k(\phi)$ = {a, b}. Thus, $\langle M_k, a \rangle \Vdash \Box\phi$; but not $\langle M_k, b \rangle \Vdash \Box\phi$. So, it is not the case that $\langle M_k, a \rangle \Vdash \Box\Box\phi$; so transitivity does not hold in the model. Scharp's semantics for his ADT theory would thus appear to have the resources both to generate, and to solve, the sorites paradox.[5]

In the remainder of the essay, I examine six issues for ADT.

13.4 Issues for ADT

13.4.1 Issue 1: Revenge Paradoxes

- Descent

$\delta \iff$ '¬D(δ) ∨ ¬S(δ)'

(i) Suppose that D(δ).

---

[4]Thanks here to Catrin Campbell-Moore, for a correction.

[5]Scharp suggests that the truth predicate is contextually invariant, although assessment-sensitive (9.4). A second means by which the proposal could be extended in order to account for vagueness is via its convergence with the interest-relative approach advanced by Fara (2000; 2008).

Then, $D['\neg D(\delta) \lor \neg S(\delta)']$.
So, $[\neg D(\delta) \lor \neg S(\delta)]$, contrary to the supposition.
(ii) Suppose that $[\neg D(\delta) \lor \neg S(\delta)]$.
So, $D['\neg D(\delta) \lor \neg S(\delta)']$. So, $D(\delta)$, contrary to the supposition.

- Ascent

$\alpha \iff \text{`}\neg A(\alpha) \lor \neg S(\alpha)\text{'}$
(i) Suppose that $A(\alpha)$.
Then, $A[\neg A(\alpha) \lor \neg S(\alpha)]$
So, $[\neg A(\alpha) \lor \neg S(\alpha)]$, contrary to the supposition.
(ii) Suppose that $A[\neg A(\alpha) \lor \neg S(\alpha)]$
Then $A(\alpha)$, contrary to the supposition.

Similarly to the response to the alethic paradoxes, Scharp avers that α and δ are unsafe, and so they are Ascending True although not Descending True.

However, Scharp concedes that ADT does not invalidate all unsafe sentences, because some theorems of ADT are not Descending True (see p. 154). Crucially, then, Scharp's response, both to the alethic paradoxes and to the revenge sentences which are generated using only the resources of his own theory, fails to generalize. Because some theorems of ADT are not Descending True, some theorems of ADT are unsafe, and therefore Scharp's proposed restriction to safe predicates in order to avoid paradox serves only, as it were, to temper the flames on one side of the room, while they flare throughout the remainder.

A second maneuver exploits the fact that some unsafe sentences are derivable in ADT.

A sentence, γ, comprising the singleton U^+ is *positively unsafe* iff it is derivable in ADT.

A sentence, γ, comprising the singleton U^- is *negatively unsafe* iff it is not derivable in ADT.

$\gamma \iff \neg D(\gamma)$ and γ is not U^+.

To show that γ is unsafe, suppose for reductio that $D(\gamma)$. Then, $D[\neg D(\gamma)$ and γ is not in $U^+]$, so $\neg D(\gamma)$ and γ is not in U^+. So, $\neg(\gamma)$, contrary to the supposition. So, by reductio, $\neg D(\gamma)$. So γ is unsafe.

Suppose for reductio that $\neg A(\gamma)$. Then, $\neg A[\neg D(\gamma)$ and γ is not in $U^+]$. So, $\neg \gamma$, i.e. either $D(\gamma)$ (by the definition of γ), or D is in U^+. If $D(\gamma)$, then

A(γ) (from the definition of Safety). Thus, by reductio, A(γ). Thus, γ is unsafe.

Suppose for reductio that γ is U^+, s.t. it is an unsafe theorem of ADT. Some sentences of ADT are not Descending True, e.g. β. So, assume that $\beta \to \gamma$. So, (a) $\neg D(\gamma)$ and (b) γ isn't U^+. Thus, by reductio, γ is not U^+. Thus, γ is U^-, i.e. unsafe and not derivable from ADT. (In order to make this proof work, Scharp needs to assume (c), i.e. that β is itself not in U^+. No argument is advanced for this. In some cases it could so be, and then the proof would be blocked.)

Scharp endeavors to minimize the crucial lacuna in his proposal to the effect that ADT validates sentences that are not Descending True. He argues:

–that a valid argument cannot take one from a D(ϕ) to a $\neg$A(ϕ);

–that – while D(ϕ) can still entail $\neg$S(ϕ) (by the construction of the paradoxes in ADT) – $\neg$S(ϕ) entails A(ϕ);

–that the Descending Liar is unsafe (caveat: the Descending Liar is provable in ADT);

–that the conjunction of the Ascending Liar and its negation is not Ascending True (caveat: the Ascending Liar is unsafe, and unsafe sentences are derivable from ADT); and

–that the axioms of ADT are at least Descending True.

13.4.2 Issue 2: Validity

Scharp mentions Field (2008)'s argument against identifying validity with necessary truth-preservation, although does not reconstruct the argument.

In order to argue against identifying validity with necessary truth-preservation, Field draws, inter alia, on Curry (1942)'s Paradox.

The argument from Curry's Paradox is such that – by (T-In) and (T-Out) – one can derive the following. If ϕ is a false sentence then,

1. $\phi \iff [T(\phi) \to \bot]$
2. $T(\phi) \iff [T(\phi) \to \bot]$
3. $T(\phi) \to [T(\phi) \to \bot]$
4. $[T(\phi) \land T(\phi)] \to \bot$ (by importation)
5. $T(\phi) \to \bot$
6. $[T(\phi) \to \bot] \to T(\phi)$
7. $T(\phi)$
8. $\bot$

So, necessary truth-preservation entails contradiction.

However, the argument need not be valid, if one preserves (T-In) and (T-Out) yet weakens the logic. One can avail of the strong Kleene (1952/1971) valuation scheme, such that $|\phi|$ is ungrounded, i.e. maps to 1/2. One can then add a Determinacy operator, such that it is not determinately true that ϕ and it is not determinately true that not ϕ; so, it is indeterminate whether ϕ.

Field argues, in virtue of the foregoing, that validity ought to be a primitive. In more recent work, Field (2015) argues that validity is primitive if and only if it is 'genuine', such that the notion cannot be identical with either its model-theoretic or proof-theoretic analyses. As an elucidation of the genuine concept, he writes that 'to regard an inference or argument as valid is to accept a constraint on belief [...; s.t.] (in the objective sense of "shouldn't") we shouldn't fully believe the premises without fully believing the conclusion' (op. cit.). (The primitivist notion is intended to hold, as well, for partial belief.)

Scharp is persuaded by Field's argument, and endorses, in turn, a primitivist notion of validity, as a primitive canon of reasoning without necessary truth-preservation. Scharp takes this to be sufficient for the retention of Condition (iii), in Montague's Lemma (151). Scharp does not provide any further account of the nature of validity in the book. In later sections of the book, he reiterates his sympathy with Field's analysis, and also avails of Kreisel's 'squeezing' argument (§8.8), to the effect that the primitive notion of validity extensionally coincides with a formal notion of validity (i.e., derivation in a first-order axiomatizable quantified logic with identity). However, one potential issue is that, in a subsequent passage, Scharp writes that: 'an argument whose premises are the members of the set G and whose conclusion is p is valid iff for every point of evaluation e [i.e., index], if all members of G are assigned tM-value [i.e., an AT- or DT-value of] 1 at e, then p is assigned tM-value 1 at e' (240); and this would appear to be a definition of validity as necessary truth-preservation.

The primitivist approach to validity is the primary consideration that Scharp explicitly avails of, when arguing that closure ought to be rejected (Condition iv, in Montague's Lemma), rather than rejecting logical tautologies as candidates for the axioms of ADT (Condition iii, in Montague's Lemma) (151). So, further remarks about the nature of validity would have been welcome. An objection to prescinding from more substantial remarks about the nature of validity might also be that Scharp exploits claims with regard to its uses. So, e.g., he writes that 'a valid argument will never take one from descending truths to something not ascending true' (177). However, that claim is itself neither a consequence of either Kreisel's squeezing argument, nor the primitivist approach to validity.

13.4.3 Issue 3: Hybrid Principles and Compositionality

- ∧-T-Imb.

  $D(\phi) \wedge D(\psi) \rightarrow A(\phi \wedge \psi)$

  v.

  $T(\phi) \wedge T(\psi) \rightarrow T(\phi \wedge \psi)$

- ∨-T-Exc.

  $D(\phi \vee \psi) \rightarrow A(\phi) \vee A(\psi)$

  v.

  $T(\phi \vee \psi) \rightarrow T(\phi) \vee T(\psi)$

  (see pp. 147, 171)

Feferman (1984)'s theory countenances a primitive truth predicate (Feferman-true, in what follows); a primitive falsity predicate; as well as a Determinacy operator (op. cit.). This is by salient contrast to Scharp's approach, on which truth is replaced with DT and AT. Scharp argues that Feferman overemphasizes the significance of the compositionality of his Determinacy operator, at the cost of not having either logical truths or the axioms of his own theory satisfy Feferman-truth. By contrast, Scharp believes that he can avail of hybrid principles, such that it is not a requirement of ADT that Descending Truth and Ascending Truth obey compositionality (157).

One objection to this maneuver is that AT and DT are separated, in the hybrid principles, between the antecedent and consequent of the conditional.[6] So, it is unclear whether Scharp's hybrid principles are sufficient to redress the failure of compositionality in ADT; i.e., there being truth-conditions for sentences whose component semantic values are, alternatively, DT and AT.

A further objection is that the foregoing might be in tension with Scharp's repeated mention of natural-language semantics, in order to argue against

---

[6]This objection is owing to Stephen Read.

competing proposals. If natural-language semantics were to vindicate principles of compositionality, then this would provide a challenge to the empirical adequacy of Scharp's ADT theory, and thereby the viability of his replacement concepts for the traditional alethic predicate.

13.4.4 Issue 4: ADT and Indeterminacy

This issue concerns whether ADT might generalize, in order to account for other philosophical issues that concern indeterminacy. Whether ADT can be so extended to other issues, such as vagueness and types of indeterminacy, is not a necessary condition on the success of the theory. However, it might be a theoretical virtue of other accounts – e.g., classical, paracomplete, intuitionist, and supervaluational approaches – that they do so generalize; and the extensions of logic and semantics to issues in metaphysics are both familiar and legion.[7]

E.g., McGee (1991) suggests replacing the truth predicate with (i) a vague truth predicate, and (ii) super-truth. The replacement predicates are not intended for deployment in inferences implicated in reasoning, such as conditional proof and arguments by reductio (155). McGee introduces a Definiteness operator, μ, in order to yield the notion of super-truth relative to a set of precisifications. There is thus a truth predicate and a super-truth predicate. Super-truth is governed by (T-In) and (T-Out). Vague truth is governed by neither.

Thus:
–If $\mu(p)$, then μ'T(p)'
(If p is definitely true, then 'p is true' is definitely true)
–If $\mu(\neg p)$, then $\mu\neg$'T(p)'
(If p is definitely not true, then 'p is true' is definitely not true)
–If p is vague, then 'T(p)' is vague
(vagueness here is secured by availing of the strong Kleene valuation scheme, such that p is ungrounded, i.e. maps neither to true nor false, and rather to .5)

McGee endeavors to avoid Revenge, by arguing that
'u' := 'u is false or u is vague'

---

[7]See Williamson (2017), for an argument for the retention of classical logic despite the semantic paradoxes, based on the abductive strength of its generalization.

collapses to u is vague. So, u is not definitely true, and not definitely vague. Further, u is not derivable within McGee's supervaluationist theory, nor within a separate, fixed-point theory that he also advances.

Scharp raises several issues for the supervaluational approach. One issue is that vague sentences cannot be precisified via supervaluation – i.e., rendered determinately true – on pain of Revenge (156).

Scharp argues that Descending Truth and Ascending Truth obey (T-Out) and (T-In), respectively, whereas – according to McGee – vague sentences do not. So, McGee's replacement restricts expressivity, whereas Scharp argues that there are no expressive restrictions on his proposal. Scharp notes, as well, that some of the axioms for McGee's theory are not definitely true – and are thus vague and not governed by T-Out or T-In – which would appear to be a considerable objection.

However, Field (2008)'s approach – K3 plus a Determinacy operator, with a multi-valued semantics for the conditional – would appear to remain a viable proposal. Extensions of Field's proposal can be to an explanation of vagueness (Field, op. cit: ch. 5) and to the logic of doxastic states (see Caie, 2012).

With regard, e.g., to the extension of Field's treatment of the paradoxes to the logic of doxastic states, Caie demonstrates that – rather than rejecting the Liar sentence – it would no longer be the case, by K3 and indeterminacy, that one could believe the Liar, and it would no longer be the case that one ought not to believe the Liar. In addition to this proposal, I provide in section **13.4.7** an epistemicist approach to Curry (1942)'s paradox which is able to retain both classical logic and the truth introduction and elimination rules.

On the supervaluational treatment of the paradoxes, the approach can more generally be extended in order to account, e.g., for the metaphysical issues surrounding fission cases and indeterminate survival. Approaches which avail of a supervaluational response to fission scenarios, and similar issues at the intersection of nonclassical logic, metaphysical indeterminacy, and decision theory, can be found, e.g., in Williams (2014).

Thus, while it is not a necessary condition on the success of treatments to the alethic paradoxes that their proposals can generalize – in order, e.g, to aid in the resolution of other philosophical issues such as epistemic and metaphysical indeterminacy – there are viable proposals which can be so extended. The competing approaches thus satisfy a theoretical virtue that might ultimately elude ADT.

13.4.5 Issue 5: Descending Truth, Ascending Truth, and Objectivity

Scharp claims that considerations of space do not permit him to elaborate on the interaction between Descending Truth, Ascending Truth, and objectivity (§8.3). Suppose that – depending on the target domain of inquiry – the truth-conditions of sentences might be sensitive to the reality of the objects and properties that the sentences concern. So, e.g., second-order implicit definitions for the cardinals might be true only if the terms therein refer to abstract entities. By contrast, what is said in sentences about humor might be true, if and only if the sentence satisfies deflationary conditions such as the T-schema.

Another objection to the replacement strategy, and of Scharp's candidate replacements in particular, is that it is unclear how – in principle – either Descending Truth or Ascending Truth can be deployed in order to capture the foregoing distinctions.

13.4.6 Issue 6: Paradox, Sense, and Signification

One final objection concerns the general methodology of the book. Scharp proceeds by endeavoring to summarize all of the extant approaches to the alethic paradoxes in the literature, and to marshal at least one issue adducing against their favor. However, there is at least one approach to the paradoxes that Scharp overlooks. This approach targets the notion of what is said by an utterance, i.e. the properties of sense and signification that a sentence might express. One such proposal is inspired by Bradwardine (c.1320/2010) and pursued by Read (2009). According to the proposal, if a sentence such as the Liar does not wholly signify that it is true, then one invalidates T-Introduction for the sentence. In a similar vein, Rumfitt (2014) argues that paradoxical sentences are a type of Scheingedanken, i.e. mock thoughts that might have a sense, although take no value; so, T-Introduction is similarly restricted.

Scharp takes it to be a virtue of his account that he can retain the disquotational principles, even though they get subsequently divided among his replacement concepts. He might then reply to the foregoing proposal by suggesting that they similarly induce expressive restrictions in a manner that his approach can circumvent.

However, there are other approaches which avail of what I shall refer to as the sense and signification strategy, and which eschew neither T-Elimination nor T-Introduction. Modulo a semantics for the conditional, K3 and indeterminacy at all orders ensures not only that hyper-determinacy – and therefore an assignment of classical values to the paradoxical sentences – is circumvented; but, furthermore, that revenge sentences cannot be derived either. Against this approach, Scharp reiterates his concern with regard to restrictions on expression. He writes, e.g., that 'Field avoids revenge only by an expressive limitation on his language' (107). However, a virtue of the approach is that, as in xeno semantics for ADT, T-Elimination and T-Introduction are preserved. Against ADT theory, K3+indeterminacy does not arbitrarily select the alethic principles that the semantic theory should satisfy. Crucially, moreover, the approach does not say more than one should like it to, as witnessed, e.g., by the derivability in ADT of both the DT and AT Liars and their revenge analogues. Rather, the language of paracompleteness and indeterminacy demonstrates that – without the loss of the foundational principles governing the alethic predicate – there are propositions which can satisfy the values in an abductively robust semantic theory.

13.4.7 Epistemic Closure and Logical Deduction

Read (2010) notes that there are at least three forms of Curry (1942)'s paradox. One form, attributed to Albert of Saxony, is conjunctive: for any sentence A, 'ϕ' := $[A \wedge F(\phi)]$. A second form, attributed both to Albert of Saxony and Bradwardine, is disjunctive: for any sentence A, 'ϕ' := $[A \vee F(\phi)]$. A third form, owing to Löb (1955) and Field is such that, for any sentence A: 'ϕ' := $[T(\phi) \to A]$, and is the version at issue in this book. Read (op. cit.) argues that T-introduction is the culprit in the first three forms of the paradox. A fourth variation on Curry's paradox can be found in Beall and Murzi (2012), who replace the truth predicate with a validity predicate.

For all sentences, ϕ,
1. $\phi \iff [T(\phi) \to \bot]$
2. $T(\phi) \iff [T(\phi) \to \bot]$
3. $T(\phi) \to [T(\phi) \to \bot]$
4. $[T(\phi) \wedge T(\phi)] \to \bot$ (by importation)
5. $T(\phi) \to \bot$ (by contraction)
6. $[T(\phi) \to \bot] \to T(\phi)$
7. $T(\phi)$
8. $\bot$

Rather than weaken the logic in a type-free setting (Field, 2008); eschew contraction (Beall and Murzi, op. cit.); or eschew T-introduction in virtue of defective properties of signification (Read, 2009), the current approach argues that Curry's paradox is classically sound; that the truth introduction and elimination rules can yet be retained; and that the paradox is problematic only because it exhibits an instance of epistemic indeterminacy.

The epistemic indeterminacy entrained by Curry's paradox occurs because step 5 invalidates axiom K on its epistemic interpretation: '$K(\phi \to \psi) \to (K\phi \to K\psi)$'.

This provides a counter-example to *epistemic closure* for reductio proofs. To see this, let ϕ denote '$T(\phi)$' and ψ denote '$\bot$'. Then:

(*) $K(T\phi \to \bot) \to [K(T\phi) \to K(\bot)]$.

In (*), $K(\bot)$ is false, because – by reflexivity – only truths can be known. However, $K(T\phi)$ can be known, because it is an instance of T-introduction. Thus, the conditional in the consequent of (*) has itself a true antecedent and false consequent, and is thus false. Finally, the conditional in the antecedent of (*), '$K(T\phi \to \bot)$' is true, in virtue of the proof of Curry's paradox. So the instance of K expressed in (*) has a true antecedent and false consequent, providing a counter-instance to K; and so K is not a valid axiom in our system of epistemic logic.

Taking stock: One of the things which is problematic about Curry's paradox is that – despite being classically sound and entailing contradiction via truth introduction and elimination rules – at least one of the steps in its derivation is epistemically indeterminate, by invalidating axiom K, i.e. epistemic closure. The invalidity of modal axiom K, i.e. epistemic closure, generalizes for reductio proofs.

13.5 Concluding Remarks

In this essay, I have outlined Scharp's theory of ADT and its semantics. I then proffered four novel extensions of the theory, and detailed six issues that the theory faces.

Chapter 14

Intention: Hyperintensional Semantics and Decision Theory

14.1 Introduction

Formal treatments of imperatival notions have been pursued both logically and semantically. In the logical setting, deontic claims have been interpreted as types of a modal operator, where a condition holding across the points of a space abbreviates the property of obligation, and its dual abbreviates the property of permissibility.[1] In the twentieth century, research in deontic logic has examined the validity of the rule of necessitation ($\vdash \phi \rightarrow \vdash \Box \phi$) (von Wright, 1981); modal axiom 4 ($\Box \phi \rightarrow \Box\Box\phi$) (see Barcan Marcus, 1966); and modal axiom GL $[\Box(\Box\phi \rightarrow \phi) \rightarrow \Box\phi]$ (see Smiley, 1963). The semantic approach has been inspired by the works of Kratzer (1977, 2012), Stalnaker (1978), and Veltman (1996), arguing that there are modal operators on a set of points which are not straightforwardly truth-conditional, instead recording an update on that set which is taken to be pragmatic (see Yalcin, 2012). The types of obligation have proliferated, as variations on the 'ought'-operator – e.g., what one ought to do relative to a time and one's states of information, by contrast to what one ought to do relative to the facts – have been codified by differences in the array of intensional parameters relative to which the operator receives a semantic value (see Yalcin, op. cit.; Cariani, 2013; Dowell, 2013; et al).

---

[1]Deontic logic dates from at least as early as the fourteenth century, in the writings of Ockham, Holcot, and Rosetus. See Knuuttila (1981) for further discussion.

This essay aims to provide a theory of the structural content of the types of intention via a similar modal analysis; to explain the role of intention in practical reasoning; and to answer thereby what I will call the unification problem: i.e., the inquiry into how the various types of intention comprise a unified mental state.[2] The general significance of the present contribution is that it will provide some foundational structure to the topic, where the previous lack thereof has served only to exacerbate its intransigence.[3] I will argue that – similarly to the case of deontic judgment – the foregoing types of intention can be countenanced as modal operators. The defining contours of the contents of the states may thus be targeted via their intensional-semantic profile. The types of intention on which I will focus include (i) the notion of 'intention-in-action', as evinced by cases in which agents act intentionally; (ii) the notion of 'intention-with-which', where an agent's intentions figure as an explanation of their actions; and (iii) the notion of 'intention-for-the-future', as evinced by an agent's plans to pursue a course of action at a future time.

I will argue that the unification problem has at least two, consistent solutions. The first manner in which the operations of intention are unified is that they are defined on a single space, whose points are states of information or epistemic possibilities. I argue, then, that the significance of examining how the state of intention interacts with practical reasoning is that it provides a second means by which to account for the unity of intention's types. Although each type of intention has a unique formal clause codifying its structural content, the notions of 'intention-in-action', 'intention-with-which', and 'intention-for-the-future' are nevertheless unified, because each is directed toward the property of expected utility. Thus, acting intentionally, acting because of an intention, and intending to pursue a course of action

---

[2]The unification problem is first examined in Anscombe (1963), and has been pursued in contemporary research by, inter alia, Bratman (1984) and Setiya (2014).

[3]Compare the aims and methods pursued in the research projects of Fine (1981) and Williamson (2014b): 'The relevance of the undertaking [...] consists mainly in the general advantages that accrue from formalizing an intuitive theory. First of all, one thereby obtains a clearer view of its primitive notions and truths. This is no small thing in a subject [...] that is so conspicuously lacking in proper foundations' (Fine, op. cit.: 177); 'The aim is to gain insight into a phenomenon by studying how it works under simplified, rigorously described conditions that enable us to apply mathematical or quasi-mathematical reasoning that we cannot apply directly to the phenomenon as it occurs in the wild, with all its intractable complexity. We can then cautiously transfer our insight about the idealized model back to the phenomenon in the wild' (Williamson, op. cit.).

at a future time, are mental states whose unification consists in that each type aims toward the satisfaction of the value of an outcome – the value of which is the product of a partial belief conditional on one's acts by the utility thereof. The dissociation between an agent's intention to pursue an action and the causal relevance of the action's outcome adduces in favor of the characterization of expected utility in the setting of evidential, rather than causal, decision theory. The proposal that the content of intention is expected utility has, furthermore, the virtue of generalizing, in order to explain the nature of the intentions of non-human organisms. The contents of non-human organisms' intentions can here be understood as the value intended by their actions, as sensitive to both their prediction that the outcome will occur and the utility of its occurrence. Finally, because the aim of intention is expected utility, a precise account can be provided of how intention relates to the notions of belief and desire, while yet retaining its status as a unique mental state.

In Sections **2-3**, I delineate the intensional-semantic profiles of the types of intention, and provide a precise account of how the types of intention are unified in virtue of both their operations in a single epistemic modal space and their role in practical reasoning, i.e., evidential decision theory. I endeavor to provide reasons adducing against the proposal that the types of intention are reducible to the mental states of belief and desire, where the former state is codified by subjective probability measures and the latter is codified by a utility function. Section **4** provides concluding remarks.

14.2   The Modes of Intention

The epistemic modal space of an agent can be defined via a frame, comprised of a set of points, and a relation of accessibility thereon (see Kripke, 1963; Blackburn et al, 2001). The points in the frame are here interpreted as an agent's states of information, while the relation of accessibility can receive various interpretations. A state of information is possible, just if there is at least one point relative to which it is true, if and only if it is not necessary for the formula to be false. One of the states of information is necessary, just if it is true everywhere, i.e. relative to all the other points in the space, if and only if it is impossible for it to be false. The distinctly epistemic interpretation of possibility comes in at least two guises, defined as the dual of epistemic necessity ('$\Diamond \phi$' iff '$\neg \Box \neg \phi$'): The truth of a formula is epistemically possible, just if the formula is believed by an agent, or is conceivable to the agent. The epistemic interpretation of necessity can itself come in at least two guises: The truth of a formula is epistemically invariant or necessary, just if the truth of the formula is known by an agent, or if it is inconceivable for the formula to be false (dually, epistemically necessary), and is thus in one sense apriori.

When an agent intends to ϕ, their intention may fall into three distinct types. One type of intention concerns the intentional pursuit, by the agent, of a course of action. A second type of intention can be witnessed, when the agent cites an intention as an explanation of her pursuit of a course of action. Finally, a third type of intention can be witnessed, when the agent intends to pursue a course of action at a future time.

14.2.1 Intention-in-Action

If the agent acts intentionally, then her intention can be understood as an operation relative to her states of information. The agent acts intentionally, just if there is a world and a unique array of intensional parameters relative to which her intention is realized and receives a positive semantic value. The array of intensional parameters is two-dimensional, because the value of intending to ϕ relative to one of the parameters will constrain the value of intending to ϕ relative to the subsequent parameters. Thus, we can say that an agent intends to ϕ, if and only if she acts intentionally, only if there is both a world and array of intensional parameters, relative to which her intention is realized, i.e. receives a positive value. The intensional parameters include a context comprised of a time and location, and a pair of indices on which spaces of the agent's acts and of the outcomes of her actions are built. So, the agent's intention-in-action receives a positive semantic value only if there is at least one world in her epistemic modal space at which – relative to the context of a particular time and location, which constrains the admissibility of the actions as defined at a first index, and which subsequently constrains the outcome thereof as defined at a second index – the intention is realized.

- $[\![\text{Intenton-in-Action}(\phi)]\!]_w = 1$ only if $\exists w' [\![\phi]\!]^{w',c(=t,l),a,o} = 1$

14.2.2 Intention-as-Explanation

If the agent refers to an intention, in order to explain her pursuit of a course of action, then her intention can similarly be understood as an operation relative to her states of information. In this case, the agent intends to ϕ, just if there is a pair of formulas defined at points in her epistemic modal space, where one of the states is realized because it holds in virtue the other state being realized. Informally, the foregoing explanation can be referred to as the intention-with-which she acts. Thus, we can say that an agent intends to ϕ, if and only if her intention is an explanation for her action, only if she acts in pursuit of ψ because she intends to ϕ. In order to capture the notion of one formula holding in virtue, or because, of a distinct formula, we define grounding operators on the agent's epistemic modal space. Thus, the agent intends to ϕ because, there is an intention in virtue of which her action, ψ so as to realize ϕ, receives a positive value.

- $[\![\text{Intention-with-which}(\phi)]\!]_w = 1$ only if $\exists w'[[\![\psi]\!]^{w'} = 1 \wedge [\![G(\phi,\psi)]\!] = 1]$,

where G(x,y) is a grounding operator encoding the explanatory connection between ϕ and ψ.[4]

14.2.3 Intention-for-the-Future

Finally, an agent can intend to ϕ, because she intends to pursue a course of action at a future time. In this case, the intensional-semantic profile which records the parameters relative to which her intention receives a positive semantic value converges with a future directed modal operator to the effect that the agent *will* ϕ. Thus, an agent realizes an intention-for-the-future only if there is a possible world and a future time, relative to which the possibility that ϕ is realized can be defined. Thus:

- $[\![\text{Intention-for-the-future}(\phi)]\!]_w = 1$ only if $\exists w' \forall t \exists t'[t < t' \wedge [\![\phi]\!]^{w',t'} = 1]$.[5]

A multi-hyperintensional semantics for the types of intention can be provided as well. In this case, the worlds figuring in the parameters of the foregoing semantic clauses for the modal operators are replaced by topic- (i.e. subject matter) sensitive truthmakers. The truthmakers can be interpreted epistemically, and as being parts of worlds rather than whole worlds

---

[4]For further properties of ground operators, see **ch. 7**.

[5]See Rao and Georgeff (1991), for the suggestion that operators in a multi-modal logic can model the notion of goal-oriented intention. The foregoing intensional semantics is consistent with the logic that they proffer.

themselves. Thus epistemic topic-sensitive truthmakers figure in, and thus inform, the parameters relative to which intentions are satisfied.

This section has endeavored to accomplish two aims. The first was to provide a precise delineation of the structural content of, and therefore the distinctions between, the types of intention. Intention was shown to be a modal mental state, whose operations have a unique intensional profile, and whose values are defined relative to an agent's space of states of information. The second aim was to secure one of the means by which the unity of the distinct types of intention can be witnessed. Despite that each of the types of intention has a unique structural content, the contents of those types are each defined in a single, encompassing space; i.e, relative to the agent's space of epistemic possibilities.

14.3 Intention in Decision Theory

In Section 2, I suggested that intention is a unified, modal mental state, the contents of which are defined relative to an agent's states of information. This section examines the proposal that intentions have a dual profile (see Bratman, op. cit.), because intentions figure constitutively in practical reasoning. I argue that, because expected utility theories are the only axiomatized theories of practical reasoning, an account must be provided of the role that intention plays therein. The account will illuminate a precise relationship – which I argue is not identity – between the types of intention and the mental states of belief and desire. The account will further serve to provide a second explanation for the unity of intention's types, given the uniform role that the types of intention play in decision theory.

A model of decision theory can be understood as a tuple $\langle A, O, K, V \rangle$, where A is a set of acts; O is a set of outcomes; K encodes a set of counterfactual conditionals, where an act from A figures in the antecedent of the conditional and O figures in the conditional's consequent; and V is a function assigning a real number to each outcome. The real number is a representation of the value of the outcome. The expected value of the outcome is calculated as the product of (i) the subjective probability – i.e., the agent's partial belief or credence – that the outcome will occur, as conditional on her act, and (ii) the value or utility which she assigns to the outcome's occurrence. The agent can prefer one assignment of values to the outcome's occurrence over another. (Which preference axioms ought to be adopted is a contentious issue, and will not here be examined. see von Neumann and Morgenstern, 1944; Savage, 1954; Jeffrey, 1983; and Joyce, 1999.) In evidential decision theory, the expected utility of an outcome is calculated as the product of the agent's credence conditional on her action, by the utility of the outcome. In causal decision theory, the expected utility of an outcome is calculated as the product of the agent's credence, conditional on both her action and the causal efficacy thereof, by the utility of the outcome. Expected utility can further be augmented by a risk-weighting function: If the agent's expected utility diminishes with the order of the bets she might pursue – such that expected utility is sensitive to the agent's propensity to take risks relative to the total ordering of the gambles – then she might have a preference for a sure-gain of .5 units of value, rather than prefer a bet with a 50 percent chance of winning either 0 or 1 units of value (see Buchak, 2014).

If intention plays a constitutive role in practical reasoning, and decision theories provide the most tractable models thereof, then what is the role of intention in decision theory? The parameters in the axiomatizations of decision theory encode variables for credences, actions, outcomes, assignments of utility, background states of information pertinent to the causal relevance of actions on outcomes, and the agent's preferences. Expected utility is derived, as noted, by the interaction between an agent's credences, actions, and utility assignments. Which, then, of these parameters do an agent's intentions concern?

There are dissociations between intention and belief and between intention and desire. An agent can have a partial belief that the sun will rise, without intending to pursue any course of action. Conversely, an agent can intend to pursue a course of action, yet appreciate that there are, unfortunately, reasons for her to disbelieve that the act will obtain. An agent can desire that the sun rises, without the intention to entrain the sun's rising as consequence. Conversely, a vegetarian can intend to consume meat, if it is the only available source of protein and they are in dire need thereof, while yet desire a distinct and orthogonal outcome.

There are dissociations between intention and preference. An agent can prefer the sun's rising to the prevalence in her life of unprovoked antagonists, without either acting intentionally, possessing an intention as an explanation for some course of action, or intending to pursue any particular course of action in the future. Conversely, whether an agent's intention to pursue an action mandates a preference for the value of the outcome of that action will depend on one's preference axioms. One such axiom might be maximin, according to which the best of the worst outcomes among a set of options should be preferred, while a distinct rule might be maximax, according to which one ought to prefer and pursue the maximally valuable outcome among a set of options. Thus, intending to ϕ is not sufficient for determining whether ϕ ought to be preferred.

There are, finally, dissociations between intention and action. One might intend to calculate the value of a formula, yet not be able so to act, because their attention might be allocated elsewhere.

Acting intentionally, intending to pursue a course of action in the future, and citing an intention as an explanation for one's course of action are each, however, in some way related to the value of a course of action. When an agent acts intentionally, she acts in such a way so as to obtain an outcome that she values. When an agent pursues a course of action, and refers to her intention so to act as the explanation for that action, the intention explains the value, for the agent, in which the action and its outcomes are supposed to consist. Finally, when an agent intends to pursue a course of action in the future, her intention is similarly guided by the value of the outcome that her action will hopefully entrain. The value of the outcome will not be her bare assessment of the utility of the outcome, because – in the setting of decision theory – utility functions codify desires, such that her intention would thereby be elided with her desire for the outcome.

Because the types of intention are all directed toward the value of an outcome of a course of action – while being irreducible to, because dissociable from, the states of belief and desire – the remaining and most suitable candidate for the role of the mental state of intention in decision theory is the aim of expected utility; i.e., the value of an outcome, as arising by the interaction between the agent's partial belief or expectation that the outcome will occur as conditional on her act, and the utility that she associates with the outcome's occurrence. Because of the dissociation between an agent's intention to pursue a course of action in the future and the action's occurrence – let alone the dissociation between the intention to act in future, and the causal efficacy of the action were it to obtain – the role of intention in practical reasoning appears to be more saliently witnessed in the setting of evidential decision theory.

That the types of intention are each directed toward expected utility evinces how an agent's intentions can be sensitive to her beliefs and desires, without being reducible to them. Crucially, moreover, that the types of intention are each directed toward expected utility provides a second explanation of the way that the types of intention comprise a unified mental state.

Theoretical advantages accruing to the foregoing proposal include that it targets a foundational role for intention in decision theory. The proposal might be foundational, because it targets a basic role for intention in practical reasoning, which is consistent with the possible augmentation of the proposal with other approaches which assume a more cognitively demanding role for intention's aims. Such approaches include proposals to the effect (i) that the most fundamental type of intention is intention-with-which, such that intention's role as an explanation can be elided with its causal efficacy (see Anscombe, op. cit.; Davidson, 1963); (ii) that the content of intention is the diachronic satisfaction of self-knowledge (see Velleman, 1989); and (iii) that the role of intention in practical reasoning ought to be understood as an evaluative constraint, as determined by the virtuous traits of an agent's character (see Setiya, 2007).

14.4 Concluding Remarks

I have argued that the unification problem for the types of intention can be solved in two, consistent ways. The types of intention can be modeled as modal operators, where the unity of the operations consists, in the first instance, in that their values are defined relative to a single, encompassing, epistemic modal space. The second manner by which the unity of intention's types can be witnessed is via intention's unique role in practical reasoning. I argued that each of the types of intention – i.e., intention-in-action, intention-as-explanation, and intention-for-the-future – has as its aim the value of an outcome of the agent's action, as derived by her partial beliefs and assignments of utility, and as codified by the value of expected utility in evidential decision theory. A precise account was thereby provided of the role of epistemic modality in the unification of the types of a unique, modal mental state, whose value figures constitutively in decision-making and practical reason.

Bibliography

Adámek, J. 1974. Free Algebras and Automata Realizations in the Language of Categories. *Commentationes Mathematicae Universitatis Carolinae*, Vol. 15, No. 4.

Adámek, J., and J. Rosicky. 1994. *Locally Presentable and Presentable Categories*. Cambridge University Press.

Adámek, J., S. Milius, L. Moss. 2018. Fixed Points of Functors. *Journal of Logical and Algebraic Methods in Programming*, 95.

Adámek, J., S. Milius, and L. Moss. Forthcoming. *Initial Algebras and Terminal Coalgebras: The Theory of Fixed Points of Functors*. Cambridge University Press.

Adriaans, P. 2020. Information. *The Stanford Encyclopedia of Philosophy* (Fall 2020 Edition), E.N. Zalta (ed.), URL = <https://plato.stanford.edu/archives/fall2020/entries/information/>.

Alberucci, L. 2002. The Modal μ-Calculus and Logics of Common Knowledge. PhD Thesis, Universität Bern.

Angell, R. 1989. Deducibility, Entailment, and Analytic Entailment. In J. Norman and R. Sylvan (eds.), *Directions in Relevant Logic*. Kluwer Academic Publishers.

Anscombe, G.E.M. 1957/1963. *Intention*. Harvard University Press.

Aristotle. 1987a. *De Caelo*, tr. J.L. Stocks, revised J. Barnes (Revised Oxford Aristotle), text: D.J. Allan (Oxford Classical Texts, 1936). In J.L. Ackrill (ed.), *A New Aristotle Reader*. Oxford University Press.

Aristotle. 1987b. *Prior Analytics I*, tr. P.T. Geach (Clarendon Aristotle Series), text: W.D. Ross (Oxford Classical Texts, 1949). In J.L. Ackrill (ed.), *A New Aristotle Reader*. Oxford University Press.

Aristotle. 1987c. *Physics*, tr. E. Hussey (Clarendon Aristotle Series, 1983), text: W.D. Ross (Oxford Classical Texts, 1950). In J.L. Ackrill (ed.), *A New Aristotle Reader*. Oxford University Press.

Aristotle. 1987d. *De Interpretatione*, tr. J.L. Ackrill (Clarendon Aristotle Series), text: L. Minio-Paluello (Oxford Classical Texts, 1956). In J.L. Ackrill (ed.), *A New Aristotle Reader*. Oxford University Press.

Arntzenius, F. 2012. *Space, Time, and Stuff*. Oxford University Press.

Artemov, S. 1995. Operational Modal Logic. Technical Report MSI 95–29. Cornell University.

Artemov, S., and M. Fitting. 2019. *Justification Logic*. Cambridge University Press.

Avigad, J. 2004. Forcing in Proof Theory. *Bulletin of Symbolic Logic*, 10:3.

Avigad, J. 2021. Foundations. https://arxiv.org/pdf/2009.09541.pdf

Awodey, S. 2004. An Answer to Hellman's Question: 'Does Category Theory Provide a Framework for Mathematical Structuralism?'. *Philosophia Mathematica*, 12:3.

Awodey, S. 2019. Intensionality, Invariance, and Univalence. slides.

Awodey, S, Á. Pelayo, and M. Warren. 2013. Voevodsky's Univalence Axiom in Homotopy Type Theory. *Notices of the American Mathematical Society*, 60:9.

Azzouni, J. 2004. Proof and Ontology in Euclidean Mathematics. In T. Kjeldsen, S. Pedersen, and L. Sonne-Hansen (eds.), *New Trends in the History and Philosophy of Mathematics*. University Press of Southern Denmark.

Azzouni, J. 2013. The Compulsion to Believe: Logical Inference and Normativity. In G. Preyer and G. Peter (eds.), *Philosophy of Mathematics: Set Theory, Measuring Theories, and Nominalism*. De Gruyter.

Bagaria, J., N. Castells, and P. Larson. 2006. An Ω-logic Primer. *Trends in Mathematics: Set Theory*. Birkhäuser Verlag.

Bagaria, J., P. Koellner, and W.H. Woodin. 2019. Large Cardinals beyond Choice. *Bulletin of Symbolic Logic*, 25:3.

Baker, A., and M. Colyvan. 2011. Indexing and Mathematical Explanation. *Philosophia Mathematica* (3), Vol. 19, No. 3.

Balaguer, M. 1998. *Platonism and Anti-Platonism in Mathematics*. Oxford University Press.

Balaguer, M. 2001. A Theory of Mathematical Correctness and Mathematical Truth. *Pacific Philosophical Quarterly*, 82.

Balog, K. 1999. Conceivability, Possibility, and the Mind-Body Problem. *Philosophical Review*, 108:4.

Baltag, A. 2003. A Coalgebraic Semantics for Epistemic Programs. *Electronic Notes in Theoretical Computer Science*, 82:1.

Baltag, A. 2016. To Know is to Know the Value of a Variable. In L. Beklemishev, S. Demri, and A. Mate (eds.), *Advances in Modal Logic, Vol. 1*. CSLI Publications.

Baltag, A., and B. Renne. 2016. Dynamic Epistemic Logic. *The Stanford Encyclopedia of Philosophy* (Winter 2016 Edition), E.N. Zalta (ed.), URL = <https://plato.stanford.edu/archives/win2016/entries/dynamic-epistemic/>.

Baltag, A., and S. Smets. 2020. Learning What Others Know. In L. Kovacs and E. Albert (eds.), *LPAR23 proceedings of the International Conference on Logic for Programming AI and Reasoning*, EPiC Series in Computing, Volume 73.

Barcan, R. 1946. A Functional Calculus of First Order Based on Strict Implication. *Journal of Symbolic Logic*, 11.

Barcan, R. 1947. The Identity of Individuals in a Strict Functional Calculus of Second Order. *Journal of Symbolic Logic*, 12.

Barcan Marcus, R. 1966. Iterated Deontic Modalities. *Mind*, 75.

Barcan Marcus, R. 1993. *Modalities*. Oxford University Press.

Barnes, E., and J.R.G. Williams. 2011. A Theory of Metaphysical Indeterminacy. In K. Bennett and D. Zimmerman (eds.), *Oxford Studies in Metaphysics, Volume 6*. Oxford University Press.

Baron, S., M. Colyvan, and D. Ripley. 2020. A Counterfactual Approach to Explanation in Mathematics. *Philosophia Mathematica* (3), Vol. 28, No. 1.

Bar-On, D. 2004. *Speaking My Mind: Expression and Self-Knowledge*. Oxford University Press.

Bar-On, D. 2012. Expression, Truth, and Reality. In A. Coliva (ed.), *Mind, Meaning, and Knowledge*. Oxford University Press.

Bar-On, D., and C. Wright. 2023. *Expression and Self-Knowledge*. John Wiley and Sons.

Bayne, T., and M. Montague (eds.). 2011. *Cognitive Phenomenology*. Oxford University Press.

Bealer, G. 1982. *Quality and Concept*. Oxford University Press.

Bealer, G. 1993. A Solution to Frege's Puzzle. *Philosophical Perspectives*, 7.

Beall, JC, and J. Murzi. 2012. Two Flavors of Curry's Paradox. *Journal of Philosophy*, 120.

Beddor, B., and S. Goldstein. 2022. A Question-sensitive Theory of Intention. *Philosophical Quarterly*, pqac031, https://doi.org/10.1093/pq/pqac031.

Beiser, F. 1987. *The Fate of Reason*. Harvard University Press.

Benacerraf, P. 1965. What Numbers Could Not Be. *Philosophical Review*, 74.

Benacerraf, P. 1973. Mathematical Truth. *Journal of Philosophy*, 70.

Bengson, J. 2015. The Intellectual Given. *Mind*, 124:495.

van Benthem, J. 1983. *Modal Logic and Classical Logic*. Bibliopolis.

van Benthem, J. 1984/2003. Correspondence Theory. In D. Gabbay and F. Guenthner (eds.), *Handbook of Philosophical Logic, 2nd Edition, Volume 3*. Kluwer.

van Benthem, J. 2010. *Modal Logic for Open Minds*. CSLI.

Bernays, P. 1942. A System of Axiomatic Set Theory: Part IV. General Set Theory. *Journal of Symbolic Logic*, 7:4.

Bernoulli, J. 1713/2006. *The Art of Conjecturing*, tr. E.D. Sylla. Johns Hopkins University Press.

Berry, S. 2022. *A Logical Foundation for Potentialist Set Theory*. Cambridge University Press.

Berto, F. 2014. On Conceiving the Inconsistent. *Proceedings of the Aristotelian Society, New Series*, Vol. 114.

Berto, F. 2018. Aboutness in Imagination. *Philosophical Studies*, 175.

Berto, F. 2019. The Theory of Topic-Sensitive Intentional Modals. In I. Sedlar and M. Blicha (eds.), *The Logica Yearbook, 2018*. College Publications.

Berto, F. 2022. *Topics of Thought*. Oxford University Press.

Berto, F., and P. Hawke. 2021. Knowledge relative to Information. *Mind*, 130: 517.

Berto, F., and A. Özgün. 2021. Dynamic Hyperintensional Belief Revision. *The Review of Symbolic Logic*, Vol. 14, Issue 3.

Berto, F., and T. Schoonen. 2018. Conceivability and Possibility: Some Dilemmas for Humeans. *Synthese*, 195.

Bjerring, J.C. 2012. Problems in Epistemic Space. *Journal of Philosophical Logic*, 43(1).

Blackburn, P., and J. van Benthem. 2007. Modal Logic: A Semantic Perspective. In Blackburn, van Benthem, and F. Wolter (eds.), *Handbook of Modal Logic*. Elsevier.

Blackburn, P., M. de Rijke, and Y. Venema. 2001. *Modal Logic*. Cambridge University Press.

Blackburn, S. 1984. *Spreading the Word*. Oxford University Press.

Block, N. 2006. Max Black's Objection to Mind-Body Identity. In D. Zimmerman (ed.), *Oxford Studies in Metaphysics, Volume 3*. Oxford University Press.

Block, N. 2015. The Canberra Plan Neglects Ground. In T. Horgan, M. Sabates, and D. Sosa (eds.), *Qualia and Mental Causation in a Physical World: Themes from the Philosophy of Jaegwon Kim*. Cambridge University Press.

Boghossian, P. 2008. Epistemic Rules. *Journal of Philosophy*, 105:9.

Boghossian, P. 2012. Blind Rule-Following. In Coliva (ed.), *Mind, Meaning, and Knowledge: Themes from Crispin Wright*. Oxford University Press.

Boghossian, P., and T. Williamson. 2020. *Debating the A Priori*. Oxford University Press.

Boh, I. 1993. *Epistemic Logic in the Later Middle Ages*. Routledge.

Bolzano, B. 1810/2004. *Contributions to a Better-grounded Presentation of Mathematics*. In Bolzano, *The Mathematical Works of Bolzano*, tr. and ed. Russ. Oxford University Press.

BonJour, L. 1998. *In Defense of Pure Reason*. Cambridge University Press.

Boole, G. 1854. *The Laws of Thought*. London: Walton and Maberly. Cambridge: MacMillan.

Boolos, G. 1971. The Iterative Conception of Set. *Journal of Philosophy*, 68.

Boolos, G. 1984. To Be is to Be the Value of a Variable (or to Be some Values of some Variables). *Journal of Philosophy*, 81.

Boolos, G. 1985. Nominalist Platonism. *Philosophical Review*, 94:3.

Boolos, G. 1987/1998. The Consistency of Frege's *Foundations of Arithmetic*. In J.J. Thomson (ed.), *On Being and Saying*. MIT Press. Reprinted in Boolos, *Logic, Logic, and Logic*, ed. R. Jeffrey. Harvard University Press.

Boolos, G. 1990/1998. The Standard of Equality of Numbers. In Boolos (ed.), *Meaning and Method*. Cambridge University Press. Reprinted in Boolos (1998).

Boolos, G. 1993. *The Logic of Provability*. Cambridge University Press.

Boolos, G. 1997. Is Hume's Principle Analytic? In R. Heck (ed.), *Language, Thought, and Logic*. Oxford University Press.

Bozianu, R., C. Dima, and C. Enea. 2013. Model checking an Epistemic μ-calculus with Synchronous and Perfect Recall. In B. Schipper (ed.), *Proceedings of the 14th Conference on Theoretical Aspects of Rationality and Knowledge – TARK 2013*.

Bradwardine, T. c.1320/2010. *Insolubilia*, tr. and ed. S. Read. Dallas Medieval Texts and Translations, 10.

Brandom, R. 2008. *Between Saying and Doing*. Oxford University Press.

Brandom, R. 2014. *From Empiricism to Expressivism*. Harvard University Press.

Bratman, M. 1984. Two Faces of Intention. *Philosophical Review*, 93:3.

Brouwer, L.E.J. 1911. Über Abbildungen von Mannigfaltigkeiten. *Mathematische Annalen*, 71.

Buchak, L. 2014. *Risk and Rationality*. Oxford University Press.

Bueno, O. 2017. The Epistemology of Modality and the Epistemology of Mathematics. In B. Fischer and F. Leon, *Modal Epistemology after Rationalism*. Springer.

Bueno, O., and S. Shalkowski. 2015. Modalism and Theoretical Virtues: Toward an Epistemology of Modality. *Philosophical Studies*, 172.

Bull, R., and K. Segerberg. 2001. Basic Modal Logic. In D. Gabbay and F. Guenther (eds.), *Handbook of Philosophical Logic, 2nd ed., Vol. 3*. Kluwer Academic Publishers.

Bulling, N., and W. Jamroga. 2011. Alternating Epistemic Mu-Calculus. *IJCAI'11: Proceedings of the Twenty-Second international joint conference on Artificial Intelligence - Volume One*.

Burali-Forti, C. 1897/1967. A Question on Transfinite Numbers (tr. J. van Heijenoort). In van Heijenoort (ed.), *From Frege to Gödel*. Harvard University Press.

Burge, T. 2010. *Origins of Objectivity*. Oxford University Press.

Burgess, J. 2004/2008. *E Pluribus Unum*: Plural Logic and Set Theory. *Philosophia Mathematica*, 12:3. Reprinted in Burgess (2008), *Mathematics, Models, and Modality*. Oxford University Press.

Burgess, J. 2014. Intuitions of Three Kinds in Gödel's Views on the Continuum. In J. Kennedy (ed.), *Interpreting Gödel*. Cambridge University Press.

Burgess, J., and G. Rosen. 1997. *A Subject with No Object*. Oxford University Press.

Buridan, J. 2001. *Summulae de Dialectica*, tr. G. Klima. Yale University Press.

Caie, M. 2011. Belief and Indeterminacy. *Philosophical Review*, 121:1.

Caie, M. 2013. Rational Probabilistic Incoherence. *Philosophical Review*, 122:4.

Cameron, R. 2008. Truthmakers and Ontological Commitment. *Philosophical Studies*, 140.

Camp, E. 2007. Thinking with Maps. *Philosophical Perspectives*, 21.

Canavotto, I., F. Berto, and A. Giordani. 2022. Voluntary Imagination: a Fine-grained Analysis. *Review of Symbolic Logic*, 15:2.

cantor, G. 1874/1996. "On a Property of the Set of Real Algebraic Numbers", tr. W. Ewald. In Ewald (ed.), *From Kant to Hilbert: A Source Book in the Foundations of Mathematics, Vol. II*. Oxford University Press.

cantor, G. 1882/1996. "The Early Correspondence between Cantor and Dedckind: Halle a. S., 5th Nov. 1882", tr. W. Ewald. In Ewald (1996).

Cantor, G. 1883/1996. *Foundations of a General Theory of Manifolds*, tr. W. Ewald. In Ewald (1996).

Cantor, G. 1891/1996. "On an Elementary Question in the Theory of Manifolds", tr. W. Ewald. In Ewald (1996).

Cantor, G. 1895/2007. *Contributions to the Founding of the Theory of Transfinite Numbers*, tr. P. Jourdain. Cosimo.

cantor, G. 1897a/1996. "Cantor's Late Correspondence with Dedekind and Hilbert: Halle, 26 Sept. 1897", tr. W. Ewald. In Ewald (1996).

cantor, G. 1897b/1996. "Cantor's Late Correspondence with Dedekind and Hilbert: Halle, 2 October 1897", tr. W. Ewald. In Ewald (1996).

cantor, G. 1899/1996. "Cantor's Late Correspondence with Dedekind and Hilbert: Halle, 3 August 1899", tr. W. Ewald. In Ewald (1996).

Cantor, G. 1899/1967. "Letter to Dedekind", tr. S. Bauer-Mengelberg and J. van Heijenoort. In van Heijenoort (ed.), *From Frege to Gödel*. Harvard University Press.

Cantù, P. 2020. Grassmann's Concept Structuralism. In E. Reck and G. Schiemer (eds.), *The Pre-History of Mathematical Structuralism*. Oxford University Press.

Cappelen, H. 2012. *Philosophy without Intuitions*. Oxford University Press.

Cappelen, H. 2018. *Fixing Language: An Essay on the Foundations of Conceptual Engineering*. Oxford University Press.

Cappelen, H. 2020. Conceptual Engineering: The Master Argument. In H. Cappelen, D. Plunkett, and A. Burgess (eds.), *Conceptual Engineering and Conceptual Ethics*. Oxford University Press.

Cariani, F. 2013. 'Ought' and Resolution Semantics. *Nous*, 47:3.

Carlson, T. 2016. Collapsing Knowledge and Epistemic Church's Thesis. In L. Horsten and P. Welch (eds.), *Gödel's Disjunction*. Oxford University Press.

Carnap, R. 1937. *The Logical Syntax of Language*, tr. A. Smeaton. Routledge.

Carnap, R. 1945. On Inductive Logic. *Philosophy of Science*, 12:2.

Carnap, R. 1947. *Meaning and Necessity*. University of Chicago Press.

Carnap, R. 1950. Empiricism, Semantics, and Ontology. *Revue Internationale de Philosophie*, 4.

Carreiro, F., and Y. Venema. 2014. PDL inside the μ-calculus. In R. Gore, B. Kooi, and A. Kurucz (eds.), *Advances in Modal Logic, Vol. 10*. CSLI Publications.

Catren, G. 2023. Abstraction, Equality and Univalence. *Philosophical Transactions of the Royal Society A*, (eds.), G. Catren and F. Holik.

Chalmers, D. 1990. Syntactic Transformations on Distributed Representations. *Connection Science*, 2: 1.

Chalmers, D. 1996. *The Conscious Mind*. Oxford University Press.

Chalmers, D. 2002. Does Conceivability Entail Possibility? In T. Gendler and J. Hawthorne (eds.), *Conceivability and Possibility*. Oxford University Press.

Chalmers, D. 2004. Epistemic Two-dimensional Semantics. *Philosophical Studies*, 118.

Chalmers, D. 2006a. The Foundations of Two-dimensional Semantics. In M. Garcia-Carpintero and J. Macia (eds.), *Two-dimensional Semantics*. Oxford University Press.

Chalmers, D. 2006b. Two-dimensional Semantics. In E. Lepore and B. Smith (eds.), *Oxford Handbook of Philosophy of Language*. Oxford University Press.

Chalmers, D. 2009. Ontological Anti-realism. In Chalmers, D. Manley, and M. Wasserman (eds.), *Metametaphysics*. Clarendon Press.

Chalmers, D. 2010a. *The Character of Consciousness*. Oxford University Press.

Chalmers, D. 2010b. Inferentialism and Analyticity.

Chalmers, D. 2011. The Nature of Epistemic Space. In A. Egan and B. Weatherson (eds.), *Epistemic Modality*. Oxford University Press.

Chalmers, D. 2012. *Constructing the World*. Oxford University Press.

Chalmers, D. 2013. The Contents of Consciousness. *Analysis*, 73:2.

Chalmers, D. 2014a. Strong Necessities and the Mind-Body Problem. *Philosophical Studies*, 167:3.

Chalmers, D. 2014b. Intensions and Indeterminacy. *Philosophy and Phenomenological Research*, 89:1.

Chalmers, D. 2014c. Frontloading and Fregean Sense. *Analysis*, 74:4.

Chalmers, D. 2020. What is Conceptual Engineering and What Should It Be? *Inquiry*, doi: 10.1080/0020174X.2020.1817141.

Chalmers, D. 2021. Inferentialism, Australian-style. *Proceedings and Addresses of the American Philosophical Association*, 92.

Chalmers, D. 2023. The Computational and the Representational Language-of-Thought Hypotheses. *Behavioral and Brain Sciences*, 43, e269. DOI: https://doi.org/10.1017/S0140525X23001796.

Chalmers, D., and B. Rabern. 2014. Two-dimensional Semantics and the Nesting Problem. *Analysis*, 74:2.

Chihara, C. 1990. *Constructibility and Mathematical Existence*. Oxford University Press.

Chihara, C. 2004. *A Structural Account of Mathematics*. Oxford University Press.

Chudnoff, E. 2013. *Intuition*. Oxford University Press.

Church, A. 1936. An Unsolvable Problem of Elementary Number Theory. *American Journal of Mathematics*, 58:2.

Church, A. 1954. Intensional Isomorphism and the Identity of Belief. *Philosophical Studies*, 5:5.

Church, A. 1993. A Revised Formulation of the Logic of Sense and Denotation. *Nous*, 27:2.

Clark, P. 2007. Frege, Neo-Logicism, and Applied Mathematics. In R. Cook (ed.), *The Arché Papers on the Mathematics of Abstraction*. Springer.

Clarke-Doane, J. 2016. What is Benacerraf's Problem? In F. Pataut (ed.), *Truth, Objects, Infinity*. Springer.

Clarke-Doane, J. 2021. Metaphysical and Absolute Possibility. *Synthese*, 198.

Clifford, W.K. 1877. The Ethics of Belief. *Contemporary Review*, 29.

Coliva, A, (ed.). 2012. *Mind, Meaning, and Knowledge: Themes from the Philosophy of Crispin Wright*. Oxford University Press.

Cook, R. 2014a. *The Yablo Paradox*. Oxford University Press.

Cook, R. 2014b. Possible Predicates and Actual Properties. *Synthese*, DOI 10.1007/s11229-014-0592-1.

Cook, R. 2016. Necessity, Necessitism, and Numbers. *Philosophical Forum*, 47: 3-4.

Cook, R., and P. Ebert. 2005. Abstraction and Identity. *Dialectica*, 59:2.

Copeland, B., and O. Shagrir. Turing versus Gödel on Computability and the Mind. In Copeland, C. Posy, and Shagrir (eds.), *Computability*. The MIT Press.

Correia, F., and B. Schnieder (eds.). 2012. *Metaphysical Grounding*. Cambridge University Press.

Cotnoir, A. 2014. Does Universalism Entail Extensionalism? *Nous*, DOI: 10.1111/nous.12063.

Cotnoir, A. ms. Are Ordinary Objects Abstracta?

Cousot, P., and R. Cousot. 1979. Constructive Versions of Tarski's Fixed Point Theorem. *Pacific Journal of Mathematics*, Vol. 82, No. 1.

Crusius, C.A. 1745. *Entwurf der nothwendigen Vernunftwahrheiten* (Sketch of the Necessary Truths of Reason). Leipzig.

Curry, H. 1942. The Inconsistency of Certain Formal Logics. *Journal of Symbolic Logic*, 7(3).

Daston, L. 1988. *Classical Probability in the Enlightenment*. Princeton University Press.

Daston, L. 1994. How Probabilities came to be Objective and Subjective. *Historia Mathematica*, 21.

Davies, M. 2004. Reference, Contingency, and the Two-Dimensional Framework. *Philosophical Studies*, 118.

Davies, M., and L. Humberstone. 1980. Two Notions of Necessity. *Philosophical Studies*, 38.

Davidson, D. 1963. Actions, Reasons, and Causes. *Journal of Philosophy*, 23.

Dean, W. 2016. Algorithms and the Mathematical Foundations of Computer Science. In L. Horsten and P. Welch (eds.), *Gödel's Disjunction*. Oxford University Press.

Dean, W. 2021. Computational Complexity Theory. *The Stanford Encyclopedia of Philosophy* (Fall 2021 Edition), E.N. Zalta (ed.), URL = <https://plato.stanford.edu/archives/fall2021/entries/computational-complexity/>.

Dedekind, R. 1872/1996. 'Continuity and Irrational Numbers', tr. Ewald. In Ewald (1996).

Dedekind, R. 1888/1996. *Was sind und was sollen die Zahlen?*, tr. Ewald. In Ewald (1996).

de Finetti, B. 1937/1964. *Foresight: Its Logical Laws, Its Subjective Sources*, tr. H. Kyburg. In Kyburg and H. Smokler (eds.), *Studies in Subjective Probability*. Wiley.

De Florio, C. and L. Zanetti. 2020. On the Schwartzkopff-Rosen Principle. *Philosophia* 48 (1).

deRosset, L., and Ø. Linnebo. 2023. Abstraction and Grounding. *Philosophy and Phenomenological Research*, https://doi.org/10.1111/phpr.13036.

DeRose, K. 1991. Epistemic Possibilities. *Philosophical Review*, Vol. 100, Issue 4.

Dima, C., B. Maubert, and S. Pinchinat. 2014. The Expressive Power of Epistemic μ-Calculus. arXiv.

Donaldson, T. 2017. The (Metaphysical) Foundations of Arithmetic? *Nous* 51 (4).

Dorr, C. 2003. Vagueness without Ignorance. *Philosophical Perspectives*, 17.

Dorr, C., and J. Goodman. 2019. Diamonds are Forever. *Nous*, https://doi.org/10.1111/nous.12271.

Dowell, J.L. 2013. Flexible Contextualism about Deontic Modals. *Inquiry*, 56.

Dummett, M. 1959. Truth. *Proceedings of the Aristotelian Society, Supplementary Volume*, 59.

Dummett, M. 1963/1978. The Philosophical Significance of Gödel's Theorem. In Dummett (1978), *Truth and Other Enigmas*. Harvard University Press.

Dummett, M. 1978. *Truth and Other Enigmas*. Harvard University Press.

Dummett, M. 1991. *Frege: Philosophy of Mathematics*. Duckworth.

Dummett, M. 1996. What is Mathematics about? In Dummett, *The Seas of Language*. Oxford University Press.

Eagle, A. 2008. Mathematics and Conceptual Analysis. *Synthese*, 161:1.

Easwaran, K. 2014. Decision Theory without Representation Theorems. *Philosophers' Imprint*, 14:27.

Easwaran, K., and B. Fitelson. 2015. Accuracy, Coherence, and Evidence. In T. Gendler and J. Hawthorne (eds.), *Oxford Studies in Epistemology, Volume 5*. Oxford University Press.

Edgington, D. 2004. Two Kinds of Possibility. *Proceedings of the Aristotelian Society, Supplementary Volumes*, Vol. 78.

Eklund, M. 2006. Neo-Fregean Ontology. *Philosophical Perspectives*, 20.

Eklund, M. 2009. Carnap and Ontological Pluralism. In Chalmers, D. Manley, and M. Wasserman (eds.), *Metametaphysics*. Clarendon Press.

Eklund, M. 2016. Hale and Wright on the Metaontology of Neo-Fregeanism. In P. Ebert and M. Rossberg (eds.), *Abstractionism: Essays in Philosophy of Mathematics*. Oxford University Press.

Elohim, D. 2018. Grounding, Conceivability, and the Mind-Body Problem. *Synthese*, 195:2.

Elohim, D. 2019. Hyperintensional Ω-Logic. In M.V. D'Alfonso and D. Berkich (eds.), *On the Cognitive, Ethical, and Scientific Dimensions of Artificial Intelligence*. Springer Verlag.

Enqvist, S., F. Seifan, and Y. Venema. 2019. Completeness for μ-calculi: A Coalgebraic Approach. *Annals of Pure and Applied Logic*, 170.

Evans, G. 1979. Reference and Contingency. *The Monist*, 62:2.

Evans, G. 1982. *The Varieties of Reference*, ed. J. McDowell. Oxford University Press.

Evans, G. 1985. *Collected Papers*. Oxford University Press.

Evans, G. 2004. Comment on 'Two Notions of Necessity'. *Philosophical Studies*, 118:1/2.

Ewald, W. (ed.). 1996. *From Kant to Hilbert: A Source Book in the Foundations of Mathematics, Vol. II*. Oxford University Press.

Fagin, R., J. Halpern, Y. Moses, and M. Vardi. 1995. *Reasoning about Knowledge*. MIT Press.

Fara, D.G. 2000. Shifting Sands. *Philosophical Topics*, 28. Originally published under the name 'Delia Graff'.

Fara, D.G. 2008. Profiling Interest Relativity. *Analysis*, 68:4.

Feferman, S. 1984. Toward Useful Type-free Theories, I. *Journal of Symbolic Logic*, 49:1.

Feferman, S. 1986. Gödel's Life and Work. In Gödel (1986), *Collected Works, Volume I*, eds. S. Feferman, J. Dawson, S. Kleene, G. Moore, R. Solovay, and J. van Heijenoort. Oxford University Press.

Ferrari, M., and S. Adam-Day. 2017. Handout. 'Homotopy Type Theory, Logic, Sets and n-Types'. 11am, Wednesday 4th October 2017.

Field, H. 1980/2016. *Science without Numbers*, 2nd ed. Oxford University Press.

Field, H. 1989. *Realism, Mathematics, and Modality*. Basil Blackwell.

Field, H. 2001. *Truth and the Absence of Fact*. Oxford University Press.

Field, H. 2004. The Consistency of the Naive Theory of Properties. *Philosophical Quarterly*, 54:214.

Field, H. 2008. *Saving Truth from Paradox*. Oxford University Press.

Field, H. 2015. What is Logical Validity? In C. Caret and O. Hjortland (eds.), *Foundations of Logical Consequence*. Oxford University Press.

Fine, K. 1974. Logics containing K4: Part I. *Journal of Symbolic Logic*, 39:1.

Fine, K. 1981. First-order Modal Theories I – Sets. *Nous*, 15:2.

Fine, K. 1985. *Reasoning with Arbitrary Objects*. Blackwell Publishing.

Fine, K. 1985. The Logic of Essence. *Journal of Philosophical Logic*, 24:3.

Fine, K. 1994. Essence and Modality. *Philosophical Perspectives*, 8.

Fine, K. 2000. Semantics for the Logic of Essence. *Journal of Philosophical Logic*, 29.

Fine, K. 2001. The Question of Realism. *Philosophers' Imprint*, 1:1.

Fine, K. 2002. *The Limits of Abstraction*. Oxford University Press.

Fine, K. 2005a. *Modality and Tense*. Oxford University Press.

Fine, K. 2005b. Our Knowledge of Mathematical Objects. In T. Gendler and J. Hawthorne (eds.), *Oxford Studies in Epistemology, Volume 1*. Oxford University Press.

Fine, K. 2005c. Class and Membership. *Journal of Philosophy*, 102:11.

Fine, K. 2006. Relatively Unrestricted Quantification. In A. Rayo and G. Uzquiano (eds.), *Absolute Generality*. Oxford University Press.

Fine, K. 2007. Response to Weir. *Dialectica*, 61:1.

Fine, K. 2012a. A Difficulty for the Possible Worlds Analysis of Counterfactuals. *Synthese*, DOI 10.1007/s11229-012-0094-y.

Fine, K. 2012b. Counterfactuals without Possible Worlds. *Journal of Philosophy* 109:3.

Fine, K. 2012c. The Pure Logic of Ground. *Review of Symbolic Logic*, 5:1.

Fine, K. 2012d. Guide to Ground. In F. Correia and B. Schnieder (eds.), *Metaphysical Grounding*. Cambridge University Press.

Fine, K. 2013. Truth-maker Semantics for Intuitionistic Logic. *Journal of Philosophical Logic*, 43.

Fine, K. 2015a. Angellic Content. *Journal of Philosophical Logic*, DOI 10.1007/s10992-015-9371-9.

Fine, K. 2015b. Identity Criteria and Ground. *Philosophical Studies*, DOI 10.1007/s11098-014-0440-7.

Fine, K. 2015c. Unified Foundations for Essence and Ground. *Journal of the American Philosophical Association*, 1:2.

Fine, K. 2017a. A Theory of Truthmaker Content I: Conjunction, Disjunction, and Negation. *Journal of Philosophical Logic*, 46:6.

Fine, K. 2017b. A Theory of Truthmaker Content II: Subject-matter, Common Content, Remainder, and Ground. *Journal of Philosophical Logic*, 46:6.

Fine, K. 2017c. Truthmaker Semantics. In B. Hale, C. Wright, and A. Miller (eds.), *A Companion to Philosophy of Language*. Blackwell.

Fine, K. 2020. Comments on Steven T. Kuhn's "Necessary, Transcendental, and Universal Truth", in M. Dumitru (ed.), *Metaphysics, Meaning, and Modality*. Oxford University Press.

Fine, K. 2021. Constructing the Impossible. In L. Walters and J. Hawthorne (eds.), *Conditionals, Paradox, and Probability: Themes from the Philosophy of Dorothy Edgington*. Oxford University Press.

Fine, K. 2023. An Epistemized Truthmaker Semantics for Epistemic Logic: Response to Hawke's and Özgün's "Truthmaker Semantics for Epistemic

Logic". In F. Faroldi and F. van de Putte (eds.), *Kit Fine on Truthmakers, Relevance, and Non-Classical Logic*. Springer.

Fine, K., and M. Jago. 2019. Logic for Exact Entailment. *Review of Symbolic Logic*,12:3.

Fintel, K. von, and I. Heim. 2011. *Intensional Semantics*. MIT.

Fitelson, B. 2001. *Studies in Confirmation Theory*. Ph.D. Dissertation, University of Wisconsin - Madison.

Flagg, R. 1985a. Epistemic Set Theory is a Conservative Extension of Intuitionistic Set Theory. *Journal of Symbolic Logic*, 50:4.

Flagg, R. 1985b. Church's Thesis is Consistent with Epistemic Arithmetic. In S. Shapiro (ed.), *Intensional Mathematics*. Elsevier Science Publishers.

Fodor, J. 1975. *The Language of Thought*. Harvard University Press.

Fodor. J. 1981. *Representations*. MIT Press.

Fodor, J. 1987. *Psychosemantics*. MIT Press.

Fodor, J. 2000. *The Mind Doesn't Work that Way*. MIT Press.

Fontaine, G. 2010. *Modal Fixpoint Logic*. ILLC Dissertation Series DS-2010-09.

Fontaine, G., and T. Place. 2010. Frame Definability for Classes of Trees in the μ-calculus. In P. Hlineny and A. Kucera (eds.), *Mathematical Foundations of Computer Science 2010*. Springer.

Fontaine, G., and Y. Venema. 2018. Some Model Theory for the Modal μ-Calculus. *Logical Methods in Computer Science*, Vol. 14(1:14).

Frege, G. 1884/1980. *The Foundations of Arithmetic*, 2nd ed., tr. J.L. Austin. Northwestern University Press.

Frege, G. 1893/2013. *Basic Laws of Arithmetic, Vol. I-II*, tr. and ed. P. Ebert, M. Rossberg, C. Wright, and R. Cook. Oxford University Press.

Frith, C., S.-J. Blakemore, and D. Wolport. 2000. Abnormalities in the Awareness and Control of Action. *Philosophical Transactions of the Royal Society of London B*, 355.

Fritz, P. 2016. First-order Modal Logic in the Necessary Framework of Objects. *Canadian Journal of Philosophy*, 46:4-5.

Fritz, P. 2017. A Purely Recombinatorial Puzzle. *Nous*, 51:3.

Fritz, P., and J. Goodman. 2017. Counting Incompossibles. *Mind*, 126:504.

Gendler, T. 2000. *Thought Experiments*. Routledge.

Gendler, T. 2010. *Intuition, Imagination, and Philosophical Methodology*. Oxford University Press.

Gendler, T., and J. Hawthorne (eds.). 2002. *Conceivability and Possibility*. Oxford University Press.

Gentzen, G. 1936. Die Widerspruchsfreiheit der reinen Zahlentheorie. *Mathematische Annalen*, 112. Reprinted as "The Consistency of Elementary Number Theory", in Gentzen, 1969. *The Collected Papers of Gerhard Gentzen*, tr. and ed. M. Szabo. North-Holland Publishing Company.

Gertler, B. 2021. Self-Knowledge. *The Stanford Encyclopedia of Philosophy* (Summer 2024 Edition), E.N. Zalta U. Nodelman (eds.), URL = <https://plato.stanford.edu/archives/sum2024/entries/self-knowledge/>.

Giaquinto, M. 2008. Visualizing in Mathematics. In P. Mancosu (ed.), *The Philosophy of Mathematical Practice*. Oxford University Press.

Gödel, K. 1929. *On the Completeness of the Calculus of Logic*. In Gödel (1986), *Collected Works, Volume I*, eds. S. Feferman, J. Dawson, S. Kleene, G. Moore, R. Solovay, and J. van Heijenoort. Oxford University Press.

Gödel, K. 1931. On Formally Undecidable Propositions of *Principia Mathematica* and Related Systems I. In Gödel (1986).

Gödel, K. 1932b. On Completeness and Consistency. In Gödel (1986).

Gödel, K. 1933. An Interpretation of the Intuitionistic Propositional Calculus. In Gödel (1986).

Gödel. K. 1934. *On Undecidable Propositions of Formal Mathematical Systems*. In Gödel (1986).

Gödel, K. 1938. Lecture at Zilsel's. In Gödel (1995), *Collected Works, Volume III*, eds. S. Feferman, J. Dawson, W. Goldfarb, C. Parsons, and R. Solovay. Oxford University Press.

Gödel, K. 193?. Undecidable Diophantine Propositions. In Gödel (1995).

Gödel, K. 1940. *The Consistency of the Axiom of Choice and of the Continuum Hypothesis with the Axioms of Set Theory*. In Gödel (1990), *Collected Works, Volume II*, eds. S. Feferman, J. Dawson, S. Kleene, G. Moore, R. Solovay, and J. van Heijenoort. Oxford University Press.

Gödel, K. 1946. Remarks before the Princeton Bicentennial Conference on Problems in Mathematics. In Gödel (1990).

Gödel, K. 1947/1964. What is Cantor's Continuum Problem? In Gödel (1990).

Gödel, K. 1951. Some Basic Theorems on the Foundations of Mathematics and their Implications. In Gödel (1995).

Gödel, K. 1953/9-III. Is Mathematics Syntax of Language? In Gödel (1995).

Gödel, K. 1953/9-V. Is Mathematics Syntax of Language? In Gödel (1995).

Gödel, K. 1961. The Modern Development of the Foundations of Mathematics in the Light of Philosophy. In Gödel (1995).

Gödel, K. 1963. Tillich ca. June $\overline{63}$. In Gödel Papers 3b/188, 012868.5. Firestone Library, Princeton.

Gödel, K. 1972a. On an Extension of Finitary Mathematics which has not yet been used. In Gödel (1990).

Gödel, K. 1972b. Some Remarks on the Undecidability Results. In Gödel (1990).

Gödel. 1986. *Collected Works, Volume I*, eds. S. Feferman, J. Dawson, S. Kleene, G. Moore, R. Solovay, and J. van Heijenoort. Oxford University Press.

Gödel. 1990. *Collected Works, Volume II*, eds. S. Feferman, J. Dawson, S. Kleene, G. Moore, R. Solovay, and J. van Heijenoort. Oxford University Press.

Gödel. 1995. *Collected Works, Volume III*, eds. S. Feferman, J. Dawson, W. Goldfarb, C. Parsons, and R. Solovay. Oxford University Press.

Goldblatt, R. 1987. *Logics of Time and Computation*. CSLI.

Goldblatt, R. 1993. *Mathematics of Modality*. CSLI Publications.

Goldblatt, R. 2006. Mathematical Modal Logic: A View of its Evolution. In D. Gabbay and J. Woods (eds.), *Handbook of the History of Logic, Volume 7: Logic and the Modalities in the Twentieth Century*. Elsevier.

Gómez-Torrente, M. 2002. Vagueness and Margin for Error Principles. *Philosophy and Phenomenological Research*, 64:1.

Goodman, N.D. 1985. A Genuinely Intensional Set Theory. In Shapiro (ed.), *Intensional Mathematics*. Elsevier Science Publishers.

Goodman, N.D. 1990. Topological Models of Epistemic Set Theory. *Annals of Pure and Applied Mathematics*, 46.

Grassmann, H. 1844/1995. *Linear Extension Theory*. In Grassmann, *A New Branch of Mathematics: The 'Ausdehnungslehere' of 1844, and Other Works*, tr. L. Kannenberg. Open Court Publishing Company.

Gurevich, Y. 1999. The Sequential ASM Thesis. *Bulletin of the European Association for Theoretical Computer Science*, 67.

Haas-Spohn, U. 1995. *Versteckte Indexikalität und subjektive Bedeutung*. Akademie Verlag.

Halbach, V., and A. Visser. 2014. Self-reference in Arithmetic I. *Review of Symbolic Logic*, 7:4.

Halbach, V., and P. Welch. 2009. Necessities and Necessary Truths. *Mind*, 118.

Halbach, V., H. Leitgeb, and P. Welch. 2003. Possible-Worlds Semantics for Modal Notions Conceived as Predicates. *Journal of Philosophical Logic*, 32, 2.

Hale, B. 1987. *Abstract Objects*. Basil Blackwell.

Hale, B. 2000a. Reals by Abstraction. *Philosophia Mathematica*, 3:8. Reprinted in Hale and Wright (2001).

Hale, B. 2000b. Abstraction and Set Theory. *Notre Dame Journal of Formal Logic*, 41:4.

Hale, B. 2013a. *Necessary Beings*. Oxford University Press.

Hale, B. 2013b. Properties and the Interpretation of Second-order Logic. *Philosophia Mathematica*, 3:21.

Hale, B. 2020. CCCP. In A. Miller (ed.), *Logic, Language, and Mathematics: Themes from the Philosophy of Crispin Wright*. Oxford University Press.

Hale, B., and C. Wright. 2001. *The Reason's Proper Study*. Oxford University Press.

Hale, B., and C. Wright. 2002. Benacceraf's Dilemma Revisited. *European Journal of Philosophy*, 10:1.

Hale, B., and C. Wright. 2005. Logicism in the Twenty-first Century. In S. Shapiro (ed.), *The Oxford Handbook of Philosophy of Mathematics and Logic*. Oxford University Press.

Hale, B., and C. Wright. 2009. The Metaontology of Abstraction. In D. Chalmers, D. Manley, and R. Wasserman (eds.), *Metametaphysics*. Oxford University Press.

Hamblin, C. 1958. Questions. *Australasian Journal of Philosophy*, 36:3.

Hamblin, C. 1973. Questions in Montague English. *Foundations of Language*, 10.

Hamkins, J. 2012. The Set-theoretic Multiverse. *Review of Symbolic Logic*, 5:3.

Hamkins, J., and B. Löwe. 2008. The Modal Logic of Forcing. *Transactions of the American Mathematical Society*, 360:4.

Hamkins, J., and B. Löwe. 2013. Moving Up and Down in the Generic Multiverse. In K. Lodaya (ed.), *Logic and Its Applications*. Springer.

Hamkins, J., and Ø. Linnebo. 2022. The Modal Logic of Set-theoretic Potentialism and the Potentialist Maximality Principles. *Review of Symbolic Logic*, 15:1.

Hardy, G., and E.M. Wright. 1979. *An Introduction to the Theory of Numbers*, 5th ed. Oxford University Press.

Haslanger, S. 2000. Gender and Race. *Nous*, 34.

Haslanger, S. 2021. How not to change the Subject. In T. Marques and A. Wikforss (eds.), *Shifting Concepts*. Oxford University Press.

Haslanger, S. 2012. *Resisting Reality*. Oxford University Press.

Haugeland, J. 1978. The Nature and Plausibility of Cognitivism. *Behavioral and Brain Sciences*, 2.

Hawke, P. 2024. Modal Knowledge for Expressivists. *Journal of Philosophical Logic*, 53:4.

Hawke, P., and S. Steinert-Threlkeld. 2021. Semantic Expressivism for Epistemic Modals. *Linguistics and Philosophy*.

Hawke, P., and A. Özgün. 2023. Truthmaker Semantics for Epistemic Logic. In F. Faroldi and F. van de Putte (eds.), *Kit Fine on Truthmakers, Relevance, and Non-Classical Logic*. Springer.

Hawley, K. 2007. Neo-Fregeanism and Quantifier Variance. *Proceedings of the Aristotelian Society Supplementary Volume*, 81.

Hawthorne, J. 2002. Deeply Contingent A Priori Knowledge. *Philosophy and Phenomenological Research*, Vol. LXV, No. 2.

Hawthorne, J. 2005. Vagueness and the Mind of God. *Philosophical Studies*, 122.

Hazen, A. 1985. Review of Crispin Wright's *Frege's Conception of Numbers as Objects*. *Australasian Journal of Philosophy*, 63:2.

Heck, R. 1992. On the Consistency of Second-order Contextual Definitions. *Nous*, 26.

Heck, R. 2011. *Frege's Theorem*. Oxford University Press.

Heim, I. 1992. Presupposition Projection and the Semantics of Attitude Verbs. *Journal of Semantics*, 9.

Hellman, G. 1990. Toward a Modal-Structural Interpretation of Set Theory. *Synthese*, 84.

Hellman, G. 1993. *Mathematics without Numbers*. Oxford University Press.

Helmholtz, H. von. 1878/1977. The Facts in Perception. In Helmholtz, *Epistemological Writings*, tr. M. Lowe, eds. R. Cohen and Y. Elkana. D. Reidel Publishing.

Henkin, L. 1952. A Problem concerning Provability. *Journal of Symbolic Logic*, 17:2.

Henkin, L., J.D. Monk, and A. Tarski. 1971. *Cylindric Algebras*, Part I. North-Holland.

Hilbert, D. Letter from Hilbert to Frege, 29.12.1899. In G. Frege, *Philosophical and Mathematical Correspondence*, tr. H. Kaal, ed., G. Gabriel, H. Hermes, F. Kambartel, C. Thiel, and A. Veraart. Basil Blackwell.

Hilbert, D. 1923/1996. 'The Logical Foundations of Mathematics', tr. Ewald. In Ewald (1996).

Hill, C. 1997. Imaginability, Conceivability, Possibility, and the Mind-Body Problem. *Philosophical Studies*, 87:1.

Hill, C. 2006. Modality, Modal Epistemology, and the Metaphysics of Consciousness. In S. Nichols (ed.), *The Architecture of the Imagination*. Oxford University Press.

Hilpinen, R. (ed.). 1981. *New Studies in Deontic Logic*. D. Reidel Publishing.

Hintikka, J. 1962. *Knowledge and Belief.* Cornell University Press.

Hjortland, O. 2016. Anti-exceptionalism about Logic. *Synthese*, DOI 10.1007/s11098-016-0701-8.

Hoare, C. 1969. An Axiomatic Basis for Computer Programming. *Communications of the Association for Computing Machinery*, 12.

Hodes, H. 1984. Logicism and the Ontological Commitments of Arithmetic. *Journal of Philosophy*, 81:3.

Hodges, W. 1997. Compositional Semantics for a Language of Imperfect Information. *Logic Journal of the IGPL*, 5:4.

Holliday, W. 2021. Possibility Semantics. In M. Fitting (ed.), *Selected Topics from Contemporary Logics*. College Publications.

Holliday, W., and M. Mandelkern. 2024. The Orthologic of Epistemic Modals. *Journal of Philosophical Logic*, 53:4.

Hornsby, J. 1997. The Presidential Address: Truth: The Identity Theory. *Proceedings of the Aristotelian Society, New Series*, Vol. 97.

Horsten, L. 1994. Modal-Epistemic Variants of Shapiro's System of Epistemic Arithmetic. *Notre Dame Journal of Formal Logic*, Vol. 35, No. 2.

Horsten, L. 1998. In Defense of Epistemic Arithmetic. *Synthese*, Vol. 116, No. 1.

Horsten, L. 2010. Perceptual Indiscriminability and the Concept of a Color Shade. In R. Dietz and S. Moruzzi (eds.), *Cuts and Clouds*. Oxford University Press.

Horsten, L., and P. Welch (eds.). 2016. *Gödel's Disjunction*. Oxford University Press.

Horty, J. 2001. *Agency and Deontic Logic*. Oxford University Press.

Horty, J., and N. Belnap. 1995. The Deliberative Stit. *Journal of Philosophical Logic*, 24:6.

Huemer, M. 2007. Epistemic Possibility. *Synthese*, Vol. 156, No. 1.

Hughes, J. 2001. *A Study of Categories of Algebras and Coalgebras*. Ph.D. Dissertation, Carnegie Mellon University.

Hume, D. 1739-1740/2007. *A Treatise of Human Nature*, eds. D.F. Norton and M.J. Norton. Oxford University Press.

Husserl, E. 1921/2001. *Logical Investigations, Vol. II*, trans. J.N. Findlay, ed. D. Moran. Routledge.

Husserl, E. 1929/1999. *Formal and Transcendental Logic*, tr. D. Cairns. In D. Welton (ed.), *The Essential Husserl*. Indiana University Press.

Incurvati, L., and J. Schlöder. 2020. Epistemic Multilateral Logic. *Review of Symbolic Logic*, https://doi.org/10.1017/S1755020320000313.

Incurvati, L., and J. Schlöder. 2021. Inferential Expressivism and the Negation Problem. *Oxford Studies in Metaethics*, 16.

Jacobi, F. 1787. *David Hume on Faith, or Idealism or Realism, a Dialogue*. In Jacobi (1994), *The Main Philosophical Writings and the Novel* Allwill, tr. and ed. G. di Giovanni. McGill-Queen's University Press.

Jackson, F. 1982. Epiphenomenal Qualia. *Philosophical Quarterly*, 32.

Jackson, F. 2011. Possibilities for Representation and Credence. In A. Egan and B. Weatherson (eds.), *Epistemic Modality*. Oxford University Press.

Jackson, F., K. Mason, and S. Stich. 2008. Folk Psychology and Tacit Theories. In D. Braddon-Mitchell and R. Nola (eds.), *Conceptual Analysis and Philosophical Naturalism*. MIT Press.

Jago, M. 2009. Logical Information and Epistemic Space. *Synthese*, 167(2).

Jago, M. 2014. The Problem of Rational Knowledge. *Erkenntnis*, 79.

James, W. 1896. The Will to Believe. *The New World*, 5.

Janin, D., and I. Walukiewicz. 1996. On the Expressive Completeness of the Propositional μ-Calculus with respect to Monadic Second-order Logic. In U. Montanari and V. Sasone (eds.), *Proceedings CONCUR, '96*. Springer.

Jech, T. 2003. *Set Theory*, 3rd Millennium ed. Springer.

Jeffrey, R. 1965/1990. *The Logic of Decision*. University of Chicago Press.

Johnson-Laird, P.N. 2004. The History of Mental Models. In K. Manktelow and M. Chung (eds.), *Psychology of Reasoning*. Psychology Press.

Jónsson, B., and A. Tarski, 1951. Boolean Algebras with Operators. Part I. *American Journal of Mathematics*, Vol. 73, No. 4.

Jónsson, B., and A. Tarski. 1952. Boolean Algebras with Operators. *American Journal of Mathematics*, Vol. 74, No. 1.

Joyce, J. 1998. A Non-pragmatic Vindication of Probabilism. *Philosophy of Science*, 65:4.

Joyce, J. 1999. *The Foundations of Causal Decision Theory*. Cambridge University Press.

Joyce, J. 2009. Accuracy and Coherence. In F. Huber and C. Schmidt-Petri (eds.), *Degrees of Belief*. Springer.

Joyce, J. 2011. The Development of Subjective Bayesianism. In D. Gabbay, S. Hartmann, and J. Woods (eds.), *Handbook of the History of Logic, Volume 10: Inductive Logic*. Elsevier.

Kalocinski, D., and M. Godziszewski. 2018. Semantics of the Barwise Sentence: Insights from Expressiviness, Complexity and Inference. *Linguistics and Philosophy*, Vol. 41, No. 4.

Kamp, H. 1967. The Treatment of 'Now' as a 1-Place Sentential Operator. multilith, UCLA.

Kanamori, A. 2007. Gödel and Set Theory. *Bulletin of Symbolic Logic*, 13:2.

Kanamori, A. 2008. Cohen and Set Theory. *Bulletin of Symbolic Logic*, 14:3.

Kanamori, A. 2009. *The Higher Infinite*, 2nd ed. Springer-Verlag.

Kanamori, A. 2012a. Set Theory from Cantor to Cohen. In D. Gabbay, A. Kanamori, and J. Woods (eds.), *Handbook of the History of Logic, Volume 6: Sets and Extensions in the Twentieth Century*. Elsevier.

Kanamori, A. 2012b. Large Cardinals with Forcing. In D. Gabbay, A. Kanamori, and J. Woods (eds.), *Handbook of the History of Logic, Volume 6: Sets and Extensions in the Twentieth Century*. Elsevier.

Kant, I. 1763/1992. *Attempt to introduce the Concept of Negative Magnitudes into Philosophy*, tr. and ed. D. Walford and R. Meerbote. In Kant, *Theoretical Philosophy, 1755-1770*, tr. and ed. D. Walford and R. Meerbote. Cambridge University Press.

Kant, I. 1787/1998. *Critique of Pure Reason*, tr. and ed. P. Guyer and A. Wood. Cambridge University Press.

Kaplan, D. 1975. How to Russell a Frege-Church. *Journal of Philosophy*, 72.

Kaplan, D. 1979. On the Logic of Demonstratives. *Journal of Philosophical Logic*, 8:1.

Kaplan, D. 1995. A Problem in Possible-World Semantics. In W. Sinnott-Armstrong, D. Raffman, and N. Asher (eds.), *Modality, Morality, and Belief: Essays in Honor of Ruth Barcan Marcus*. Cambridge University Press.

Katz, J. 1998. *Realistic Rationalism*. MIT Press.

Kennedy, J. 2013. On Formalism Freeness: Implementing Gödel's 1946 Princeton Bicentennial Lecture. *Bulletin of Symbolic Logic*, 19:3.

Kleene, S. 1952/1971. *Introduction to Metamathematics*. Woltors-Noordhoff Publishing and North-Holland Publishing.

Kleyko, D., Davies, M., Frady, E. P., Kanerva, P., Kent, S. J., Olshausen, B. A., Osipov, E., and Rabaey, J. M. 2022. Vector Symbolic Architectures as a Computing Framework for Emerging Hardware. *Proceedings of the IEEE*, 110: 1538.

Kment, B. 2012. Haecceitism, Chance, and Counterfactuals. *Philosophical Review*, 121:4.

Knaster, B. 1928. Un Théorème sur les Fonctions d'Ensembles. *Annales Polonici Mathematici*, 6.

Kneale, M. 1938. Logical and Metaphysical Necessity. *Proceedings of the Aristotelian Society*, 38, 1.

Knuuttila, S. 1981. The Emergence of Deontic Logic in the Fourteenth Century. In Hilpinen (1981).

Koch, S. 2023. Why Conceptual Engineers should not worry about Topics. *Erkenntnis*, 88.

Koellner, P. 2006. On the Question of Absolute Undecidability. *Philosophia Mathematica*, (III), 14.

Koellner, P. 2009. On Reflection Principles. *Annals of Pure and Applied Logic*, 157.

Koellner, P. 2010. On Strong Logics of First and Second Order. *Bulletin of Symbolic Logic*, 16:1.

Koellner, P. 2014. Large Cardinals and Determinacy. *The Stanford Encyclopedia of Philosophy* (Spring 2014 Edition), E.N. Zalta (ed.), URL = <https://plato.stanford.edu/archives/spr2014/entries/large-cardinals-determinacy/>.

Koellner, P. 2016. Gödel's Disjunction. In L. Horsten and P. Welch (eds.), *Gödel's Disjunction*. Oxford University Press.

Koellner, P. 2018a. On the Question of Whether the Mind Can Be Mechanized, I: From Gödel to Penrose. *The Journal of Philosophy*, 115, 7.

Koellner, P. 2018b. On the Question of Whether the Mind Can Be Mechanized, II: Penrose's New Argument. *The Journal of Philosophy*, 115, 9.

Koellner, P., and W.H. Woodin. 2010. Large Cardinals from Determinacy. In M. Foreman and A. Kanamori (eds.), *Handbook of Set Theory, Volume 3*. Springer.

Koopman, B. 1940. The Axioms and Algebra of Intuitive Probability. *Annals of Mathematics*, Second Series, 41:2.

Korbmacher, J. 2016. *Properties grounded in Identity: A Study of Essential Properties*. PhD Dissertation, LMU München: Faculty of Philosophy, Philosophy of Science and the Study of Religion.

Kratzer, A. 1977/2012. What 'Must' and 'Can' Must and Can Mean. *Linguistics and Philosophy*, 1:3. Reprinted in Kratzer, *Modals and Conditionals*. Oxford University Press.

Kratzer, A. 1979. Conditional Necessity and Possibility. In R. Bäuerle, U. Egli, and A.v. Stechow (eds.), *Semantics from Different Points of View*. Springer.

Kratzer, A. 1981/2012. Partition and Revision: the Semantics of Counterfactuals. *Journal of Philosophical Logic*, 10:2. Reprinted in Kratzer (2012).

Kremer, P. 2009. Dynamic Topological S5. *Annals of Pure and Applied Logic*, 160.

Kriegel, U. 2015. *The Varieties of Consciousness*. Oxford University Press.

Kripke, S. 1963. Semantical Considerations on Modal Logic. *Acta Philosophica Fennica*, 16.

Kripke, S. 1965. Semantical Analysis of Intuitionistic Logic I. In J. Crossley and M. Dummett (eds.), *Formal Systems and Recursive Functions*. North Holland.

Kripke, S. 1972. Naming and Necessity. In D. Davidson and G. Harman (eds.), *Semantics of Natural Language*, 2nd ed. D. Reidel.

Kripke, S. 1980. *Naming and Necessity*. Harvard University Press.

Kuhn, S. 1980. Quantifiers as Modal Operators. *Studia Logica*, 39:2-3.

Kuhn. S. 2020. Necessary, Transcendental, and Universal Truth. In M. Dumitru (ed.), *Metaphysics, Meaning, and Modality*. Oxford University Press.

Kurz, A., and A. Palmigiano. 2013. Epistemic Updates on Algebras. *Logical Methods in Computer Science*, 9:4:17.

Lambek, J. 1968. A Fixpoint Theorem for Complete Categories. *Mathematische Zeitschrift*, Vol. 103.

Lange, M. 2017. *Because without Cause*. Oxford University Press.

Laplace, P. 'Memoir on the Probability of the Causes of Events', tr. S. Stigler. *Statistical Science*, 1:3.

Lappin, S. 2014. Intensions as Computable Functions. *Linguistic Issues in Language Technology*, Vol. 9.

Lawvere, F.W. 2005. An Elementary Theory of the Category of Sets. *Reprints in Theory and Applications of Categories*, 11.

Leach-Krouse, G. ms. Ω-Consequence Interpretations of Modal Logic.

Leech, J. 2019. Martha Kneale on Why Metaphysical Necessities are not A Priori. *Journal of the American Philosophical Association*, 5:4.

Leibniz, G.W. 1704/1981. *New Essays on Human Understanding*, tr. and ed. P. Remnant and J. Bennett. Cambridge University Press.

Leibniz, G.W. 1714/1998. *Monadology*, tr. R. Franks and R.S. Woolhouse. In Franks and Woolhouse (eds.), *G.W. Leibniz: Philosophical Texts*. Oxford University Press.

Leitgeb, H. 2009. On Formal and Informal Provability. In O. Bueno and Ø. Linnebo (eds.), *New Waves in Philosophy of Mathematics*. Palgrave Macmillan.

Leitgeb, H., and R. Pettigrew. 2010. An Objective Justification of Bayesianism I: Measuring Inaccuracy. *Philosophy of Science*, 77.

Lemmon, E.J. (with D. Scott). 1966/1977. Intensional Logic/*An Introduction to Modal Logic*, ed. K. Segerberg. In N. Rescher (ed.), *American Philosophical Quarterly Monograph Series*, Vol. 11. Basil Blackwell.

Leng, M. 2007. What's There to Know? In Leng, A. Paseau, and M. Potter (eds.), *Mathematical Knowledge*. Oxford University Press.

Leng, M. 2009. 'Algebraic' Approaches to Mathematics. In O. Bueno and Ø. Linnebo (eds.), *New Waves in Philosophy of Mathematics*. Palgrave Macmillan.

Leng, M. 2010. *Mathematics and Reality*. Oxford University Press.

Lessing, G.E. 1777/2005. "On the Proof of Spirit and of Power", tr. and ed. H. Nisbett. In Lessing, *Philosophical and Theological Writings*, tr. and ed. Nisbett. Cambridge University Press.

Levine, J. 1983. Materialism and Qualia: The Explanatory Gap. *Pacific Philosophical Quarterly*, 64.

Lewis, C.I. 1923. A Pragmatic Conception of the Apriori. *Journal of Philosophy*, 20:7.

Lewis, C.I. 1932. Alternative Sytems of Logic. *The Monist*, Vol. 42, No. 4.

Lewis, D. 1969. Review of *Art, Mind, and Religion*. *Journal of Philosophy*, 66:1.

Lewis, D. 1972. Psychophysical and Theoretical Identifications. *Australasian Journal of Philosophy*, 50:3.

Lewis, D. 1973. *Counterfactuals*. Blackwell.

Lewis, D. 1979. Attitudes De Dicto and De Se. *Philosophical Review*, 88:4.

Lewis, D. 1980. Index, Context, and Content. In S. Kanger and S. Öhman (eds.), *Philosophy and Grammar*. D. Reidel Publishing. Reprinted in Lewis (1998), *Papers in Philosophical Logic*. Cambridge University Press.

Lewis, D. 1980/1987. A Subjectivist's Guide to Objective Chance. In Lewis (1987), *Philosophical Papers, Volume 2*. Oxford University Press.

Lewis, D. 1986. *On the Plurality of Worlds*. Basil Blackwell.

Lewis, D. 1988/1998. Statements Partly about Observation. *Philosophical Papers*, 17:1. Reprinted In Lewis (1998).

Linnebo, Ø. 2006. Sets, Properties, and Unrestricted Quantification. In A. Rayo and G. Uzquiano (eds.), *Absolute Generality*. Oxford University Press.

Linnebo, Ø. 2007. Burgess on Plural Logic and Set Theory. *Philosophia Mathematica*, 3:15.

Linnebo, Ø. 2008. Structuralism and the Notion of Dependence. *Philosophical Quarterly*, 58:230.

Linnebo, Ø. 2009. Bad Company Tamed. *Synthese*, 170.

Linnebo, Ø. 2010. Some Criteria on Acceptable Abstraction. *Notre Dame Journal of Formal Logic*, 52:3.

Linnebo, Ø. 2013. The Potential Hierarchy of Sets. *Review of Symbolic Logic*, 6:2.

Linnebo, Ø. 2018a. *Thin Objects: An Abstractionist Account.* Oxford University Press.

Linnebo, Ø. 2018b. Putnam on Mathematics as Modal Logic. In G. Hellmann and R. Cook (eds.), *Hilary Putnam on Logic and Mathematics*. Springer.

Linnebo, Ø. 2018c. Dummett on Indefinite Extensibility. *Philosophical Issues*, 28.

Linnebo, Ø. 2023. Platonism in the Philosophy of Mathematics. *The Stanford Encyclopedia of Philosophy* (Summer 2023 Edition), E.N. Zalta and U. Nodelman (eds.). URL = <https://plato.stanford.edu/archives/sum2023/entries/platonism-mathematics/>.

Linnebo, Ø., and R. Pettigrew. 2014. Two Types of Abstraction for Structuralism. *Philosophical Quarterly*, doi: 10.1093/pq/pqt044.

Linnebo, Ø., and A. Rayo. 2012. Hierarchies, Ontological and Ideological. *Mind*, 121:482.

Linnebo, Ø., and S. Shapiro. 2020. Realizability as a Kind of Truth-Making. In M. Szatkowski (ed.)., *Quo Vadis, Metaphysics?*. De Gruyter.

Linnebo, Ø., and G. Uzquiano. 2009. What Abstraction Principles Are Acceptable? Some Limitative Results. *British Journal for the Philosophy of Science*, 60.

Liu, T., J. Abrams, and M. Carrasco. 2009. Voluntary Attention Enhances Contrast Appearance. *Psychological Science*, 20:3.

Löb, M.H. 1955. Solution of a Problem of Leon Henkin. *Journal of Symbolic Logic*, 20:2.

Lowe, E.J. 2008. Two Notions of Being: Entity and Essence. *Royal Institute of Philosophy Supplements*, Volume 62.

Lowe, E.J. 2012. What is the Source of Our Knowledge of Modal Truths? *Mind*, Volume 121, Issue 484.

Lucas, J.R. 1961. Minds, Machines, and Gödel. *Philosophy*, Vol. 36, No. 137.

MacFarlane, J. 2005. Making Sense of Relative Truth. *Proceedings of the Aristotelian Society*, 105.

MacFarlane, J. 2011. Epistemic Modals are Assessment-Sensitive. In A. Egan and B. Weatherson (eds.), *Epistemic Modality*. Oxford University Press.

Machery, E. 2005. You Don't Know How You Think: Introspection and the Language of Thought. *British Journal for the Philosophy of Science*, 56:3.

Machery, E. 2006. Two Dogmas of Neo-Empiricism. *Philosophy Compass*, 1/4.

Machery, E. 2009. *Doing without Concepts*. Oxford University Press.

Mackie, P. 2006. *How Things Might Have Been: Individuals, Kinds, and Essential Properties*. Oxford University Press.

Macpherson, F. 2011. Individuating the Senses. In Macpherson (ed.), *The Senses*. Oxford University Press.

Maddy, P. 1988a. Believing the Axioms I. *Journal of Symbolic Logic*, 53:2.

Maddy, P. 1988b. Believing the Axioms II. *Journal of Symbolic Logic*, 53:3.

Maddy, P. 1990. *Realism in Mathematics*. Oxford University Press.

Mamassian, P., M. Landy, and L. Maloney. 2002. Bayesian Modelling of Visual Perception. In R. Rao and M. Lewicki (eds.), *Probabilistic Models of the Brain*. MIT Press.

Mancosu, P. 2008. Mathematical Explanation: Why it Matters. In Mancosu (ed.), *The Philosophy of Mathematical Practice*. Oxford University Press.

Marcus, G. 2001. *The Algebraic Mind: Integrating Connectionism and Cognitive Science*. MIT Press.

Mares, E. 2019. Propositional Functions. The Stanford Encyclopedia of Philosophy (Winter 2019 Edition), Edward N. Zalta (ed.), URL = <https://plato.stanford.edu/archives/win2019/entries/propositional-function/>.

Mares, E. 2024. Relevance Logic. The Stanford Encyclopedia of Philosophy (Summer 2024 Edition), Edward N. Zalta Uri Nodelman (eds.), URL = <https://plato.stanford.edu/archives/sum2024/entries/logic-relevance/>.

Marfori, M.A., and L. Horsten. 2016. Epistemic Church's Thesis and Absolute Undecidability. In L. Horsten and P. Welch (eds.), *Gödel's Disjunction*. Oxford University Press.

Maudlin, T. 2013. The Nature of the Quantum State. In A. Ney and D. Albert (eds.), *The Wave Function*. Oxford University Press.

McDowell, J. 1994. *Mind and World*. Harvard University Press.

McFetridge, I. 1990. *Logical Necessity*, eds. J. Haldane and R. Scruton. Aristotelian Society Series, Volume 11.

McGee, V. 1991. *Truth, Vagueness, and Paradox*. Hackett Publishing.

McGee, V. 1992. Two Problems with Tarski's Theory of Logical Consequence. *Proceedings of the Aristotelian Society*, New Series, 92.

McGee, V. 1996. Logical Operations. *Journal of Philosophical Logic*, 25:6.

McLarty, C. 2008. Introduction to Categorical Foundations of Mathematics.

McLarty, C. 2009. What Does It Take to Prove Fermat's Last Theorem? Grothendieck and the Logic of Number Theory. *Bulletin of Symbolic Logic*.

Mendelssohn, M. 1764/1997. *On Evidence in Metaphysical Sciences*, tr. and ed. D. Dahlstrom. In Mendelssohn, *Philosophical Writings*, tr. and ed. Dahlstrom. Cambridge University Press.

Menger, K. 1930/1979. On Intuitionism. In Menger, *Selected Papers in Logic and Foundations, Didactics, Economics*, ed. H. Mulder. D. Reidel Publishing.

Meyer, J.J., and W. van der Hoek. *Epistemic Logic for AI and Computer Science*. Cambridge University Press.

Montague, R. 1960/1974. Logical Necessity, Physical Necessity, Ethics, and Quantifiers. In Montague, *Formal Philosophy*, ed. R. Thomason. Yale University Press.

Montague, R. 1963. Syntactical Treatments of Modality, with Corollaries on Reflexion Principles and Finite Axiomatizability. *Acta Philosophica Fennica*, 16.

Montero, B. 2010. A Russellian Response to the Structural Argument against Physicalism. *Journal of Consciousness Studies*, 17:3-4.

Moore, G.E. c.1941-1942/1962. *Commonplace Book, 1919-1953*, ed. C. Lewy. George Allen and Unwin.

Moschovakis, Y. 1998. On the Founding of a Theory of Algorithms. In H. Dales and G. Olivieri (eds.), *Truth in Mathematics*. Clarendon Press.

Moschovakis, Y. 2006. A Logical Calculus of Meaning and Synonymy. *Linguistics and Philosophy*, 29.

Mostowski, M., and D. Wojtyniak. 2004. Computational Complexity of the Semantics of Some Natural Language Constructions. *Annals of Pure and Applied Logic*, 127.

Mostowski, M., and J. Szymanik. 2012. Semantic Bounds for Everyday Language. *Semiotica*, 188.

Moss, S. 2015. On the Semantics and Pragmatics of Epistemic Vocabulary. *Semantics and Pragmatics*, vol. 8, no. 5.

Moss, S. 2018. *Probabilistic Knowledge*. Oxford University Press.

Mount, B.M. 2017. *The Kinds of Mathematical Objects*. PhD Thesis, University of Oxford.

Muskens, R. 2005. Sense and the Computation of Reference. *Linguistics and Philosophy*, 28.

Myhill, J. 1985. Intensional Set Theory. In Shapiro (ed.), *Intensional Mathematics*. Elsevier Science Publishers.

Nado, J. 2021. Conceptual Engineering, Truth, and Efficacy. *Synthese*, 198.

Nagel, J. 2007. Epistemic Intuitions. *Philosophy Compass*, 2/6.

Nagel, J. 2013a. Defending the Evidential Value of Epistemic Intuitions. *Philosophy and Phenomenological Research*, 87:1.

Nagel, J. 2013b. Knowledge as a Mental State. In T. Gendler and J. Hawthorne (eds.), *Oxford Studies in Epistemology, Volume 4*. Oxford University Press.

Nagel, J. 2014. Intuition, Reflection, and the Command of Knowledge. *Proceedings of the Aristotelian Society*, 88.

Neitz, J., and G. Jacobs. 1986. Polymorphism of the Long-wavelength Cone in Normal Human Colour Vision. *Nature*, 323.

Neitz, M., and J. Neitz. 2000. Molecular Genetics of Color Vision and Color Vision Defects. *Archives of Ophthalmology*, 118.

Newell, A. 1980. Physical Symbol Systems. *Cognitive Science*, 4.

Ney, A. 2013. Ontological Reduction and the Wave Function Ontology. In Ney and D. Albert (eds.), *The Wave Function*. Oxford University Press.

Nichols, S. (ed.). 2006. *The Architecture of the Imagination*. Oxford University Press.

Nida-Rümelin, M. 2002. Phenomenal Concepts and Phenomenal Properties. Presented at the NEH 'Institute on Consciousness and Intentionality', July 2002. <http://consc.net/neh/papers/nidarumelin.htm>.

Nida-Rümelin, M. 2003. Phänomenale Begriffe. In U. Haas-Spohn, ed., *Intentionalität zwischen Subjektivität und Weltbezug*. Mentis.

Nolan, D. 1996. Recombination Unbound. *Philosophical Studies*, 84:2-3.

Nolan, D. 2002. *Topics in the Philosophy of Possible Worlds*. Routledge.

Nozick, R. 1981. *Philosophical Explanations*. Harvard University Press.

O'Keefe, J., and L. Nadel. 1978. *The Hippocampus as a Cognitive Map*. Oxford University Press.

Oliver, A., and T. Smiley. 2013. *Plural Logic*. Oxford University Press.

Pacherie, E. 2012. Action. In K. Frankish and W. Ramsey (eds.), *The Cambridge Handbook of Cognitive Science*. Cambridge University Press.

Papineau, D. 2015. Can We Really See a Million Colors? In P. Coates and S. Coleman (eds.), *Phenomenal Qualities*. Oxford University Press.

Paris, J.B. 2001. A Note on the Dutch Book Method. In *Proceedings of the Second International Symposium on Imprecise Probabilities and Their Applications, ISIPTA*. Shaker.

Parsons, C. 1964. Frege's Theory of Number. In M. Black (ed.), *Philosophy in America*. Allen and Unwin.

Parsons, C. 1977/1983. What is the Iterative Conception of Set? In P. Benacerraf and H. Putnam (eds.), *Philosophy of Mathematics*, 2nd ed. Cambridge University Press.

Parsons, C. 1979-1980. Mathematical Intuition. *Proceedings of the Aristotelian Society*, New Series, 80.

Parsons, C. 1983. *Mathematics in Philosophy*. Cornell University Press.

Parsons, C. 1990. Introductory Note to [Gödel] 1946. In Gödel (1990), *Collected Works, Volume II*, eds. S. Feferman, J. Dawson, S. Kleene, G. Moore, R. Solovay, and J. van Heijenoort. Oxford University Press.

Parsons, C. 1993. On Some Difficulties concerning Intuition and Intuitive Knowledge. *Mind*, 102:406.

Parsons, C. 1997. What Can We Do 'In Principle'? In M. Chiara, K. Doets, D. Mundici, and J. van Benthem (eds.), *Logic and Scientific Methods*. Kluwer Academic Publishers.

Parsons, C. 2008. *Mathematical Thought and Its Objects*. Cambridge University Press.

Parsons, C. 2014. Analyticity for Realists. In J. Kennedy (ed.), *Interpreting Gödel*. Cambridge University Press.

Parry, W. T. 1968. The Logic of C.I. Lewis. In P. Schlipp (ed.), *The Philosophy of C.I. Lewis*. Cambridge University Press.

Parry, W. T. 1989. Analytic Implication: Its History, Justification and Varieties. In Norman and Sylvan (eds.), *Directions in Relevant Logic*. Kluwer Academic Publishers.

Pascal, B. 1654/1959. 'Fermat and Pascal on Probability', tr. V. Sanford. In D. Smith (ed.), *A Source Book in Mathematics*. Dover.

Peacocke, C. 2000. Explaining the A Priori: The Programme of Moderate Rationalism. In P. Boghossian and Peacocke (eds.), *New Essays on the A Priori*. Oxford University Press.

Peano, G. 1889/1967. The Principles of Arithmetic, Presented by a New Method (tr. J. van Heijenoort). In van Heijenoort (ed.), *From Frege to Gödel*. Harvard University Press.

Pearl, J. 2009. *Causality: Models, Reasoning, and Inference*, 2nd ed. Cambridge University Press.

Peirce, C.S. 1933. *Collected Papers of Charles Sanders Peirce: Vol. IV, The Simplest Mathematics*, ed., C. Hartshorne and P. Weiss. Harvard University Press.

Penrose, R. 1989. *The Emperor's New Mind*. Oxford University Press.

Penrose, R. 1994. *Shadows of the Mind*. Oxford University Press.

Perry, J. 2001. *Knowledge, Possibility, and Consciousness*. MIT Press.

Pettigrew, R. 2012. Accuracy, Chance, and the Principal Principle. *Philosophical Review*, 121:2.

Pettigrew, R. 2014. Accuracy and Evidence. *Dialectica*, 67:4.

Pettigrew, R. ms. An Introduction to Toposes.

Plantinga, A. 1976. Actualism and Possible Worlds. *Theoria*, 42.

iantadosi, S. T. 2021. The computational origin of representation. *Minds and Machines*, 31

Piccinini, G. 2004. Functionalism, Computationalism, and Mental States. *Studies in History and Philosophy of Science*, 35.

Pincock, C. 2012. *Mathematics and Scientific Representation*. Oxford University Press.

Pinder, M. 2020. On Strawson's Critique of Explication as a Method in Philosophy. *Synthese*, 197:3.

Pitt, D. 2004. The Phenomenology of Cognition. *Philosophy and Phenomenological Research*, 69:1.

Pohlová, V. 1973. On Sums in Generalized Algebraic Categories. *Czechoslovak Mathematical Journal*, Vo. 23, No. 2.

Potter, M. 2009. The Logic of the *Tractatus*. In D. Gabbay and J. Woods (eds.), *Handbook of the History of Logic, Volume 5: Logic from Russell to Church*. Elsevier.

Pratt, V. 1976. Semantical Considerations on Floyd-Hoare Logic. In *Proceedings of the 17th Annual Symposium on Foundations of Computer Science*. IEEE.

Prelević, D. 2021. The Chalmers Trilemma Re-examined. *Journal of Philosophical Research*, 46.

Price, H. 2013. *Expressivism, Pragmatism and Representationalism*. Cambridge University Press.

Priest, G. 2006. *In Contradiction*, 2nd ed. Oxford University Press.

Priest, G. 2019. Some Comments and Replies. In C. Baskent and T. Ferguson (eds.), *Graham Priest on Dialetheism and Paraconsistency*. Springer.

Prinzing, M. 2018. The Revisionist's Rubric: Conceptual Engineering and the Discontinuity Objection. *Inquiry*, 61:8.

Prior, A. 1968/2003. *Papers on Time and Tense*, ed. P. Hasle, P. Øhrstrøm, T. Braüner, and J. Copeland. Oxford University Press.

Pryor, J. 1999. Immunity to Error through Misidentification. *Philosophical Topics*, 26.

Putnam, H. 1967a. Mathematics without Foundations. *Journal of Philosophy*, 64.

Putnam, H. 1967b. Psychological Predicates. In W. Capitan and D. Merrill (eds.) *Art, Mind, and Religion*. University of Pittsburgh Press. Reprinted as 'The Nature of Mental States', in N. Block (ed.), *Readings in Philosophy of Psychology, Volume One*. Harvard University Press, 1990.

Putnam, H. 1980. Models and Reality. *Journal of Symbolic Logic*, 45:3.

Putnam, H. 1981. *Reason, Truth, and History*. Cambridge University Press.

Pylyshyn, Z. 1978. Computational Models and Empirical Constraints. *Brain and Behavioral Sciences*, 1.

Pylyshyn, Z. 2007. *Things and Places*. MIT Press.

Quine, W.V. 1951. Two Dogmas of Empiricism. *Philosophical Review*, 60:1.

Quine, W.V. 1963. Necessary Truth. In Quine (1976), *The Ways of Paradox*, 2nd ed. Harvard University Press.

Quine, W.V. 1967. Introductory Note to Bertrand Russell, 1908a, "Mathematical Logic as based on the Theory of Types. In van Heijenoort (ed.), *From Frege to Gödel*. Harvard University Press.

Quine, W.V. 1968. Propositional Objects. *Crtica*, 2:5.

Raffman, D. 2000. Is Perceptual Indiscriminability Non-transitive? In C. Hill (ed.), *Philosophical Topics*, 28.

Ramsey, F.P. 1926. Truth and Probability. In Ramsey (1960), *The Foundations of Mathematics*, ed. R. Braithwaite. Littlefield, Adams, and Co.

Rantala, V. 1982. Quantified Modal Logic: Non-normal Worlds and Propositional Attitudes. *Studia Logica*, 41.

Rao, A., and M. Georgeff. 1991. Modeling Rational Agents within a BDI-Architecture. In J. Allen, R. Fikes, and E. Sandewall (eds.), *KR '91: Principles of Knowledge Representation and Reasoning*. Morgan Kaufmann.

Rasiowa, H. 1963. On Modal Theories. *Acta Philosophica Fennica*, 16.

Raatikainen, P. 2022. Gödel's Incompleteness Theorems. *The Stanford Encyclopedia of Philosophy* (Spring 2022 Edition), E.N. Zalta (ed.), URL = <https://plato.stanford.edu/archives/spr2022/entries/goedel-incompleteness/>.

Rayo, A. 2003. Success by Default? *Philosophia Mathematica*, 3:11.

Rayo, A. 2006. Beyond Plurals. In Rayo and G. Uzquiano (eds.), *Absolute Generality*. Oxford University Press.

Rayo, A. 2013. *The Construction of Logical Space*. Oxford University Press.

Rayo, A. 2020. On the Open-Endedness of Logical Space. *Philosophers' Imprint*, 20:4.

Read, S. 2009. Plural Signification and the Liar Paradox. *Philosophical Studies*, 145.

Read, S. 2010. Field's Paradox and Its Medieval Solution. *History and Philosophy of Logic*, 31.

Recanati, F. 2007. *Perspectival Thought*. Oxford University Press.

Reinhardt, W. 1974a. Remarks on Reflection Principles, Large Cardinals, and Elementary Embeddings. In T. Jech (ed.), *Proceedings of Symposia in Pure Mathematics, Vol. 13, Part 2: Axiomatic Set Theory*. American Mathematical Society.

Reinhardt, W. 1974b. Set Existence Principles of Shoenfield, Ackermann, and Powell. *Fundamenta Mathematicae*, 84.

Reinhardt, W. 1986. Epistemic Theories and the Interpretation of Gödel's Incompleteness Theorems. *Journal of Philosophical Logic*, 15:4.

Reinhardt, W. 1988. Epistemic Set Theory. *Notre Dame Journal of Formal Logic*, 29:2.

Rescorla, M. 2009. Cognitive Maps and the Language of Thought. *British Journal for the Philosophy of Science*, 60.

Rescorla, M. 2013. Bayesian Perceptual Psychology. In M. Mohan (ed.), *The Oxford Handbook of the Philosophy of Perception*. Oxford University Press.

Rescorla, M. 2015. The Representational Foundations of Computation. *Philosophia Mathematica*, 23:3.

Rescorla, M. 2024. The Language of Thought Hypothesis. *The Stanford Encyclopedia of Philosophy*, (Summer 2024 Edition), E.N. Zalta and Uri Nodelman (eds.), URL = <https://plato.stanford.edu/archives/sum2024/entries/language-thought/>.

Restall, G. 2003. Just What is Full-Blooded Platonism? *Philosophia Mathematica*, Volume 11, Issue 1.

Reynolds, J., and D. Heeger. 2009. The Normalization Model of Attention. *Neuron* 61.2.

Roberts, A. 2019. Modal Expansionism. *Journal of Philosophical Logic*, 48.

Roca-Royes, S. 2011a. Conceivability and *De Re* Modal Knowledge. *Nous*, 45:1.

Roca-Royes, S. 2011b. Modal Knowledge and Counterfactual Knowledge. *Logique Et Analyse*, 54:216.

Roca-Royes, S. 2012. Essentialist Blindness Would Not Preclude Counterfactual Knowledge. *Philosophia Scientiae*, 16:2.

Roca-Royes, S. 2016. Similarity and Possibility. In B. Fischer and F. Leon (eds.), *Modal Epistemology after Rationalism*. Synthese Library.

Rodin, A. 2014. *Axiomatic Method and Category Theory*. Springer.

Rosen, G. 2010. Metaphysical Dependence: Grounding and Reduction. In B. Hale and A. Hoffman (eds.), *Modality: Metaphysics, Logic, and Epistemology*. Oxford University Press.

Rosen, G., and S. Yablo. 2020. Solving the Caesar Problem – With Metaphysics. In A. Miller (ed.), *Logic, Language, and Mathematics: Themes from the Philosophy of Crispin Wright*. Oxford University Press.

Rossi, N., and A. Özgün. 2023. A Hyperintensional Approach to Positive Epistemic Possibility. *Synthese*, 202:44.

Routley, R., and R. Meyer. 1972a. The Semantics of Entailment II. *Journal of Philosophical Logic*, 1.

Routley, R., and R. Meyer. 1972b. The Semantics of Entailment III. *Journal of Philosophical Logic*, 1.

Routley, R., and R. Meyer. 1973. The Semantics of Entailment I. In H. Leblanc (ed.), *Truth, Syntax, and Modality*. North-Holland.

Rumfitt, I. 2014. Truth and Meaning. *Proceedings of the Aristotelian Society Supplementary Volume*, 88.

Russell, B. 1903/2010. *The Principles of Mathematics*. Routledge.

Russell, B. 1905. On Some Difficulties in the Theory of Transfinite Numbers and Order Types. *Proceedings of the London Mathematical Society*, 4:14.

Russell, B. 1908a. Mathematical Logic as based on the Theory of Types. In van Heijenoort (ed.), *From Frege to Gödel*. Harvard University Press.

Russell, B. 1919. The Philosophy of Logical Atomism. *The Monist*, 29:3.

Russell, G. 2015. The Justification of the Basic Laws of Logic. *Journal of Philosophical Logic*, DOI 10.1007/s10992-015-9360-z.

Russell, J. 2013. Possible Worlds and the Objective World. *Philosophy and Phenomenological Research*, doi: 10.1111/PHPR.12052.

Rutten, J. 2019. *The Method of Coalgebra*. CWI.

Sambin, G., and S. Valentini. 1998. Building up a Toolbox for Martin-Löf's Type Theory: Subset Theory. In G. Sambin and J. Smith (eds.), *Twenty-five Years of Constructive Type Theory. Proceedings of a Congress held in Venice, October 1995*. Clarendon Press.

Savage, L. 1954/1972. *The Foundations of Statistics*. Dover.

Sawyer, S. 2018. The Importance of Concepts. *Proceedings of the Aristotelian Society*, 118:2.

Sawyer, S. 2020. Thought and Talk. In H. Cappelen, D. Plunkett, and A. Burgess (eds.), *Conceptual Engineering and Conceptual Ethics*. Oxford University Press.

Segerberg, K. 1973. Two-dimensional Modal Logic. *Journal of Philosophical Logic*, 2:1.

Scedrov, A. 1986. Embedding Sheaf Models for Set Theory into Boolean-valued Permutation Models with an Interior Operator. *Annals of Pure and Applied Logic*, 32.

Scharp, K. 2013. *Replacing Truth*. Oxford University Press.

Schiffer, S. 1981. "Truth and the Theory of Content". In H. Parret and J. Bouveresse (eds.), *Meaning and Understanding*. Walter de Gruyter.

Schoonen, T. 2020. *Tales of Similarity and Imagination*. ILLC Dissertation Series DS-2020-15.

Schroeder, M. 2011. Ought, Agents, and Actions. *Philosophical Review*, 120:1.

Schwartzkopff, R. 2011. Numbers as Ontologically Dependent Objects: Hume's Principle Revisited. *Grazer Philosophische Studien* 82 (1).

Seager, W. 1995. Consciousness, Information, and Panpsychism. *Journal of Consciousness Studies*, 2:3.

Segerberg, K. 1971. *An Essay in Classical Modal Logic*. *Filosofiska Studier*, Vol. 13. Uppsala Universitet.

Setiya, K. 2007. *Reasons without Rationalism*. Princeton University Press.

Setiya, K. 2018. Intention. *The Stanford Encyclopedia of Philosophy* (Fall 2018 Edition), E.N. Zalta (ed.), URL = <https://plato.stanford.edu/archives/fall2018/entries/intention/>.

Shagrir, O. 2014. Kripke's Infinity Argument. In J. Berg (ed.), *Naming, Necessity, and More*. Springer.

Shapiro, S. 1985. Epistemic and Intuitionistic Arithmetic. In Shapiro (ed.), *Intensional Mathematics*. Elsevier Science Publishers.

Shapiro, S. 1991. *Foundations without Foundationalism*. Oxford University Press.

Shapiro, S. 1994. Reasoning, Logic, and Computation. *Philosophia Mathematica*, 3:2.

Shapiro, S. 1997. *Philosophy of Mathematics*. Oxford University Press.

Shapiro, S. 1998. Logical Consequence: Models and Modality. In M. Schirn (ed.), *The Philosophy of Mathematics Today*. Oxford University Press.

Shapiro, S. 2000. Frege meets Dedekind: A Neologicist Treatment of Real Analysis. *Notre Dame Journal of Formal Logic*, 41:4.

Shapiro, S., and G. Hellman. 2015. Frege meets Aristotle: Points as Abstracts. *Philosophia Mathematica*, doi:10.1093/philmat/nkv021.

Shapiro, S., and Ø. Linnebo. 2015. Frege meets Brouwer (or Heyting or Dummett). *Review of Symbolic Logic*, doi:10.1017/S1755020315000039.

Shapiro, S., and A. Weir. 1999. New V, ZF, and Abstraction. *Philosophia Mathematica*, 3:7.

Shapiro, S., and C. Wright. 2006. All Things Indefinitely Extensible. In A. Rayo and G. Uzquiano (eds.), *Absolute Generality*. Oxford University Press.

Shea, N. 2013. Neural Mechanisms of Decision-making and the Personal Level. In K.W.M. Fulford, M. Davies, R. Gipps, G. Graham, J. Sadler, G. Stanghellini, and T. Thornton (eds.), *The Oxford Handbook of Philosophy and Psychiatry*. Oxford University Press.

Sher, G. 1991. *The Bounds of Logic*. MIT Press.

Sher, G. 2001. The Formal-Structural View of Logical Consequence. *Philosophical Review*, 110:2.

Sher, G. 2008. Tarski's Thesis. In D. Patterson (ed.), *New Essays on Tarski and Philosophy*. Oxford University Press.

Shoemaker, S. 1968. Self-reference and Self-awareness. *Journal of Philosophy*, 65.

Shoenfield, J. 1967. *Mathematical Logic*. Addison-Wesley.

Shulman, M. 2017. Homotopy Type Theory: A Synthetic Approach to Higher Equalities. In E. Landry (ed.), *Categories for the Working Philosopher*. Oxford University Press.

Shulman, M. 2022. Towards third generation HOTT. Part 1: Basic Syntax. slides.

Sider, T. 2007. Neo-Fregeanism and Quantifier Variance. *Proceedings of the Aristotelian Society Supplementary Volume*, 81.

Sider, T. 2009. Williamson's Many Necessary Existents. *Analysis*, 69.

Sider, T. 2011. *Writing the Book of the World*. Oxford University Press.

Sider, T. 2016. On Williamson and Simplicity in Modal Logic. *Canadian Journal of Philosophy*, 46:4-5.

Simons, P. 2016. Applications of Complex Numbers and Quaternions. In P. Ebert and M. Rossberg (eds.), *Abstractionism: Essays in Philosophy of Mathematics*. Oxford University Press.

Simpson, S. 1999. *Subsystems of Second Order Arithmetic*. Springer.

Skiles, A. 2015. Against Grounding Necessitarianism. *Erkenntnis*, 80.

Skolem, T. 1923. The Foundations of Elementary Arithmetic established by means of the Recursive Mode of Thought, without the use of Apparent Variables ranging over Infinite Domains (tr. S. Bauer-Mengelberg). In J. van Heijenoort (ed.), *From Frege to Gödel*. Harvard University Press.

Skow, B. 2005. *Once upon a Spacetime*. Ph.D. Dissertation, New York University.

Skow, B. 2008. Haecceitism, Anti-Haecceitism and Possible Worlds. *Philosophical Quarterly*, 58.

Smart, J.J.C. 1959. Sensations and Brain Processes. *Philosophical Review*, 68:2.

Smiley, T. 1963. The Logical Basis of Ethics. *Acta Philosophica Fennica*, 16.

Smithies, D. 2013a. The Nature of Cognitive Phenomenology. *Philosophy Compass*, 8:8.

Smithies, D. 2013b. The Significance of Cognitive Phenomenology. *Philosophy Compass*, 8:8.

Sorensen, R. 1988. *Blindspots*. Oxford University Press.

Sorensen, R. 1999. *Thought Experiments*. Oxford University Press.

Srinivisan, A. 2013. Are We Luminous? *Philosophy and Phenomenological Research*, doi: 10.1111/phpr.12067.

Stalnaker, R. 1968. A Theory of Conditionals. In J. Cornman (ed.), *Studies in Logical Theory*. Blackwell Publishing.

Stalnaker, R. 1978. Assertion. In P. Cole (ed.), *Syntax and Semantics, Vol. 9*. Academic Press.

Stalnaker, R. 2003. *Ways a World Might Be*. Oxford University Press.

Stalnaker, R. 2004. Assertion Revisited. *Philosophical Studies*, 118.

Stalnaker, R. 2006. On Logics of Knowledge and Belief. *Philosophical Studies*, 128.

Stalnaker, R. 2008. *Our Knowledge of the Internal World*. Oxford University Press.

Stalnaker, R. 2011. Conditional Propositions and Conditional Assertions. In A. Egan and B. Weatherson (eds.), *Epistemic Modality*. Oxford University Press.

Stalnaker, R. 2012. *Mere Possibilities*. Princeton University Press.

Stanley, J. 2014. Constructing Meanings. *Analysis*, 74.4.

Stanley, J., and T. Williamson. 2001. Knowing How. *Journal of Philosophy*, 98.

Starr, W. 2013. A Uniform Theory of Conditionals. *Journal of Philosophical Logic*, http://dx.doi.org/10.1007/s10992-013-9300-8.

Starr, W. 2016. Dynamic Expressivism about Deontic Modality. In N. Charlow and M. Chrisman (eds.), *Deontic Modality*. Oxford University Press.

Stern, J. 2016. *Toward Predicate Approaches to Modality*. Springer.

Stoljar, D. 2001. Two Conceptions of the Physical. *Philosophy and Phenomenological Research*, 62:2.

Stoljar, D. 2014. Four Kinds of Russellian Monism. In U. Kriegel (ed.), *Contemporary Debates in Philosophy of Mind*. Routledge.

Strawson, P. 1963/1991. Carnap's View on Constructed Systems versus Natural Languages in Analytic Philosophy. In P.A. Schlipp (ed.), *The Philosophy of Rudolf Carnap*. Open Court.

Strawson, G. 2006. Realistic Monism. *Journal of Consciousness Studies*, 13.

Studd, J. 2013. The Iterative Conception of Set: a (Bi-)Modal Characterisation. *Journal of Philosophical Logic*, 42.

Studd, J. 2019. *Everything, More or Less*. Oxford University Press.

Tarski, A. 1955. A Lattice-Thoretical Fixpoint Theorem and Its Applications. *Pacific Journal of Mathematics*, Vol. 5, No. 2.

Tennant, N. 1984. "Constructive logicism: an adequate theory of number", Minutes of the Cambridge University Moral Sciences Club, October 23rd.

Tennant, N. 1987. *Anti-Realism and Logic: Truth as Eternal*. Oxford University Press.

Tennant, N. 2009. Natural Logicism via the Logic of Orderly Pairing. in S. Lindström, E. Palmgren, K. Segerberg, and V. Stoltenberg-Hansen (eds.), *Logicism, Intuitionism, Formalism*. Springer Verlag.

Tennant, N. 2022. *The Logic of Number*. Oxford University Press.

Tennant, N. 2024. Frege's Class Theory and the Logic of Sets. In T. Piecha and K. Wehmeier (eds.), *Peter Schroeder-Heister on Proof-Theoretic Semantics* Springer.

Thomason, R. 1981. Deontic Logic and the Role of Freedom in Moral Deliberation. In Hilpinen (1981).

Thomasson, A. 2007. Modal Normativism and the Methods of Metaphysics. *Philosophical Topics*, 35:1-2.

Thomasson, A. 2009. Answerable and Unanswerable Questions. In Chalmers, D. Manley, and M. Wasserman (eds.), *Metametaphysics*. Clarendon Press.

Thomasson, A. 2020. A Pragmatic Method for Normative Conceptual Work. In H. Cappelen, D. Plunkett, and A. Burgess (eds.), *Conceptual Engineering and Conceptual Ethics*. Oxford University Press.

Tichy, P. 1969. Intension in Terms of Turing Machines. *Studia Logica*, 24.

Tieszen. R. 1989. *Mathematical Intuition: Phenomenology and Mathematical Knowledge.* Kluwer.

Tieszen, R. 2005. *Phenomenology, Logic, and the Philosophy of Mathematics.* Cambridge University Press.

Tieszen, R. 2013. *After Gödel: Platonism and Rationalism in Mathematics and Logic.* Oxford University Press.

Turing, A. 1936. On Computable Numbers, with an Application to the Entscheidungsproblem. *Proceedings of the London Mathematical Society*, 2:42

Turing, A. 1950. Computing Machinery and Intelligence. *Mind*, 59: 236.

Turner, J. 2014. Scrying an Indeterminate World. *Philosophy and Phenomenological Research*, 89:1.

Turner, J. 2016. *The Facts in Logical Space.* Oxford University Press.

The Univalent Foundations Program. 2013. *Homotopy Type Theory: Univalent Foundations of Mathematics.* Institute for Advanced Study.

Uzquiano, G. 2003. Plural Quantification and Classes. *Philosophia Mathematica*, 3:11.

Uzquiano, G. 2015a. Varieties of Indefinite Extensibility. *Notre Dame Journal of Formal Logic*, 58:1.

Uzquiano, G. 2015b. Modality and Paradox. *Philosophy Compass*, 10:4.

van Atten, M. 2015. *Essays on Gödel's Reception of Leibniz, Husserl, and Brouwer.* Springer.

van Fraassen, B. 1969. Facts and Tautological Entailments. *Journal of Philosophy*, 66.

van Heijenoort, J. (ed.). 1967. *From Frege to Gödel.* Harvard University Press.

Velleman, J.D. 1989. *Practical Reflection.* Princeton University Press.

Veltman, F. 1996. Defaults in Update Semantics. *Journal of Philosophical Logic*, 25:3.

Venema, Y. 2007. Algebras and Coalgebras. In P. Blackburn, J. van Benthem, and F. Wolter (eds.), *Handbook of Modal Logic.* Elsevier.

Venema, Y. 2012. Lectures on the Modal μ-Calculus.

Venema, Y. 2013. Cylindric Modal Logic. In H. Andrka, M. Ferenczi, and I. Nmeti (eds.), *Cylindric-Like Algebras and Algebraic Logic*. Jnos Bolyai Mathematical Society and Springer-Verlag.

Venema, Y. 2014. Expressiveness modulo Bisimilarity. In A. Baltag and S. Smets (eds.), *Johan van Benthem on Logic and Information Dynamics*. Springer.

Venema, Y. 2020. Lectures on the Modal μ-Calculus.

Venema, Y., and J. Vosmaer. 2014. Modal Logic and the Vietoris Functor. In G. Bezhanishvili (ed.), Leo Esakia on Duality in Modal and Intuitionistic Logics. Springer.

Vetter, B. 2013. 'Can' without Possible Worlds: Semantics for Anti-Humeans. *Philosophers' Imprint*, 13:16.

Vlach, F. 1973. *'Now' and 'Then': A Formal Study in the Logic of Tense Anaphora*. Ph.D. Dissertation, UCLA.

Voevodsky, V. 2006. A Very Short Note on the Homotopy λ-Calculus. Unpublished.

von Neumann, J. 1923/1967. On the Introduction of Transfinite Numbers (tr. van Heijenoort). In J. van Heijenoort (ed.), *From Frege to Gödel*. Harvard University Press.

von Neumann. 1925/1967. An Axiomatization of Set Theory (tr. S. Bauer-Mengelberg and D. Follesdal). In van Heijenoort (ed.), *From Frege to Gödel*. Harvard University Press.

von Neumann, J., and O. Morgenstern. 1944. *Theory of Games and Economic Behavior*. Princeton University Press.

von Wright, G.H. 1951. An Essay in Modal Logic. North-Holland Publishing Company.

Walsh, S. 2016. Fragments of Frege's *Grundgesetze* and Gödel's Constructible Universe. *Journal of Symbolic Logic*, 81:2.

Wang, H. 1974. *From Mathematics to Philosophy*. Routledge and Kegan Paul.

Wang, H. 1986. *A Logical Journey*. MIT Press.

Waterlow, S. 1982. *Passage and Possibility: A Study of Aristotle's Modal Concepts*. Oxford University Press.

Waxman, D. ms$_1$. Imagining the Infinite.

Waxman, D. ms$_2$. Did Gentzen prove the Inconsistency of Arithmetic?

Weir, A. 2003. Neo-Fregeanism: An Embarrassment of Riches. *Notre Dame Journal of Formal Logic*, 44.

Williams, J.R.G. 2012. Gradational Accuracy and Nonclassical Semantics. *Review of Symbolic Logic*, 5:4.

Williams, J.R.G. 2014. Nonclassical Minds and Indeterminate Survival. *Philosophical Review*, 123:4.

Williamson, T. 1990/2013. *Identity and Discrimination*. Wiley-Blackwell.

Williamson, T. 1994. *Vagueness*. Routledge.

Williamson, T. 1998. Indefinite Extensibility. *Grazer Philosophische Studien*, 55.

Williamson, T. 2002. *Knowledge and Its Limits*. Oxford University Press.

Williamson, T. 2007. *The Philosophy of Philosophy*. Oxford University Press.

Williamson, T. 2009. Some Computational Constraints in Epistemic Logic. In S. Rahman, J. Symons, D. Gabbay, and J. van Bendegem (eds.), *Logic, Epistemology, and the Unity of Science*. Springer.

Williamson, T. 2013a. *Modal Logic as Metaphysics*. Oxford University Press.

Williamson, T. 2013b. How Deep is the Distinction between A Priori and A Posteriori Knowledge? In A. Casullo and J. Thurow (eds.), *The Apriori in Philosophy*. Oxford University Press.

Williamson, T. 2014a. Logic, Metalogic, and Neutrality. *Erkenntnis*, 79.

Williamson, T. 2014b. A Note on Gettier Cases in Epistemic Logic. *Philosophical Studies*, DOI 10.1007/s11098-014-0357-1.

Williamson, T. 2014c. Very Improbable Knowing. *Erkenntnis*, DOI 10.1007/s10670-013-9590-9.

Williamson, T. 2016a. Absolute Provability and Safe Knowledge of Axioms. In L. Horsten and P. Welch (eds.), *Gödel's Disjunction*. Oxford University Press.

Williamson, T. 2016b. Modal Science. *Canadian Journal of Philosophy*, Volume 46, Issue 4-5.

Williamson, T. 2017. Semantic Paradoxes and Abductive Methodology. In B. Armour-Garb (ed.), *Reflections on the Liar*. Oxford University Press.

Williamson, T. 2018. Dummett on the Relation between Logics and Metalogics. In M. Frauchiger (ed.), *Truth, Meaning, Justification, and Reality: Themes from Dummett*. De Gruyter.

Williamson, T. 2023. Boghossian, Müller-Lyer, the Parrot, and the Nazi. In L.R.G. Oliveira (ed.), *Externalism about Knowledge*. Oxford University Press.

Wittgenstein, L. 1921/1974. *Tractatus Logico-Philosophicus*, tr. D.F. Pears and B.F. McGuinness. Routledge.

Wittgenstein, L. 1933-1937/2005. *The Big Typescript: TS 213*, tr. and ed. C. Luckhardt and M. Aue. Blackwell Publishing.

Wittgenstein, L. 1969. *On Certainty*, tr. D. Paul and G.E.M. Anscombe, ed. Anscombe and G.H. von Wright. Basil Blackwell.

Wittgenstein, L. 1975, 1976. *Wittgenstein's Lectures on the Foundations of Mathematics. Cambridge, 1939*, ed. C. Diamond. University of Chicago Press.

Wittgenstein, L. 1979. *Notebooks, 1914-1916*, 2nd ed., tr. G.E.M Anscombe, ed. G.H. von Wright and Anscombe. University of Chicago Press.

Wittgenstein, L. 2009. *Philosophical Investigations*, 4th ed., tr. G.E.M. Anscombe, P.M.S. Hacker, and J. Schulte. Blackwell Publishing.

Wittgenstein, L. 2010. *Remarks on the Foundations of Mathematics*, 3rd ed., tr. G.E.M. Anscombe, ed. G.H. von Wright, R. Rees, and Anscombe. Basil Blackwell.

Woodin, W.H. 1999. *The Axiom of Determinacy, Forcing Axioms, and the Nonstationary Ideal*. de Gruyter.

Woodin, W.H. 2001a. The Continuum Hypothesis, Part I. *Notices of the American Mathematical Society*; 48:6.

Woodin, W.H. 2001b. The Continuum, Hypothesis, Part II. *Notices of the American Mathematical Society*; 48:7.

Woodin, W.H. 2010. Strong Axioms of Infinity and the Search for V. In R. Bhatia, A. Pal, G. Rangarajan, V. Srinivas, and M.Vanninathan (eds.), *Proceedings of the International Congress of Mathematicians (ICM 2010), Vol. I*, https://doi.org/10.1142/7920.

Woodin, W.H. 2011a. The Realm of the Infinite. In M. Heller and Woodin (eds.), *Infinity: New Research Frontiers*. Cambridge University Press.

Woodin, W.H. 2011b. The Transfinite Universe. In M. Baaz, C. Papadimitriou, H. Putnam, D. Scott, and C. Harper (eds.), *Kurt Gödel and the Foundations of Mathematics*. Cambridge University Press.

Woodin, W.H. ms. The Ω Conjecture (slides).

Woodin, W.H. 2019. The Continuum Hypothesis (slides).

Woodward, J., and C. Hitchcock. 2003. Explanatory Generalizations, Part I. *Nous*, 37:1.

von Wright, G.H. 1957. On Double Quantification. In von Wright, *Logical Studies*. Routledge.

Wright, C. 1983. *Frege's Conception of Numbers as Objects*. Aberdeen University Press.

Wright, C. 1985. Skolem and the Sceptic. *Proceedings of the Aristotelian Society, Supplementary Volume*, 59.

Wright, C. 1987/1993. *Realism, Meaning, and Truth*. Blackwell.

Wright, C. 1995. Intuitionists Are Not (Turing) Machines. *Philosophia Mathematica*, 3:3.

Wright, C. 1997. The Philosophical Significance of Frege's Theorem. In R. Heck (ed.), *Language, Thought, and Logic*. Oxford University Press. Reprinted in Hale and Wright (2001).

Wright, C. 2000. Neo-Fregean Foundations for Real Analysis. *Notre Dame Jouranl in Formal Logic*, 41:4.

Wright, C. 2001. *Rails to Infinity: Essays on Themes from Wittgenstein's Philosophical Investigations*. Harvard University Press.

Wright, C. 2004. Intuition, Entitlement and the Epistemology of Logical Laws. *Dialectica*, 58:1.

Wright, C. 2007. On Quantifying into Predicate Position. In M. Leng, A. Paseau, and M. Potter (eds.), *Mathematical Knowledge*. Oxford University Press.

Wright, C. 2012a. Frege and Benacerraf's Problem. In R. DiSalle, M. Frappier, and D.Brown (eds.), *Analysis and Interpretation in the Exact Sciences: Essays in Honour of William Demopoulos*. Springer.

Wright, C. 2012b. Replies Part I: The Rule-Following Considerations and the Normativity of Meaning. In Coliva (2012).

Wright, C. 2012c. Replies Part II: Knowledge of Our Own Minds and Meanings. In Coliva (2012).

Wright, C. 2014. On Epistemic Entitlement II. In D. Dodd and E. Zardini (eds.), *Scepticism and Perceptual Justification*. Oxford University Press.

Wright, C. 2016. Abstraction and Epistemic Entitlement. In P. Ebert and M. Rossberg, *Abstractionism*. Oxford University Press.

Wright, C. 2020. Frege and Logicism. In In A. Miller (ed.), *Logic, Language, and Mathematics: Themes from the Philosophy of Crispin Wright*. Oxford University Press.

Yablo, S. 1993. Paradox without Self-reference. *Analysis*, 53.

Yablo, S. 2008. *Thoughts*. Oxford University Press.

Yablo, S. 2011. A Problem about Permission and Possibility. In A. Egan and B. Weatherson (eds.), *Epistemic Modality*. Oxford University Press.

Yablo, S. 2014. *Aboutness*. Oxford University Press.

Yablo, S. 2015. Parts and Differences. *Philosophical Studies*, Volume 173. https://doi.org/10.1007/s11098-014-0433-6.

Yalcin, S. 2007. Epistemic Modals. *Mind*, 116:464.

Yalcin, S. 2008. *Modality and Inquiry*. PhD Dissertation, MIT.

Yalcin, S. 2011. Nonfactualism about Epistemic Modality. In A. Egan and B. Weatherson (eds.), *Epistemic Modality*. Oxford University Press.

Yalcin, S. 2012. Bayesian Expressivism. *Proceedings of the Aristotelian Society*, doi: 10.1111/j.1467-9264.2012.00329.x

Yalcin, S. 2016. Beliefs as Question-sensitive. *Philosophy and Phenomenological Research*, doi: 10.1111/phpr.12330.

Zalta, E.N. 2001. Fregean Senses, Modes of Presentation, and Concepts. *Philosophical Perspectives*, 15.

Zardini, E. 2019. Non-transitivism and the Sorites Paradox. In S. Oms and Zardini (eds.), *The Sorites Paradox*. Cambridge University Press.

Zeimbekis, J., and A. Raftopoulos (eds.). 2015. *The Cognitive Penetrability of Perception*. Oxford University Press.

Zermelo, E. 1904/1967. Proof that Every Set Can be Well-Ordered (tr. S. Bauer-Mengelberg). In van Heijenoort (ed.), *From Frege to Gödel*. Harvard University Press.

Zermelo, E. 1908/1967. A New Proof of the Possibility of a Well-ordering (tr. S. Bauer-Mengelberg). In van Heijenoort (ed.), *From Frege to Gödel*. Harvard University Press.

Zermelo, E. 1930a/2010. On the Set-theoretic Model. In Zermelo (2010), *Collected Works, Volume 1*, tr. and ed. H.-D. Ebbinghaus, C. Fraser, and A. Kanamori. Springer.

Zermelo, E. 1930b/2010. On Boundary Numbers and Domains of Sets. In Zermelo (2010).

www.ingramcontent.com/pod-product-compliance
Lightning Source LLC
Chambersburg PA
CBHW071204240526
45470CB00018B/1447